EXPERT SYSTEMS FOR BUSINESS

Concepts and Applications

Second Edition

Featuring VP-Expert®

Version 3.1

D. V. Pigford
Greg Baur
Western Kentucky University

boyd & fraser publishing company

I(T)P An International Thomson Publishing Company

Danvers • Albany • Bonn • Boston • Cincinnati • Detroit • London • Madrid • Melbourne
Mexico City • New York • Paris • San Francisco • Singapore • Tokyo • Toronto • Washington

To Sport, Pepper Sprout, and Maggie Mae

Senior Acquisitions Editor: James H. Edwards
Production Coordinator: Patty Stephan
Production Services: Matrix Productions Inc.
Cover Design: Hannus Design Associates
Cover Photo: Myron J. Dorf/The Stock Market
Manufacturing Coordinator: Tracy Megison
Marketing Coordinator: Daphne Snow

Printed in the United States of America

This book is printed on recycled, acid-free paper that meets Environmental Protection Agency standards.

For more information, contact boyd & fraser publisher company:

boyd & fraser publishing company
One Corporate Place • Ferncroft Village
Danvers, Massachusetts 01923, USA

International Thomson Publishing Europe
Berkshire House 168-173
High Holborn
London, WC1V 7AA, England

Thomas Nelson Australia
102 Dodds Street
South Melbourne 3205
Victoria, Australia

Nelson Canada
1120 Birchmont Road
Scarborough, Ontario
Canada M1K 5G4

International Thomson Editores
Campose Eliseos 385, Piso 7
Col. Poanco
11560 Mexico D.F. Mexico

International Thomson Publishing GmbH
Konigswinterer Strasse 418
53227 Bonn, Germany

International Thomson Publishing Asia
221 Henderson Road
#05-10 Henderson Building
Singapore 0315

International Thomson Publishing Japan
Hirakawacho Kyowa Building, 3F
2-2-1 Hirakawacho
Chiyoda-ku, Tokyo 102, Japan

1 2 3 4 5 6 7 8 9 10 MT 7 6 5 4

Library of Congress Cataloging-in-Publication Data

Pigford, Darleen, 1943-
Expert systems for business : concepts and applications / D.V. Pigford, Greg Baur. -- 2nd ed.
p. cm.
Includes bibliographical references and index.
ISBN 0-87709-127-7
1. Business--Data processing. 2. Expert systems (Computer science) I. Baur, Gregory R. II. Title.
HF5548.2.P515 1995 93-4499
006.313--dc20 CIP

Text only ISBN: 0-87709-127-7
Text with 3.5″ Student Data Disk ISBN: 0-87709-512-4
Text with 3.5″ Education Version of VP-Expert 3.1 and 3.5″ Student Data Disk ISBN: 0-87709-129-3

Contents

Preface

About This Book

Expert Systems for Business: Concepts and Applications, Second Edition is a blend of theory and practice. The primary goal of this book is to give users a basic understanding of expert systems and how they are developed. The first part of the text (the chapters) presents theory and gives users a broad exposure to expert systems, a branch of applied artificial intelligence. The second part of the book (the modules) features a practical, hands-on approach to using an expert system development tool. VP-Expert Version 3.1, a product of Wordtech Systems, Inc., is featured in the applications portion of this text. VP-Expert was chosen because of its user friendliness, power, ease of use, cost, and interface with spreadsheets and databases.

We believe that the theory/practice approach is most practical. With a combined total of over fifty years of teaching experience in a variety of disciplines, we have found that a person who is actively involved in learning has a more successful experience than someone who is not.

Who Should Use This Book

Expert Systems for Business: Concepts and Applications, Second Edition is written for students enrolled in an introductory course on expert systems or artificial intelligence at the undergraduate or two-year college level. We assume that users have some experience with IBM-compatible microcomputers, but we do not assume any prior knowledge or experience with expert systems or expert system development tools. However, users with a background in these areas will be able to pursue the topics presented in greater depth.

Users with little or no background in expert systems will find that the first part of the book prepares them to deal with the process of expert system development. On the other hand, users with some background in expert systems can proceed directly to the second part of the book to begin hands-on work with VP-Expert, using the theory portion of the text as a reference, if necessary.

Changes in the Second Edition

There are several major changes in the second edition of this text. Additional features include an improved graphical trace, coverage of string functions, and expanded use of commands to enhance the graphical user interface (menus, formfields, buttons, and gauges). The featured version of VP-Expert has been updated from Version 3.0 to Version 3.1. There is a marked increase in the number of exercises in every chapter and module and a corresponding increase in the size of the test bank in the Instructor's Manual.

Organization of This Book

Part I: The Concepts

Chapter 1 describes artificial intelligence (AI) and how it compares with human intelligence. The term *intelligent computer* is defined, and the nature of applied AI and its advantages and limitations are discussed.

Chapter 2 presents an overview of expert systems technology. It includes definitions of expert systems and knowledge and a brief historical perspective. Also discussed are the components and features of expert systems and a brief description of expert systems development tools. Finally, a discussion of expert system applications and their advantages and limitations completes this chapter. Many of these topics are described in greater depth in subsequent chapters.

Chapter 3 deals with knowledge representation and knowledge characteristics. Several knowledge representation models, including production rules, frames, semantic nets, and predicate calculus, are introduced.

Chapter 4 begins with a discussion of various reasoning methods. The concept of heuristics, an integral part of expert systems, and the problems of search are discussed. Inference strategies, such as deduction, induction, and abduction, are explained. The chapter describes uncertainty and how expert systems deal with it.

Finally, forward and backward chaining, which are the primary reasoning control structures used in expert systems, are defined.

Chapter 5 presents the activities involved in creating and updating the knowledge base. The choices of an appropriate problem for expert systems development and the right development tool are the first topics of discussion. The processes of acquiring and organizing the knowledge are then examined. The final discussion in the chapter centers on updating and maintaining the knowledge base.

Chapter 6 describes activities performed by the inference engine. The inference engine's ability to deal with uncertainty and its associated problems, such as rule conflict resolution, is explained.

Chapter 7 examines the value of the third major component of the expert system, the user interface. Human factors are compared to the relative aspects of the user interface.

Chapter 8 focuses on the relationship of software engineering to expert systems development, final system development, and integrating that system into the organization. Topics of discussion include the software life cycle, the expert systems life cycle, and the software engineering life cycle. The stages of prototyping and full system development are examined. A number of difficulties that occur in expert systems development are resolved. Finally, the chapter examines how the expert system should be integrated into the organization.

Chapter 9 examines the future of expert systems and expert systems development Future trends in knowledge representation, knowledge acquisition, system development tools, and user interfaces are interpreted. Future availability and cost of expert systems technology are examined. Finally, the chapter presents the concept of intelligent systems.

Part II: The Applications

Module 1 contains a brief introduction and discusses examples of expert systems. The module's primary purpose is getting users started with VP-Expert. System requirements and initial software installation procedures are introduced.

Module 2 presents an overview of the main components of VP-Expert. The structure of the Main Menu, using expert system consultation, and obtaining help from VP-Expert are the focus of this module.

Module 3 details forming rules with VP-Expert. An expert system is created from a table, and the complete creation cycle is overviewed.

Module 4 discusses how VP-Expert deals with reasoning and uncertainty. Utilities such as TRACE and SET are discussed along with forward and backward chaining.

Module 5 emphasizes how VP-Expert links to peripheral software packages. Details of its linkage to database and spreadsheet software are discussed.

Module 6 describes how VP-Expert deals with uncertainty by means of confidence factors. Procedures for calculating confidence factors and their use with various rule forms are examined.

Module 7 discusses the VP-Expert user interface. Formatting the screen, controlling the printer, improving the run-time environment, and using a mouse are the main thrust of this module. String commands and their applications are also included, and commands to enhance the graphical user interface (menus, formfields, buttons, and gauges) are explained.

Finally, Module 8 contains a summary of VP-Expert commands and their use. It also includes a discussion of some of VP-Expert's advanced programming features.

The Appendices: Technical Information

The appendices start with a visual Hierarchy Chart of VP-Expert commands, which also appears on the inside front cover of the text. Appendix B is a table of ASCII characters and values. Appendix C contains answers to the odd-numbered chapter and module exercises. Appendix D contains a classification of all VP-Expert commands by functionality: database, graphics, logic, printing, and so on. A complete reference list follows Appendix E. It contains all text/sources referenced in the text.

How to Use the Text

The text may be used in a variety of ways, depending on the user's needs. The material can be presented in the following ways: Chapters 1 through 9 alone, Modules 1 through 8 alone, Chapters 1 through 9 followed by Modules 1 through 8, or integrating the chapters and modules (e.g., Chapter 1 followed by Module 1, Chapter 2 followed by Module 2, and so on).

For classroom use, we suggest integrating chapters and modules. We have attempted to write the chapters and modules in parallel. This way, students can study theory and follow immediately with implementation.

Ancillaries

The Instructor's Manual to accompany *Expert Systems for Business: Concepts and Applications, Second Edition* contains the following:

- A brief summary of each chapter and module

- A detailed teaching outline for each chapter and module
- Transparency masters
- A test bank with true/false, multiple-choice, and short answer questions and answers for each chapter and module
- Answers to even-numbered exercises for each chapter and module
- Answers to selected assignments

Student Data Disk

A Student Data Disk containing the Sample Files required in the text is available for use with the book.

VP-Expert Educational Software

Special packages of this book containing an educational version of VP-Expert Version 3.1 are available.

Technical Support

Technical support for VP-Expert is available from Wordtech Systems, Inc., 1590 Solano Way, Unit C, Concord, CA 94520-9300 at (510)689-1200.

Acknowledgments

We wish to thank all of the professors and other professionals who used the first edition of this book; they made it possible for the second edition to be created. Our thanks to Eric Stein of Penn State University—Grand Valley for his helpful comments and suggestions. We are grateful to Tony Kirchner and Jenny Chiles for their help with the preparation of this "many-faceted" manuscript.

We wish to acknowledge Wordtech Systems, Inc., manufacturers of VP-Expert, for their technical assistance. Their cooperation has been very valuable to our efforts.

Finally, we are especially grateful to Patty Stephan, our Production Coordinator, for her patience, cooperation, guidance, and support in the preparation of this second edition.

We sincerely hope that you find this text enjoyable and beneficial. We have enjoyed preparing it.

D.V. Pigford
Greg Baur
August 1994

Part I

THE CONCEPTS

1 The Intelligent Computer

***A**rtificial intelligence (AI)* *is a segment of computer science devoted to the development of computer hardware and software that is designed to imitate the human mind. As such, AI has as its main goal the task of solving the problem of making computers smarter; that is, the capability to do more and more things that are considered by people to require some level of intelligence.*

Computers have already demonstrated that they can perform a number of tasks faster and more accurately than human beings. They can, among other things, perform mathematical calculations, manipulate data in the forms of letters and numbers (including storage and retrieval), and make simple decisions. AI researchers are working to improve these current capabilities and to find new ways to use computers requiring even more intelligence.

In this chapter, we will discuss what is meant by intelligence, both human and artificial, and how intelligent computing compares with conventional computing. We will also examine what is meant by applied artificial intelligence and how expert systems fit into the picture. We will discuss advantages and limitations of artificial intelligence. Finally, we will examine the role of AI in the business world.

The Nature of Intelligence

A good place to begin our discussion is to answer the question: "What is intelligence?" According to *Webster's Second Edition New World Dictionary,* intelligence is "the ability to learn or understand from experience; the ability to acquire and retain knowledge; the ability to respond quickly and successfully to new situations; or the use of the faculty of reason in solving problems, directing conduct, etc., effectively."

If we pause and look at this definition for a moment, we will find nothing really startling about it: it contains terms that most of us would associate with intelligence and behaviors exhibited by human beings.

We can summarize this definition by saying that intelligence is the ability to acquire knowledge and use it. In this process, we use thinking and reasoning. This description characterizes what we commonly call human intelligence.

On the other hand, artificial intelligence attempts to imitate human characteristics through the use of computer hardware and software; the end result produces what we can call an intelligent computer. At this time, the computer's ability to imitate is limited but continues to progress.

AI systems today are good at taking knowledge given by humans and storing it for later retrieval. The body of knowledge possessed by the system is called its **knowledge base** and consists of facts, concepts, procedures, and relationships. This knowledge has been organized by a human being so that it is understandable and can be readily used in problem-solving or decision-making systems. One limiting characteristic of this knowledge is that its scope represents a particular subject or domain; it does not have the same breadth that a human being would have. For example, consider an AI system designed to diagnose lung cancer. The knowledge base could consist of facts (patient vital signs), concepts (human will to live), procedures (treatments and surgery), and relationships (incidence of smoking and lung problems). Medical professionals would have to identify and gather all relevant data for the knowledge base, and the system development team would organize and implement a usable computer system to assist doctors in diagnosing lung cancer. Obviously, this system's scope is relevant for lung cancer but not for other forms of cancer.

A second component of artificial intelligence is concerned with techniques used to allow the computer to think and reason. With these capabilities, the computer can draw inferences and make conclusions based on the knowledge it possesses. In the lung cancer example, the system implementation could make inferences and judgments to assist the doctor.

Thus, the intelligent computer is one that can do more than solve problems whose solutions require calculation, data storage and retrieval, or simple control. It is sometimes referred to as a *knowledge-based system.* In later chapters, we will discuss details of how knowledge is represented in the computer and how the computer can think and reason.

Intelligent Computing versus Conventional Computing

We have already alluded to the differences between intelligent computing and conventional computing. They can be summarized this way (at the risk of oversimplification): in conventional computing, the human gives a problem to the computer and tells it how to solve the problem; in intelligent computing, the human gives the problem to the computer but does not tell it how to solve the problem.

When creating a conventional computer program, a programmer develops a step-by-step procedure for solving the problem and converts it into a certain computer language. The computer then takes that procedure and applies it to a given set of data to produce a result. The conventional program guarantees a solution that will change only if a different set of data is used. However, for any given set of data, the program will produce consistent and reliable results. The conventional computer program is capable of solving exactly one problem at a time with the given step-by-step procedure or *algorithm;* a different problem requires the use of another algorithm. Examples of conventional computer programs include payroll, tax computation, inventory, exception reporting, and so on. Conventional programs are very efficient at doing what they are intended to do but are not capable of deviating from the problem at hand.

By contrast, a programmer creating an intelligent computer program gives the problem to the computer but makes no attempt to tell the computer how to solve it. The computer does not guarantee a solution nor does it guarantee totally consistent results. In this vein, it behaves much like a human expert attempting to solve a given problem.

The intelligent computer depends on its preprogrammed knowledge base and inferencing strategies to solve the problem. It is also capable of solving other related problems without any changes to its knowledge base or set of inferencing strategies. The intelligent computer may make use of algorithms, especially for search and pattern-matching activities with its knowledge base, but these algorithms are not the driving force behind the problem-solution process, as they are with conventional computing. Figure 1.1 summarizes the differences between intelligent and conventional computing.

We cannot state that intelligent computing is better than conventional computing or vice versa; that would be like comparing apples and oranges. Although there may be some common elements to the techniques used, they are not designed to solve the same kinds of problems. Conventional computing represents a mature technology that will only be improved with increases in computing speed and faster main memory access. Intelligent computing is currently a less mature technology but is progressing as researchers develop ways for computers to modify their knowledge bases internally; they will become much more powerful, with a greater range of possible applications.

Intelligent Computing	Conventional Computing
1. Does not guarantee a solution to a given problem.	1. Guarantees a solution to a given problem.
2. Produces results that may not be reliable or consistent.	2. Produces results that are inconsistent and unreliable.
3. Solves the given problem without specific program instructions.	3. Solves the given problem according to the programmer's exact instructions.
4. Can solve a range of problems in a given domain.	4. Can solve only one problem at a time in a given domain.

FIGURE 1.1 INTELLIGENT COMPUTING VERSUS CONVENTIONAL COMPUTING

Applied Artificial Intelligence

Interest in artificial intelligence has grown in recent years because of the development of several commercially feasible applications. More and more industrial companies are using AI technologies and have begun to hire professionals to work specifically in these areas. In this section, we wish to briefly discuss these technologies, their current stage of development, and their future directions.

To begin our discussion, consider the diagram in Figure 1.2, which gives an overall perspective of AI today. We recognize that there are many other areas of possible applications, but the areas shown in the figure have the greatest impact on the industrial world today. We will discuss each of these applied areas of AI: general problem solving, expert systems, natural language, computer vision, and robotics.

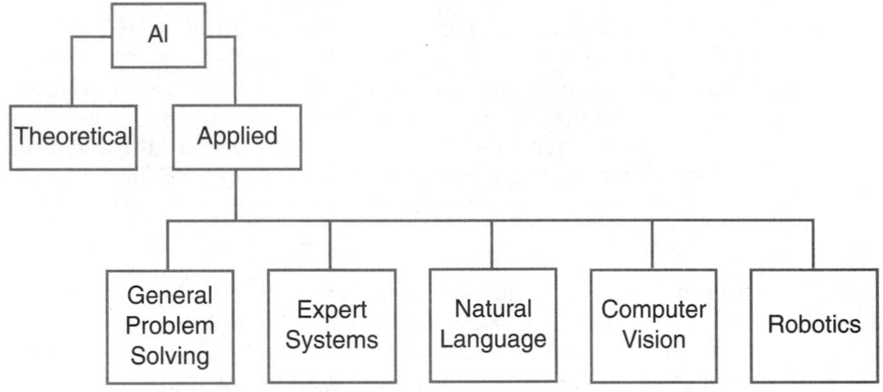

FIGURE 1.2 AN OVERVIEW OF AI TODAY

General Problem Solving

One of the goals of AI research has been to develop techniques to help people such as scientists, engineers, and mathematicians solve problems. Efforts in the 1950s and 1960s were focused on the development of a model that would solve a wide range of complex problems. The results were quite frustrating to AI researchers because as the range of problem-solving capability grew, the power of the model diminished. In the early 1970s, AI researchers finally recognized that, in order to develop a model powerful enough to be useful, the range of problems to be solved had to be narrowed to specific subject domains. This recognition was the catalyst for a number of advances in problem-solving capabilities.

Planning and scheduling are two problems of special interest to business and manufacturing professionals. A plan is a step-by-step sequence of actions designed to achieve some goal. The task is to come up with an optimal plan, one that maximizes efficiency and minimizes cost.

Existing AI programs have been developed to attain an optimal plan. The steps and situation constraints are given to the program, which examines the steps and puts them in sequence, combining some if necessary, to produce the final result.

Scheduling is another major problem for many companies. It may involve personnel, production, inventory, or any number of other areas. Again, the goal is to maximize efficiency and minimize cost. AI programs that can deal with the complexity of scheduling problems are available.

A third type of problem is one that is often found in basic research and system design development. AI techniques of *theorem proving* and *formula reduction* are sometimes used to verify the results of work done in the laboratory or the system development area.

As AI techniques become more sophisticated, the number of applications in which they will be used will increase dramatically. Previously unsolvable problems will meet their match.

Expert Systems

A second area of applied artificial intelligence is **expert systems (ES).** Expert systems are designed to simulate the behavior of a human expert as he or she attempts to solve a complex problem in a particular domain.

The key to the success of an expert system is the validity and completeness of the system's knowledge. When this knowledge is combined with various AI inferencing techniques, the outcome is a system that can solve problems and obtain results that sometimes exceed the performance of a human expert.

Any area in which human experts solve problems is a potential area for the use of expert systems. There are a number of issues regarding problem selection and system development; we will discuss them in detail throughout this book.

Natural Language

Since the beginning of computer production, communication between computers and people has been unnatural in a human sense. With the ever-increasing number of people now using computers, more attention is focused on improving human/computer communication.

Many people believe that the ideal way to communicate with a computer is via the English language in an everyday form through the use of keyboard, voice, or bar code formats, to name a few. Notable research is taking place to improve the use of *natural language*, but only with moderate success.

The problem is a complex one. The English language is not an easy language to understand—just ask a person who has English as a second language! The English language contains a number of words that have multiple meanings; meanings also depend on the context in which words or phrases are used. Systems developed utilizing natural language thus far have had limited success in realizing a general understanding of English.

Natural-language understanding systems basically use one of two techniques: key word or syntactic/semantic analysis. A system using the **key words technique** attempts to infer a meaning from the communication through the meaning of key words. Such a system is not capable of dealing with words that have multiple meanings.

The **syntactic/semantic analysis technique** holds the most promise for the development of a commercially feasible understanding. This technique breaks down a given sentence into its parts of speech (subject, verb, and so on for diagramming sentences) and attempts to infer a meaning from this analysis. Systems developed so far have had limited vocabulary capability because of the almost infinite possible combinations of sentence structure. These systems also require a great deal of computer memory for storage of the *lexicon* (dictionary) for the words that the system can deal with.

It is also difficult for the natural-language system to generate appropriate responses to the communication. Current systems, such as Eliza, usually have preprogrammed phrases or sentences for response purposes, but these are not always appropriate. Systems do not yet really possess the capability of constructing sentences word by word.

Speech-recognition systems are natural-language understanding systems that accept voice instead of keyboard input. The voice signal must be translated into a digital format that the computer can process as input. Most of these systems are voice dependent; that is, they can recognize only one human voice. Because each person speaks differently, the system must match the spoken word to a unique voice print (like a fingerprint) of that word against some word form stored in memory in order to recognize it. This recognition process must be completed before the understanding process can begin.

Speech synthesis (computer-generated speech) is in its rudimentary stages. Techniques that can adjust pitch, frequency, and inflection—characteristics expected by humans listening to oral communication—are not yet well developed.

Natural-language understanding systems have improved since their inception; however, it will be some time before we can truly communicate in natural language (English) with the computer.

Computer Vision

A fourth area of applied AI is computer vision. **Vision systems** are used when the computer requires vision capabilities to perform certain tasks.

Through the use of *computer vision,* a computer can identify features of real objects or images of objects (as in a photograph) and use this information to solve a problem. The problem-solving system uses techniques of search and pattern matching to process the visual information. Because the techniques are exhaustive, a vision system may find details that the human eye misses. Applications exist in several fields, including the military (examination of photographs for intelligence information) and robots (determination of a robot's location in a problem-solving situation).

A vision system uses a camera (two cameras for stereoptic applications) to obtain a picture or image. This image is in the form of an *analog signal*, much like the picture on a television screen. It must then be converted to a *digital signal* so that the image can be processed by the computer. The vision system runs the image through a series of steps (processing, analysis, and understanding) to obtain the desired information.

Vision systems are very complex and require a great deal of computer resources. At this point, they have limited use in real-time (immediate) situations; improved processing speed and memory-management techniques will make vision systems more useful.

Robotics

Science fiction has been instrumental in making people aware of *robots.* In reality, robots are not capable of what they are shown doing in the movies.

However, robots do have a number of capabilities and can perform useful tasks, especially in manufacturing applications. Robots can perform many tedious and repetitive tasks—a clear advantage. They can be put into situations that would be dangerous or unhealthy for human beings (mining, handling explosives, and so on).

Most robots have an arm with several joints that is moved by actuators, with a master system controller preprogrammed to perform one or several tasks in sequence. Robots are either fixed or mobile, depending on the need.

A goal of robotics researchers is to embed AI software in the robot. AI software will enable the robot to perform tasks requiring some degree of intelligence. The ultimate intelligent robot might be one that utilizes expert systems, speech recognition/synthesis, and vision technologies.

Thus, all these areas—general problem solving, expert systems, natural language, computer vision, and robotics—constitute the area commonly called applied artificial intelligence.

Advantages and Limitations of Artificial Intelligence

Thus far in this chapter, we have tried to present a perspective of what artificial intelligence is and what it is not. A number of interesting possibilities exist for the use of artificial intelligence to solve problems, but AI, as such, is not a panacea.

On the positive side, we have seen that AI techniques can be used to solve a number of complex problems. Computers will become much more useful as AI techniques are incorporated in both hardware and software to accomplish certain tasks more efficiently and at lower cost. Problems that were unsolvable before may become solvable through the use of AI techniques. Possible benefits include increased productivity, continuation of acquired expertise, and safer work environments for humans.

On the negative side, the development of AI software has been difficult at best. Development work is expensive and time-consuming, particularly if a computer language such as LISP or PROLOG has to be learned. There are not many commercial AI systems, although this situation is changing. Many research breakthroughs are a long way from happening. Added to this is a lack of trained personnel to work together in teams to develop and implement AI systems.

Even though progress will be slow, AI clearly has had a major impact on computer technology. This impact will increase in some application areas faster than in others. The intelligent computer is here to stay and will grow steadily as systems with high paybacks in productivity are implemented.

The Role of AI in Business

Historically, computers have played a major role in the business world and will continue to expand that role as advances in hardware and software are realized. The evolution of the computer's role began with data processing systems (DPS) and has progressed through the stages of management information systems (MIS), decision support systems (DSS), and expert systems (ES). The role of artificial intelligence really had its beginnings with the development of expert systems, but it is useful to describe the characteristics of the other systems.

Data processing systems (DPS) use predetermined procedures to manipulate huge volumes of data. Examples of such systems include payroll and inventory systems and, in general, involve transaction processing. Those systems are used as tools to help solve problems such as "Should the out-of-stock item be backordered?" or "Should the customer be given credit for the purchase of a large quantity of item Y?" The DPS manipulates the available data so that the decision maker can interpret the result to solve the problem (answer the question). However, because of the huge volume of data generated in the business world, more sophistication was needed.

Management information systems (MIS) were the next step. The purpose of an MIS is to summarize and select data from the massive DP files so as to give meaning to the computer output. MIS can be characterized by the terms *exception reporting* or *management by exception*.

In an MIS, managers embed data-selection criteria, based on content relevance, into the computer program, which will operate on a DP file. A typical MIS might deal with inventory control or purchasing for replacement of stock. One method might assign minimum levels of stock inventory. An MIS program could determine if the level is below the established minimum, and if so, report this exception to the manager; the process of reordering could then begin.

MIS are very good at dealing with situations that have a known problem structure and a predictable set of solutions that are usually based on past experience.

Unfortunately, as problems become larger and more complex, MIS often is inadequate, especially when used by an individual who does not have any relevant experience in solving the problem. Decision support systems can be used in these situations.

Decision support systems (DSS) are designed to help managers make more effective and efficient decisions. The manager ultimately makes the decision, but the DSS allows for the use of judgment as it supports the manager's decision process. The DSS is not designed to generate solutions or recommendations, but to help managers deal with problems in which their problem-solving abilities are below a certain threshold. For example, in the inventory control problem we discussed, a manager who has had little experience with inventory issues could use a DSS.

DSS developers are attempting to build intelligent features into the DSS. Natural-language processing and expert systems are of special interest because of the desire to make DSS more powerful and a concern for the ease of use of the human/machine interface.

Even the brief descriptions in this chapter of DSS and ES show their similarities. They also have some major differences. Let us first examine the similarities.

DSS and ES evolved from similar disciplines. DSS evolved from organizational behavior and information systems, while ES evolved from cognitive psychology and information systems. Both involve the use of the computer to make better decisions.

There are, however, some distinct differences. In general, the goal of an expert system is to replace the human expert by capturing his or her knowledge and expertise. On the other hand, the general goal of a decision support system is to assist rather than replace an expert.

This distinction, however, is not as clear-cut as it might appear. Even though the expert system replaces a human expert, its user still employs the system to help make a decision. While the expert system suggests a solution or makes recommendations, the user is not obligated to implement the expert system's proposal. In that sense, there is little difference between the expert system and the decision support system. We will discuss the features of DSS and ES in Chapter 9.

So far, we have discussed how AI is used in the business world to help people make more efficient and effective decisions and solve problems. AI is already playing, and will increasingly play, a major role in a number of other business areas.

The area of manufacturing is especially fertile. Robotics is already well established in a number of applications, such as the automated factory and assembly-line tasks that are boring or dangerous for humans. Vision techniques are being used in some applications involving robots. Natural-language processing is being used more and more as an integral part of the human/machine interface. We are now only at the tip of the AI iceberg.

Key Points

- Artificial intelligence (AI) is a branch of computer science devoted to the study of computer hardware and software designed to imitate the thinking and reasoning powers of the human mind.
- Human intelligence is the ability of humans to understand from experience, to acquire new knowledge, and to integrate experience and knowledge to solve problem situations successfully.
- An AI system consists of two parts:
  - Knowledge base (facts, concepts, procedures, and relationships)
  - Software and hardware to manipulate the knowledge base to produce thinking and reason as output
- An AI system can determine many ways to solve a given problem, whereas a conventional computer program solves a problem using step-by-step code. If the given problem is altered, the AI system may be able to handle the situation, but the conventional computer program will not.
- Another name for an AI system is an intelligent computer.
- AI is divided into theory and applications.
- Applied AI systems include:
  - General problem solving
  - Expert systems
  - Natural language
  - Computer vision
  - Robotics
- General problem solving is an applied AI system that includes planning, scheduling, and system design and development.
- An expert system is an applied AI system that simulates the behavior of a human expert in a narrowly defined or restricted domain.
- Natural language is an applied AI area that allows a user to communicate orally with the system in ordinary English sentences (as opposed to specific commands or predetermined menus).
- Computer vision is an applied AI system that allows visual input and interpretation from 2-D objects (pictures) or 3-D objects (real world).
- Robotics is an applied AI system consisting of fixed or moving mechanical arm(s) to perform given tasks (paint or assemble a car, plant explosives, explore a damaged nuclear plant, and so on).

- Advantages of AI systems include:
  - Solution of complex problems
  - Increase in productivity
  - Prevention of the loss of human expertise
  - Removal of human workers from selected dangerous environments
- Disadvantages of AI systems include:
  - Large overhead in time, money, and resources
  - Time required to learn new computer languages
  - Lack of inexpensive, usable, and reliable AI development tools to implement needed systems
  - Lack of qualified professionals to work on teams to implement AI systems
- Artificial intelligence techniques will play an increasingly greater role in the business world, especially in decision-making and problem-solving situations.
- Decision-making and problem-solving aids have evolved over the years through:
  - Data processing systems (DPS)
  - Management information systems (MIS)
  - Decision support systems (DSS)
  - Expert systems (ES)
- Robotics, computer vision, and natural-language processing have already had a major impact on many manufacturing applications.

Exercises

1. Consult your dictionary for a definition of intelligence.
2. Define artificial intelligence. How does it differ from human intelligence?
3. Name five conventional computer systems that exist in your educational or business organization. Define the purpose of each.
4. Name the five applied areas of artificial intelligence (AI).
5. What are the advantages and limitations of AI systems?
6. What is the role of AI in the business world?
7. Contrast data processing systems, management information systems, decision support systems, and expert systems.
8. What are the parts of an AI system?
9. What is another name for an AI system?
10. What is GPS (General Problem Solving)?
11. What is an expert system?

Assignments

1. A number of periodicals describe new technology and developments in artificial intelligence, including the five areas of applied AI. Magazines such as *Byte, PC World* and *AI Expert*, among others, and publications such as *Computerworld, Infoworld,* and *MacWorld*, contain articles about such developments. Consult your school or local library to locate these periodicals. Find and summarize an article about each of the five areas of applied AI described in the chapter. The *Reader's Guide to Periodical Literature* will also be helpful.
2. In the same periodicals used in the previous assignment, find ads from vendors that claim to have AI-type hardware/software products. Write for information about these products and request samples or demos. What appears to be the quality of the products? Compare your findings with a classmate or friend.
3. Consult periodicals/textbooks in your major field of interest. What discussion of intelligent computers and/or AI products do you find? Examine the periodicals/textbooks published over the last five years. What trends in the use of AI technology do you find? Share your findings with a classmate or friend.
4. What examples of AI hardware/software are there in your school or business? If you find some examples, talk with the people who have used them. What are their reactions to the product(s)? Are there plans for your school or business to purchase AI hardware/software products or to build an applied AI system in the near future? If so, what products/systems are being considered? What will be their purpose? Who will use them? Why?
5. Robotics have been highly publicized in recent movies. If you have seen one or more of these movies, use what you have learned in this chapter and critique the movie(s) for technical accuracy about the capability of robots. In your own business, what routine activity could a robot do for you? In your household, what tasks could a robot do?
6. Whether your major activity is educational, business, or engineering, what expert system would be useful to you in your work?
7. Computer vision is an area that has permeated both AI and expert systems development. The August 1991 issue of *IEEE Expert* has several interesting articles on vision-based expert systems. Pick one or two of the articles and summarize them. Share your summaries with a classmate or friend.

Expert Systems Technology

Over the past decade or so, artificial intelligence (AI) has been recognized as viable for research and commercial activities. Using computers to perform tasks previously done by human beings has resulted in the development of new and varied applications and has sparked a great deal of interest among professionals both in and out of computer science.

One of the newest applications for the computer is expert systems. In this chapter, we will discuss basic concepts of expert systems, including descriptions of expert systems and knowledge, a historical perspective, component parts, development tools, desirable system features, general applications, specific examples, and advantages and limitations. This discussion will give you a good foundation for the in-depth treatment of these topics in later chapters. The case study, WPHELP, will be introduced. WPHELP, an expert system, has been designed to assist in selecting word processing software and will be used throughout the text.

What Is an Expert System?

Simply put, an **expert system** is a computer program that emulates the behavior of human experts who are solving real-world problems associated with a particular domain of knowledge. For example, an expert system might be built that can diagnose a person's ailment when the symptoms of the ailment are input. Another example of an expert system might be one that can help choose a word processing package for an office.

Although the concept of an expert system seems simple and straightforward, the creation of the expert system is far more complex. Creating an expert system involves two steps: (1) extracting from the expert(s) the knowledge and methods used to solve the problem and (2) reforming the knowledge/methods into an organized form for later use. The processes of extraction and reformation are known as *knowledge acquisition* and *representation*, respectively, and are complex and time-consuming processes. The sum of the two processes is known as *knowledge engineering*.

Expert systems differ from conventional computer programs in several ways. Perhaps the most basic difference is that expert systems manipulate knowledge while conventional computer programs manipulate data. We will say more about the difference between knowledge and data in the next section.

Conventional computer programs are algorithmic in nature and are characterized by the use of repetitive processes. Thus, conventional programs guarantee a solution to the problem to be solved, such as payroll, accounts receivable, or inventory. Even a database program (which generates reports and makes requested queries) falls into this category.

On the other hand, expert systems are characterized by the use of *heuristics* (rules of thumb) and make use of *inferential processes*.

Human beings often use **heuristics** to solve a problem or make a decision. Heuristics often are based on experience or knowledge. For example, suppose that a complex piece of equipment in a manufacturing process breaks down. A repair person (an expert) is brought in to fix the equipment. From experience, the repair person knows that the breakdown was most likely caused by a burned-out bearing. Hence, diagnosing the cause of the breakdown starts with checking the bearings. Of course, the cause of the breakdown may be something other than a burned-out bearing, but heuristics gave the repair person a place to start.

Inferential processes involve the use of reasoning. Commonly used reasoning strategies in solving problems include deduction, induction, and abduction; expert systems make use of such strategies. We will discuss these strategies in more detail in Chapter 4.

An expert system does not guarantee a solution to a problem and is designed to work on problems that require non-algorithmic solutions—in other words, the kinds of problems that require human expertise.

Waterman (1986) states that expert systems must have four basic characteristics:

1. Expertise
2. Depth
3. Symbolic reasoning
4. Self-knowledge

Expert systems must have **expertise**, that is, the ability to perform, at least, at the level of a human expert. Expertise includes not only the capability to solve the given problem but to do so in as little time as possible. Expertise also means that the expert system's knowledge about the problem domain must be broad as well as deep.

Breadth of knowledge refers to the scope of what is known. An expert in a given subject area will have a greater breadth of knowledge than a novice. Experience is likely to be a contributing factor to breadth.

Depth of knowledge refers to the ability to extend existing knowledge and infer new knowledge. Again, experts have greater depth of knowledge about a subject area than novices. Experience is, again, a contributing factor.

To solve real-world problems effectively, an expert system must have **depth**. Real-world problems are often very difficult and complex to solve; the expert system's knowledge must be capable of dealing with these problems.

Symbolic reasoning is a third characteristic of expert systems. When an expert system solves a problem, it does so by manipulating symbols, not just by performing mathematical calculations or solving equations.

A symbol can be thought of as a string of characters that represents a real-world concept. For example:

brake system

manufacturer

0.95

all represent symbols that may be used to represent a real-world relationship. An example of such a symbolic structure might be:

(INADEQUATE brake system)

(SOLD BY automobile manufacturer)

(EQUAL (FAULT manufacturer) 0.95)

That an expert system manipulates symbols does not mean that it does not perform mathematical calculations; such calculations are performed only as needed and are not the thrust of the system's problem-solving techniques.

Finally, an expert system possesses **self-knowledge**, which can be defined as the expert system's knowledge about its knowledge. For example, most current expert systems have an *explanation facility* that allows the system to explain to a user how or why a particular conclusion was reached.

The explanation facility is but a small part of the self-knowledge characteristic. In the future, expert systems will be able to reformulate their inference strategies to operate more efficiently and to reason new knowledge from experience. Further development of self-knowledge will allow expert systems to move closer and closer to performing like human experts. We will discuss the future in more detail in Chapter 9.

What Is Knowledge?

When the term **knowledge** is used in AI work, it means the things that a computer program needs to behave intelligently. This information often takes the form of facts and rules for inferring new facts and rules. For example, we might have the following set of facts and rules from two knowledge bases. Note that new facts can be generated.

Fact: The car is out of gas,

Rule: IF the car is out of gas,
THEN the car will not start.

New Fact: The car will not start.

Fact: Can #7 contains green solution.
Can #4 contains blue solution.
Can #1 contains yellow solution.

Rule: IF the color of the solution is green,
THEN the solution is a mixture of yellow and blue solutions.

New Fact: Can #7 is a mixture of can #4 and can #1.

In the real world, facts and rules are not always entirely true or false. There may be some uncertainty about the validity of a fact or the accuracy of a rule. For example, you might hear a human expert say:

"I'm pretty sure this valve is the cause of the difficulty."

"It doesn't seem very likely that the brakes are defective."

or

"Nine times out of ten, the battery is the cause of the problem."

When this doubt occurs in a fact or rule, the expert system quantifies it with a number. These numbers, known as **certainty factors (CF)**, indicate the degree of certainty attached to a fact or rule. For example:

Fact: The car is out of gas (CF 0.80).

Rule: IF CF (car is out of gas) > = 0.80
AND the test error is greater than 0.02,
THEN the car will not start (CF 0.95).

Fact: Can #7 contains green solution (CF 0.9).

Rule: IF the can contains green solution (CF 0.9).
THEN the can is a mixture of only yellow and blue solutions (CF 0.85).

As you might suspect, an expert system must be sensitive to uncertainty in its reasoning strategies. We will explore this uncertainty and its effect on reasoning strategies in more detail in Chapters 4 and 5.

In the first examples in this section, facts and rules were used to generate new facts but were not used to generate new rules. The generation of new rules by the expert system itself is beyond the scope of current technology but is anticipated in the next decade. Some expert systems workers prefer to call current expert systems knowledge-based systems or pseudo-expert systems, leaving the term *expert system* to the future. In this book, we will not be concerned with this distinction.

Expert Systems: Past and Present

From a historical perspective, expert systems have been around a relatively short time. They developed from the efforts of AI researchers to build computer programs that could "think," that is, solve problems in a way considered by human beings to be intelligent.

The early efforts at expert system development began in the 1950s with the work of Newell, Shaw, and Simon, who created a system called the General Problem Solver (GPS). About the same time and into the 1960s, Minsky and McCarthy also did significant work in expert system development. A third group, active in the 1960s and consisting of Ledburg, Buchanan, and Feigenbaum, worked on a system called DENDRAL, which developed a structure of molecules using mass spectrometer data.

Even though the work of the three groups produced significant results, there was no solid breakthrough in achieving the original goal. The major problem was that a computer program designed to solve a wide variety of problems did not have the same power as a program designed to work with a more narrow domain.

In the 1970s, AI researchers realized that expert systems should be built with *narrow-scoped problem domains*. Researchers also recognized then that the quality of the expert system's knowledge was at least as important as the inference techniques it used.

These two realizations, along with advances in microcomputer technology, were the catalysts to advancements in expert system technology in the late 1970s and early 1980s. Expert systems then began to move from the research environment to the commercial environment.

Today, a growing number of commercial expert systems are being developed. The use of expert system development tools, which do not require high-level language programming expertise, has aided the growth of expert systems both by expanding the base of qualified developers and by simplifying the development process.

With all of these advances in expert systems technology, there is still much to be done. Current expert systems can be described as "how to" systems. All the knowledge and control structures that a given expert system possesses must be programmed into the system initially. Today's expert systems do not have the capability of "learning" (adding to the knowledge base internally), as a human expert is able to do. Researchers are working on the problem, but no real breakthrough is expected for at least a decade or two.

Components of an Expert System

Expert systems have three basic components: the knowledge base, the inference engine, and the user interface. In this section, we will very briefly describe each component. In subsequent chapters, each component will be examined in greater depth. Figure 2.1 shows a block diagram of an expert system.

The Knowledge Base

The **knowledge base** consists of the facts and rules (or other representation) that the system possesses about a particular problem domain. New facts and rules may be added externally at any time but, with current technology, only new facts (derived from the existing rules) may be added internally during program execution. More detail about the knowledge base is given in Chapter 5.

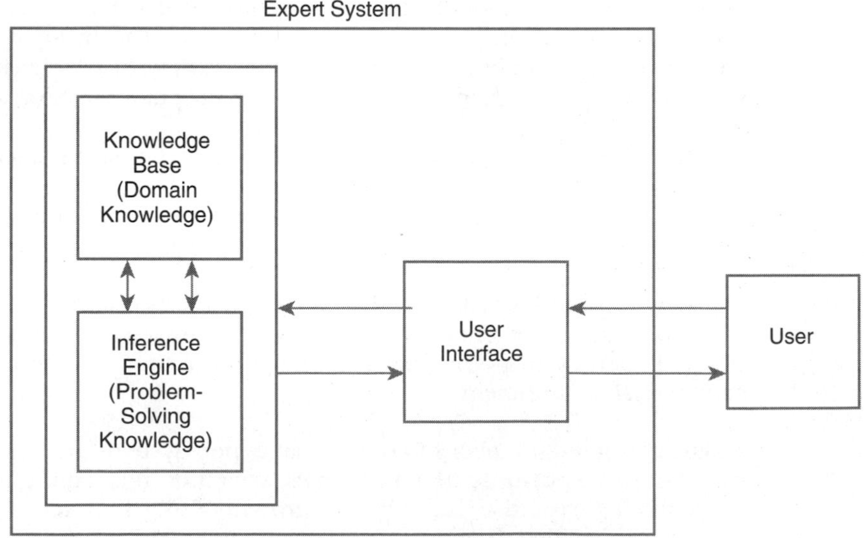

FIGURE 2.1 BLOCK DIAGRAM OF AN EXPERT SYSTEM

The Inference Engine

The **inference engine** performs the task of deciding how and when facts and rules are to be applied to make decisions (solve the problem). The structure of the inference engine is independent of the knowledge base but may be the same for similar problem structures. More detail about the inference engine is given in Chapter 6.

The User Interface

The **user interface** is the means of communication between a user wishing to solve a problem and the problem solver (expert system). This communication must be as meaningful and friendly as possible; otherwise, the expert system will be neither useful nor used. The user interface may be command-driven, menu-driven, or icon-oriented (using graphical objects). User interfaces are discussed in detail in Chapter 7.

Expert System Development Tools

Two basic types of expert systems development tools are available today: knowledge engineering languages/systems and high-level programming languages.

Knowledge engineering languages/systems (also known as *shells*) are sophisticated expert system building tools. They can be classified as skeletal or general-purpose and are composed of an expert system building language and an extensive support environment.

The advantage of the knowledge engineering language/system is that it is easier for the inexperienced expert system builder to apply. The **shell** essentially dictates how knowledge is to be represented and contains a preconstructed inference engine for accessing the knowledge base. Its disadvantage is that, because its inference engine is already structured, the range of problems that can be solved in a given domain is limited to the constraints enforced by the shell. Although the cost of good knowledge engineering shells can be relatively high (thousands of dollars), they probably do not require as much personnel and system development time, which also carry high costs. The names of knowledge engineering shells are not as well known as those of standard high-level programming languages; however, there are a number of excellent ones, such as **VP-Expert**, and more are continually being developed.

High-level programming languages used for expert system applications include *symbolic-manipulation languages* such as LISP and PROLOG and *procedure-oriented languages* such as Pascal and FORTRAN. There are advantages and disadvantages to the use of these high-level programming languages.

The main advantage is that high-level languages give the expert system builder a great deal of flexibility. Because of this flexibility, an expert system built with these tools can usually solve a wider variety of problems within the same domain. The disadvantage is that these languages are more difficult to apply because they give little or no guidance as to how knowledge should be represented or how mechanisms for accessing the knowledge base should be designed.

Desirable Features of Expert Systems

Expert systems should have several desirable features to aid the user in operating the expert system as a tool. In this section, we will describe dealing with uncertainty, explanation, ease of modification, transportability, and adaptive learning—all desirable features.

Dealing with Uncertainty

Whenever we set about to solve a problem, being able to count on our knowledge (facts and rules) as totally true or false is an advantage. We could then be certain that our derived solution was correct. Unfortunately, this desirable characteristic is seldom present. In reality, we must often deal with incomplete or less-than-accurate information.

One of an expert system's best features is its ability to deal with knowledge that is incomplete or not completely accurate. This ability allows us to obtain a solution even though we may not be 100 percent confident of its correctness.

As we noted earlier in this chapter, an expert system deals with uncertainty through the use of certainty factors. We will discuss certainty factors and uncertainty in greater detail in Chapter 6.

Certainty factors are assigned to facts or rules in the knowledge base and indicate the degree of correctness of a fact or the accuracy of a rule. We again illustrate the concept of certainty factors with several examples in Figure 2.2.

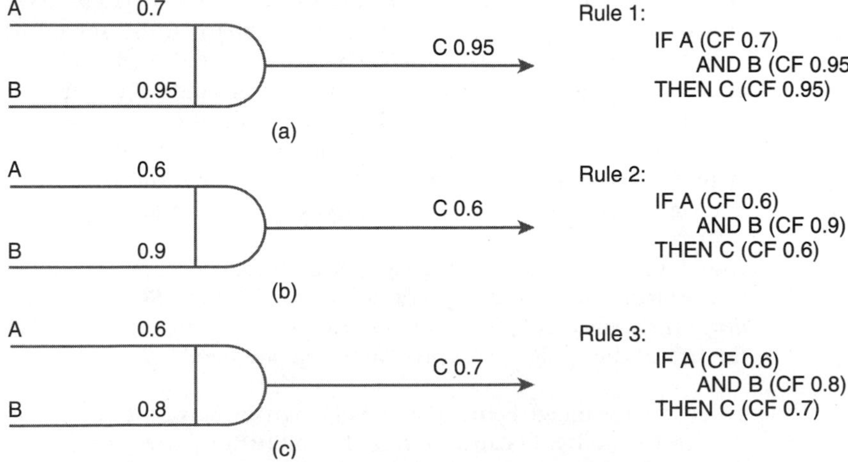

FIGURE 2.2 THREE METHODS OF DEALING WITH UNCERTAINTY

As you can see from the figure, there are several methods for calculating the certainty factor of the THEN component. Figure 2.2(a) uses the greater of the two certainty factors given in the IF component while in Figure 2.2(b), the lesser value is used. In Figure 2.2(c), the final certainty factor is the average of the two certainty factors in the IF component. Thus, three different ways to specify the certainty factor are maximum, minimum, and average of the certainty factors of the input (A, B).

Other techniques for assigning a final certainty-factor value have been developed and used. However, there is no general agreement on which method is best.

Another situation arises when several rules are chained together. In this case, the certainty factor at the end of the chain will be affected by certainty factors of elements in the chain. This situation is illustrated in Figure 2.3.

All of the certainty factors in the chain were determined by using the average method.

Explanation

A second desirable feature of expert systems is the capability to explain how and why a decision or solution was reached. The user of the expert system thus can gain access to the reasoning behind the actions of the expert system. We can illustrate this capability with Figure 2.4.

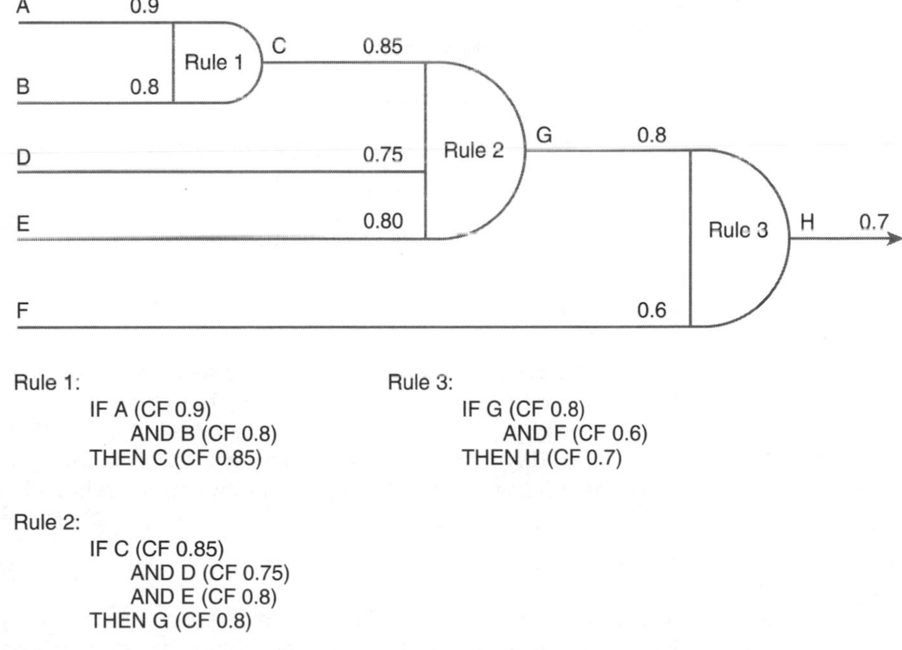

FIGURE 2.3 EFFECT OF CERTAINTY FACTORS ON AN INFERENCE CHAIN

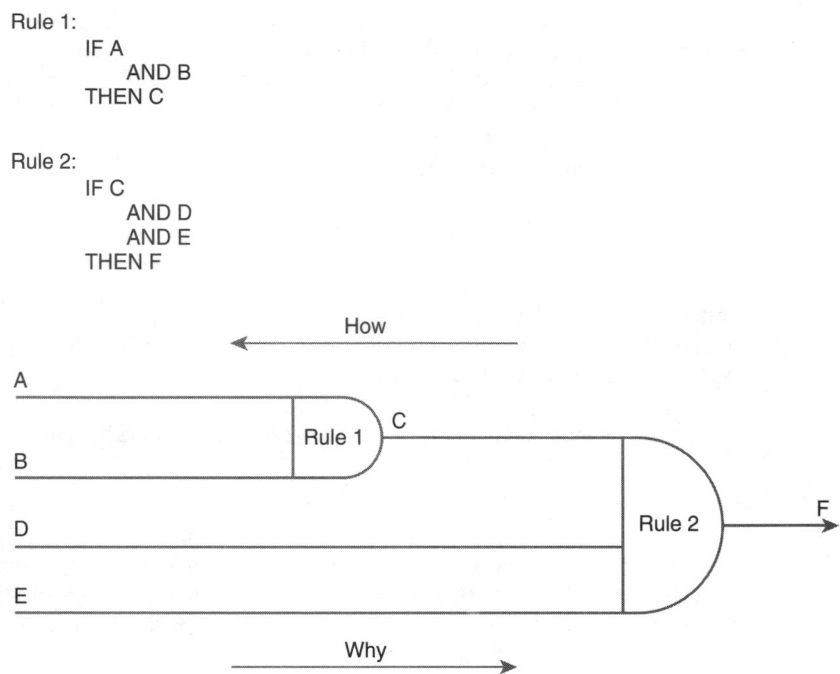

FIGURE 2.4 HOW THE EXPLANATION FACILITY IS IMPOSED ON AN INFERENCE CHAIN

The following dialogues might occur:

User: Why did you conclude F?

System: Because A and B gave me C and because C, D, and E gave me F.

User: Why did you conclude C?

System: I knew C because I knew A and B.

Ease of Modification

As new facts and rules become available or as existing facts and rules need modification, it is desirable that changes to the knowledge base be as easy to make as possible. Expert systems developed using high-level languages such as PROLOG or LISP are more difficult to modify or add to than expert systems developed with shells. Shells generally have an editing facility that makes modifications relatively easy.

Transportability

An expert system is often developed to run on only one type of computer. This is not necessarily a desirable situation because it may limit the scope of usage. If the expert system can run on more than one type of computer, a much higher level of usability can be reached.

Adaptive Learning

The **adaptive learning** feature is one possessed by few current expert systems. It involves the ability of the expert system to learn (add or modify facts and rules) on its own through experience. When this feature is available, the expert system will become much more powerful and will move closer to truly emulating human problem-solving capability. Researchers are actively working in this area, but it will be several years before this feature will be readily available. We will discuss this feature in greater detail in Chapter 9.

Applications of Expert Systems

There are so many applications for expert systems that an entire book could be filled with their descriptions. Perhaps the best way to give an overview of the different examples is to explore a generic classification scheme used by Frenzel (1987). The classifications are shown in Table 2.1.

TABLE 2.1 GENERIC CLASSIFICATION OF EXPERT SYSTEM APPLICATIONS

CONTROL	Provides intelligent automation
DEBUGGING	Recommends corrections to faults
DESIGN	Develops products to specification
DIAGNOSIS	Estimates defects
INSTRUCTION	Optimizes computer instruction
INTERPRETATION	Clarifies situations
PLANNING	Develops goal-oriented schemes
PREDICTION	Guesses outcomes intelligently
REPAIR	Automates diagnosis, debugging, planning, and fixing

An explanation of each type of application follows.

Control

Control, a very common application of computers, uses the computer to govern operation of a single component, several components, or the entire system. For example, a system that makes parts for an automobile, or tool and die, milling system could be controlled by a computer.

Many applications in this area involve the use of feedback data gathered by sensors. Examples of this kind of data might be temperature, water level, speed, or any other factor that might affect system performance.

An algorithmic computer program could be used with these applications, but an expert system would be a good choice when, say, an algorithm does not exist. The expert system is also useful when a decision about system performance needs to be made based on feedback data.

Debugging

In **debugging** applications, a system is found to be operating at less than an expected or specified level. An expert system may be used to locate the source(s) of the difficulty and then determine a solution to correct the difficulty.

Design

Design applications may involve the use of an expert system to collect information about constraints (specifications) relative to a particular product. The system then can be used as the design tool or to assist another design tool.

Several companies have used expert systems to design electronic and computer equipment and entire computer systems. They have also been used to supplement CAD/CAM systems in appliance and automobile manufacturing.

Diagnosis

It is probably fair to say that more successful expert systems have been built for **diagnostic** applications than for any of the others. In this type of application, the expert system is used to locate the source(s) of the system difficulty. It differs from an expert system that debugs in that the diagnostic expert system does not determine corrective action(s).

These expert systems have found wide use in manufacturing applications and the detection of system faults. The medical profession has begun to use expert systems to aid and verify the diagnostic process.

Instruction

Computers have been used in **instruction** applications for some time. A student sits down with a computer program and follows a certain format as he or she attempts to learn a given concept or skill.

A major problem with computer-assisted instruction (CAI) has been the lack of instructional strategy flexibility. The strategy(ies) embedded in the computer programs may not optimize the student's learning situation.

An expert system could correct this problem by determining the nature of the background knowledge each student brings to the learning situation. It could then prescribe a possible strategy and, with feedback from the student, alter the presentation of the material. In this sense, the expert system could be said to be like a debugging expert system except that its corrective action and diagnostic components possibly would be much more complex.

Interpretation

Interpretation applications involve collecting various forms of input data and then inferring a meaning or situation from that data. Interpretation is designed to help the user understand or clarify what has been recorded. Certain expert systems have been built that analyze military intelligence information; others have been used for image or signal analysis.

Planning

Applications involving a variety of **planning** techniques are those that plan a method or strategy for achieving a desired goal. Planning is necessary in the solution of complex problems. With the appropriate knowledge and input, an expert system can perform the planning phase quite well. Some examples of planning applications include military tactics, project management, scheduling, and problem-solving strategies.

Prediction

Predictive applications involve the use of input to infer possible outputs. A prediction expert system may examine all possible outputs and then choose the one that best fits the input. As input is varied, a user can determine what changes will occur in the output. Among others, weather forecasting is an example of a predictive application.

Repair

Repair applications involve an extension of debugging applications. A repair expert system takes the prescribed solution generated by a debugging expert system and implements it. A repair system, of course, must have access to the tools needed for repair. It is fairly easy to see that a repair system must have diagnostic, debugging, and planning capabilities. Examples of applications involve fault-tolerant systems, which actually may or may not be able to repair the difficulty; if not, the system is kept operational until repairs can be made by external sources.

Some Examples of Expert Systems

There are many examples of successful expert systems; in this section, we will briefly describe some of the more well-known systems.

MYCIN System

MYCIN is perhaps the most well known of the expert systems currently in use. The system was developed in the mid-1970s by Edward Feigenbaum at Stanford University. The purpose of the system was to medically diagnose a bacterial infection and prescribe antibiotic treatment. The system has proven to be very effective in assisting physicians who have had little or no experience with such infections. It also allows physicians to check the accuracy of their diagnostic/prescriptive process. MYCIN has never been a commercial success.

XCON/XSEL System

XCON/XSEL is an expert system developed jointly by Digital Equipment Corporation (DEC) and Carnegie Mellon University in the late 1970s. XCON/XSEL's purpose is to configure a computer system in the DEC VAX 11/780 series to meet specific customer orders. Large computer systems such as the VAX have a great number of different component parts designed to perform different tasks. There are thousands of combinations of these parts; each combination represents a unique computer system tailored to a specific customer. XCON/XSEL takes the given customer system specifications and produces a plan for assembling the final system.

DELTA System

DELTA is an expert system designed and developed by the General Electric Company to assist maintenance personnel in locating difficulties with electric diesel locomotive engines. These engines are very complex, and the task of troubleshooting problems is, at best, difficult. DELTA has proven to be extremely helpful to the maintenance personnel in performing the diagnostic function on the engines.

DENDRAL System

DENDRAL is an expert system also developed at Stanford University in the late 1970s. The system has knowledge of molecular structures and chemistry and is used to help identify the molecular structure of unknown chemical compounds. This assistance is very helpful to researchers as they work with existing unknown compounds for a variety of applications.

PROSPECTOR System

PROSPECTOR is an expert system designed by the Sheffield Research Institute in the late 1970s and early 1980s. The system was developed to assist geologists in locating ore deposits. The system analyzes data input by a geologist and estimates the likelihood of a particular ore deposit at a given site. In addition, PROSPECTOR can also suggest further explanations, tests, or the need for more data collection. To date, PROSPECTOR has been credited with the location of several large and valuable ore deposits in the United States.

YESMVS System

YESMVS is an expert system designed by IBM in the early 1980s. It was developed to help computer operators monitor and control the MVS (multiple virtual storage) operating system. Among other features, YESMVS runs in real time, monitors system operation, schedules large batch jobs, and alerts the operator about possible network link problems.

ACE System

ACE is an expert system designed and developed by AT&T Bell Laboratories in the early 1980s to perform troubleshooting on large telephone cable systems. The system analyzes maintenance activity data and generates reports referring to the physical location of the problem and characteristics of the cable network or the problem spot. ACE does not interact directly with a user. Instead, it interacts with **GRAS**, which is a database management and report-generation system. The system has proven especially successful in its implementation by Southwestern Bell in Denver, Colorado.

It should be fairly clear from these brief descriptions that expert systems have a wide variety of capabilities and applications. New examples are being produced almost daily by industrial companies, research organizations, and universities. A source for further information about each of the example expert systems is given in Assignment 5 at the end of this chapter.

Advantages and Limitations of Expert Systems

Existing expert systems, such as XCON/XSEL, have demonstrated that an expert system is capable of performing at or near the level of a human expert. This is important for several reasons. First, human expertise is fragile; it could be immediately lost by an organization through the resignation, retirement, or death of a key employee. An expert system is much more permanent. Second, human expertise is costly in terms of both time and money. Although initial development costs of an expert system are high, maintaining it is usually less expensive than supporting personnel. The expert system is portable and is easily shipped to remote locations. This point is especially important if human expertise is needed in more than one location at the same time. Third, an expert system is easy to document, and its results are consistent. This is not always the case with human expertise.

Expert systems do have limitations. First, they currently cannot learn from experience as a human expert can. An expert system can generate new facts but, at this point, it is incapable of generating new rules. Second, the knowledge acquisition process is a limitation. It is not an easy matter to extract knowledge from a human expert. The process of knowledge engineering is still not well understood, but its state is improving. Third, expert systems are appropriate for solving only certain kinds of problems. Problems whose solutions are not generally agreed upon by human experts are not likely to be good candidates for expert systems. Fourth, input into an expert system must have the same representation as the current knowledge base. Human experts are capable of dealing with different forms of knowledge representation, but expert systems are not.

Case Study: WPHELP

Throughout this book we use the case study WPHELP to reinforce the theory and concepts of an expert system. WPHELP is a software program that runs on IBM PCs and compatibles. The purpose of this expert system is to assist a small business manager in selecting software to process documents (letters, memos, reports, and so on) for the office. A business person can use WPHELP to obtain advice and guidance on selecting word processing software to fit the office needs of the company.

In today's office, it can cost as much as $20 to process a business letter using a typewriter. If a mistake is made, the letter must be retyped or corrected with fluid or ribbon. If 200 form letters are to be mailed, then 200 letters with different addresses must be individually typed. A *word processing* software program can eliminate these problems and reduce both the time and money involved. A user of a word processing program can operate a keyboard to type documents and then save the documents to a floppy or hard disk. The advantage is being able to retrieve, revise, and save again. A document with three misspellings can be corrected without retyping the entire document. Moreover, it is possible to create a list of addresses, say 200 customer addresses (which can be used again and again), to create one form letter (typed only one time), and to print 200 copies of that one letter with individual addresses. Clearly, using word processing software can save time and money.

WPHELP can be categorized as a design system. The expert system helps a business person plan and develop the goals for document processing for the specified business.

The system helps the user clarify and prioritize word processing needs. WPHELP, after communicating with the user through the user interface, suggests a list of possible software packages with possible steps for maximizing the resources for the office. Since WPHELP does help to interpret the word processing needs and to predict the needed outcomes in terms of software, it can be also classified as **interpretive** or **predictive**.

With more than 300 different packages ranging from $5 to $10,000 from which to choose, today's business person does not have the time and resources to research all possible word processing systems. WPHELP is designed to help the typical manager running a small business that employs as many as 20 people to clarify needs and plan for word processing in the office.

Processing documents is a vital step in maintaining a business. WPHELP's advantage is its reliable advice for determining word processing needs. On the other hand,

WPHELP has some limitations as an expert system. The nature of microcomputer hardware and software changes so fast that expert system software has to be continually updated. Rules that were true one year ago may not be true today. Constant maintenance and updating of the knowledge base are crucial. Though the domain of knowledge is limited to word processing on IBM-compatibles for a small office, document processing needs could still vary considerably. WPHELP might even advise a user to investigate another type of software.

However, even with these limitations, the advantages of WPHELP make it a viable business expert system and a useful educational case study for this text.

Key Points

- An expert system is a computer program that emulates the behavior of human experts who are solving real-world problems associated with a particular domain of knowledge.
- Knowledge engineering is the process of extracting knowledge and methods for solving problems from a human expert and then using what has been extracted to create an expert system.
- An expert system manipulates knowledge; a conventional computer program manipulates data.
- An expert system uses inferential processes and heuristics (rules of thumb); a conventional computer program uses sequential processes and algorithms.
- An expert system cannot guarantee a problem solution; a conventional computer program does guarantee a solution.
- In AI terms, knowledge is information needed by a computer program to behave intelligently. This information may take the form of facts and rules for inferring new information.
- Not all knowledge is entirely true or false. This uncertainty is quantified in an expert system with probability numbers called certainty factors (CF) or confidence ratios.
- Expert systems development work began in the 1950s but did not really move from the research environment to the commercial environment until the late 1970s and early 1980s.
- Current expert systems can be classified as "how to" systems with knowledge and inference strategies preprogrammed into the system. Future systems will be able to "learn on the fly."
- An expert system consists of three basic components:
  - Knowledge base
  - Inference engine
  - User interface
- The knowledge base contains the knowledge (facts and rules) that the expert system possesses about a particular problem domain.
- The inference engine performs the task of deciding how and when facts and rules are to be applied to make a decision (solve the problem).
- The user interface controls communication between the user and the expert system.

- There are two types of expert system development tools:
  - Knowledge engineering languages/systems (shells)
  - High-level programming languages
- Shells are easier to use by inexperienced expert system builders but are not as flexible relative to the range of potential problems to be solved because the inference engine is preconstructed.
- High-level programming languages offer greater flexibility relative to the range of problems to be solved but are more difficult to use because the inference engine must also be constructed and decisions about how knowledge is to be represented must be made.
- Expert systems have several desirable features:
  - Ability to deal with uncertainty
  - Explanation facility
  - Easy modification
  - Transportability
  - Adaptive learning
- Expert systems have been used for many types of applications including control, debugging, design, diagnosis, instruction, interpretation, planning, prediction, and repair.
- There are examples of expert systems in a wide variety of applications. Among others, MYCIN, XCON/XSEL, DELTA, DENDRAL, PROSPECTOR, YESMVS, and ACE have had notable success.
- Expert systems have several advantages:
  - The knowledge base is (usually) permanent and easy to maintain.
  - The system is relatively inexpensive to maintain and can be used in several locations at the same time.
  - The system is easy to document and produces consistent results.
- An expert system has several limitations:
  - The system cannot learn from experience.
  - The knowledge acquisition process is not well understood.
  - Not all problems are candidates for expert systems.
  - All knowledge put into the expert system must be represented in the same form.
- WPHELP is an expert system to be used throughout this text. It:
  - Is designed for business managers who run a small office with as many as 20 workers.
  - Helps a manager clarify word processing needs and suggests possible software solutions to meet these needs.
  - Has the advantages of low cost and productivity.
  - Has the disadvantages of constant revision because of changing technology, emerging new products, and varying market conditions.

Exercises

1. What is an expert system? How does it differ from a conventional computer program? How is it similar to a conventional computer program?
2. How is knowledge defined in AI terms?
3. Trace the history of expert systems. What were the significant factors that caused the current form of expert systems to be realized?

4. Describe the three components of an expert system.
5. Describe the two types of expert system development tools. What are the advantages and disadvantages of each type?
6. What are the desirable features of an expert system? How do these features enhance an expert system?
7. List the major advantages and limitations of an expert system.
8. Using the classification scheme of Table 2.1, propose an expert system of that type that would be useful to your organization.

PROPOSED EXPERT SYSTEM

CONTROL
DEBUGGING
DESIGN
DIAGNOSIS
INSTRUCTION
INTERPRETATION
PLANNING
PREDICTION
REPAIR

9. WPHELP helps a user select word processing software; define *word processing.*
10. List types of documents that a small business office might need to produce.
11. How do current expert systems differ from future expert systems?
12. What is knowledge engineering?
13. What are heuristics?

Assignments

1. Using the resources gathered in the assignments for Chapter 1, locate and summarize articles on newly created expert systems and expert systems development tools. Does there seem to be a trend toward one or more areas of applications?
2. How has expert system technology been incorporated into your major area of study/interest? What kinds of expert systems do you feel would have useful applications in your field? Do any such systems now exist? If so, gather details of their design and implementation.
3. You have read how knowledge is defined in AI terms. Compare and contrast this view of knowledge with the view of knowledge in the fields of psychology, education, and business.
4. Investigate in greater depth the projects of the three groups of early pioneers in expert system development. What specific contributions did each group make to expert systems development? What impact of each group's work has been realized in today's expert systems?

5. In the text, we briefly described some examples of working expert systems. In this assignment, you are to choose one or more of the examples and do an in-depth study of the characteristics and functions of each expert system. To help you in this assignment, we have included a source of information for each expert system; you may find other references.
   MYCIN
      Buchanan and Shortliffe, eds. *Rule-Based Expert Systems*, 295–301, 571, 596 (Menlo Park, CA: Addison-Wesley, 1984).
   XCON/XSEL
      McDermott. "Domain Knowledge and the Design Process." *Design Studies*, 3:1, January 1982.
   DELTA
      "DELTA/CATS-l." *The Artificial Intelligence Report*, 1:1, January 1984.
   DENDRAL
      Buchanan and Feigenbaum. "DENDRAL and METADENDRAL: Their Applications Dimension." *Artificial Intelligence*, vol. 11, 1978.
   PROSPECTOR
      Waterman. *A Guide to Expert Systems*, 41-60. (Menlo Park, CA: Addison-Wesley, 1986).
   YESMVS
      Gresnerietal. "YESMVS: A Continuous Real-Time Expert System." *Proceedings of the American Association for Artificial Intelligence*, 1984.
   ACE
      Wright, et al. "ACE: Going from Prototype to Product with an Expert System." *Conference Proceedings of the Association for Computing Machinery*, October 1984.
6. The forerunner of the expert system development tool (shell) was EMYCIN, which derived from MYCIN. Following is a brief description: EMYCIN is an example of an expert system development tool discussed earlier in this chapter. It is essentially the MYCIN System with the knowledge base removed. The development of EMYCIN is important because it demonstrated that a shell could be created that contained only the inference engine and the user interface. The shell could then have a number of knowledge bases inserted into it, with each knowledge base creating a unique expert system. There are a number of expert system shells available today, and they have grown in popularity for expert systems development by relatively inexperienced knowledge engineers.

   Prepare a more detailed description of EMYCIN, as you did for one of the systems in Assignment 5. This reference will help you start. Buchanan and Shortliffe, eds. *Rule-Based Expert Systems*. (Menlo Park, CA: Addison-Wesley, 1984).
7. An interesting debate on the future of expert systems can be found on pages 3-14 of the October 1991 issue of *IEEE Expert*. Read the article. What conclusions do you draw from this discussion? Summarize your findings.
8. The literature often uses the terms *expert system* and *knowledge-based system* interchangeably. There are some basic differences. Read and summarize the article "Knowledge-Based Systems: An Overview" by Mauk and Simpson, which appeared in the June 1991 issue of *IEEE Expert*. Contrast the two terms, using the article and your understanding of expert systems.

Knowledge Representation

As we mentioned in Chapter 2, an expert system contains knowledge about a particular problem domain. This knowledge represents the system's source of intelligence and is used by the inference engine to reason and draw conclusions and solutions about a particular problem in the problem domain.

The process of collecting and organizing the knowledge is perhaps the most important and difficult task involved in the creation of an expert system. We will discuss techniques for accomplishing this task in greater detail in Chapter 5. In this chapter, we will examine how the knowledge can be represented so that it can be stored and used in a computer.

Characteristics of Knowledge

To better understand what is meant by knowledge, it is useful to distinguish among three terms that are sometimes carelessly used interchangeably: data, information, and knowledge.

Data represents facts, such as raw figures (numbers) or measurements. The following are examples of data:

eight people

12 degrees

Data, by itself, may or may not answer a user's question.

Information is the refinement and use of data to answer a specific user question. Examples of information might include these:

We can fit six people in the car.

It is 32 degrees outside and the rain is turning into ice.

In addition, information can organize the data (numbers, words) so that it is meaningful to the user.

Knowledge can be thought of as refined information. It tends to be more general but also may be incomplete or fuzzy. It is more than what would be found in a conventional database. Knowledge can, of course, include facts and information and may also include concepts, procedures, models, and heuristics that can be used in problem-solving and inference systems. Knowledge can be regarded as information in context that is organized so that it is readily available for solving problems, perceptions, and learning.

Knowledge varies widely in both its content and appearance. It may be specific or general, exact or fuzzy, procedural or declarative. In the next section, we will discuss some common methods for organizing and representing knowledge: production rules, semantic networks, frames, and predicate calculus.

Knowledge Representation Models

Over the years, a number of knowledge representation models have been developed. These models have used several diverse approaches to the problem of representing knowledge, but they share two common characteristics. First, each model can be programmed with one or more existing high-level computer languages, or expert system shells, and the results can be stored in computer memory. Second, the models are designed so that their contents (facts or other types of knowledge) can be used in a reasoning process. It may be helpful to think of each model as a data structure that the inference engine can manipulate in search and pattern-matching activities.

Knowledge representation models are usually classified as either *declarative* or *procedural*. A model is said to be declarative if it is used to represent facts or assertions. A procedural model is one that is used to represent actions or procedures.

Declarative models include semantic networks, frames, and predicate calculus (logic). Procedural models include production rules. Other models do exist, but the models discussed in this section represent those most often used in the expert system development process. Table 3.1 summarizes these models.

TABLE 3.1 KNOWLEDGE REPRESENTATION MODELS

KNOWLEDGE REPRESENTATION MODEL	TYPE
Production Rules	Procedural
Semantic Networks	Declarative
Frames	Declarative
Predicate Logic (Calculus)	Declarative

The remainder of this section discusses each of these knowledge representation models or data structures in more detail.

Production Rules

Production rules (often referred to as simply rules or productions) are the most popular form of knowledge representation used in expert systems, largely because rules are relatively easy to understand and create. Rules are a good way to represent recommendations, directives, or strategies when the knowledge to be represented consists of the results of much experience in solving problems in a particular area or domain.

Rules consist of two parts, an IF and a THEN. The two parts establish a relationship that is the heart or purpose of the rule. Several pairs of terms are used to describe this IF-THEN relationship. These pairs are illustrated in Tables 3.2 and 3.3.

TABLE 3.2 IF-THEN PRODUCTION RULES

IF	THEN
antecedent	consequent
situation	action
premise	conclusion

Also, the symbol → may be used in place of IF-THEN.

TABLE 3.3 USE OF → IN IF-THEN PRODUCTION RULES

IF	THEN
antecedent →	consequent
situation →	action
premise →	conclusion

Typical rules in a rule-based expert system might look like the following:

Rule 1: IF the number of widgets is less than 60, THEN place a reorder to be delivered in 24 hours.

Rule 2: IF the battery is dead, THEN the car will not start.

These two examples represent the *situation-action* and the *premise-conclusion* forms of a rule, respectively. In these examples, the action or conclusion in the THEN part is performed or reached if the situation or premise in the IF part is true or meets certain specified criteria. We will discuss specified criteria in more detail later in this chapter and in Chapters 4 and 6.

Rules may make use of the conjunction AND or the logical OR. This use is illustrated in the following rules:

Rule 3: IF (the board appropriates the money)
AND (delivery can be assured in less than 30 days),
THEN (purchase the laser printer).

Rule 4: IF (the contractor does not meet the completion deadline),
OR (there is a cost overrun),
THEN (the contract is void).

In Rule 3, the IF part uses AND to connect the two statements (situations). To purchase the laser printer, both situations (board appropriates the money and assured less than 30-day delivery) must be true.

In the second example (Rule 4), the rule states that the contract is void if either the deadline is missed or there is a cost overrun. Both situations (past deadline and cost overrun) could occur and the contract would be voided.

Rules may contain any number of statements connected by AND or OR in the IF part. For an AND rule to be true, all of the statements connected by AND must be true. For an OR rule to be true, at least one of the statements connected by OR must be true.

To accurately represent the knowledge in a particular domain, a relatively large number of rules are needed because the rules contain the details of the knowledge. The number of rules will vary proportionately to the complexity of the knowledge. The rules are usually interrelated. For example, the action or conclusion of one rule may be the situation or premise of another rule. The final action or conclusion may involve a chaining of rules; more will be said about chaining in Chapters 4 and 5, but an example here will illustrate the point:

Rule 5: IF (the customer has an account)
AND (the account has not been used for 90 days),
THEN (mark the account as "inactive").

Rule 6: IF (the account is inactive),
THEN (call the customer to re-establish the business connection).

Rule 7: IF (the customer indicates unwillingness to do further business),
THEN (delete the customer from the database).

In this example, the three rules are interrelated. Rule 6 is invoked only if Rule 5's sequence of actions has occurred. As we shall see later, this chain may be traversed in either a forward or backward manner.

Semantic Networks

A second method of representing knowledge that has been around for several years is the **semantic network** (also known as a **semantic net**). This method is based on a network structure and is usually viewed conceptually by graphic means. A semantic net is described by points (*nodes*) connected by directed links (*arcs* or *arrows*) that show hierarchical tree relations between nodes. **Nodes** represent objects, concepts, or events. **Arcs** can be defined in a variety of ways that depend on the type of knowledge they connect.

To illustrate the use of nodes and arcs, consider Figure 3.1. In the figure, Mary, Jack, threw, red, ball, and ball-1 are all represented by nodes. Arcs between nodes show some type of relationship between the two nodes.

As we indicated, the kinds of relationships represented depend on the type of knowledge. For example, links between nodes in natural-language applications use relationships such as agent, object, and recipient. Links representing hierarchies use relationships such as "is a" (is-a), "has part" (has-part), or "instance of" (inst-of). Examples of these relationships are shown by the semantic net in Figure 3.1; this net represents the English sentence "Mary threw Jack the red ball." This sentence describes an event.

Let's examine Figure 3.1 more closely. The sentence represents an event where an action (threw) occurred between entities (Mary, Jack) concerning an object (ball-1). The action, entities, and the object are represented in separate ovals. Further, an attribute (red) and a class descriptor (ball) are also shown in separate ovals. Directed links connect pairs of ovals and are labeled with the appropriate relationship. The direction of the link is not arbitrary. The action's agent is Mary, the action's recipient is Jack, and the action's object is ball-1. In addition, the ball-1's color is red, and ball-1 is a member of (inst-of) the set (class) of objects named ball.

It may seem curious that the object is ball-1 when ball was used in the sentence. The object (ball-1) represents a particular ball in the class of all balls, and so the relationship between ball-1 and ball indicates a hierarchical relationship. However, in the sentence, just the word *ball* was used.

It also may seem that this distinction between ball-1 and ball is unimportant. After all, we human beings know the difference because we have common sense, which is not a property of computers. The computer may have many objects named ball in its memory. If it is asked to search for a ball, which ball should it look for? There also may be two red balls (a big and a small one). Hence, we must use a more specific label than just ball or red ball; the label ball-1 gives us the specificity we need.

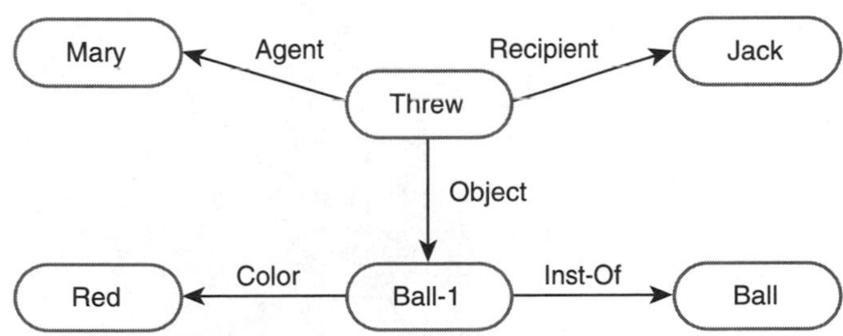

FIGURE 3.1 A SEMANTIC NET REPRESENTING THE ENGLISH SENTENCE "MARY THREW JACK THE RED BALL."

One of the appealing features of a semantic net is that it can show the **inheritance** of properties. That is, if whatever is represented by a node is part of a hierarchy, then all properties of the higher level node are also properties of any connected node at a lower level. For example, consider Figure 3.2.

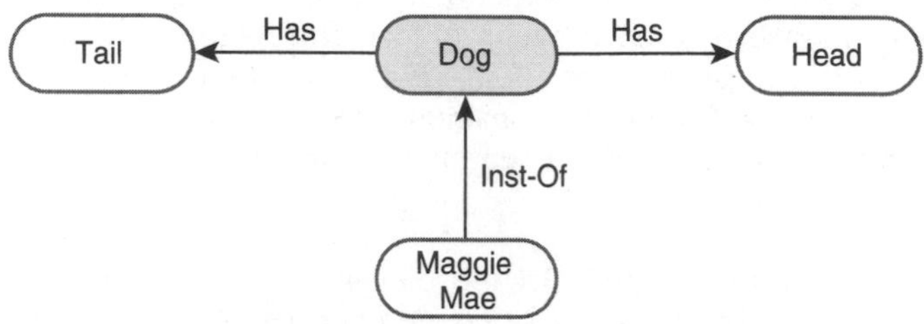

FIGURE 3.2 INHERITANCE OF PROPERTIES

Since Maggie Mae is an instance of a dog, Maggie Mae inherits the properties of having a head and a tail. We could not have a color description of dog, as we had a color description of a specific ball-1 in Figure 3.1, because not all dogs are the same color. We could, however, have a color descriptor for Maggie Mae.

Let us now examine the notion of *hierarchical representation* in greater depth. To do this, consider Figure 3.3. Note that the figure does not show all possible properties or hierarchies.

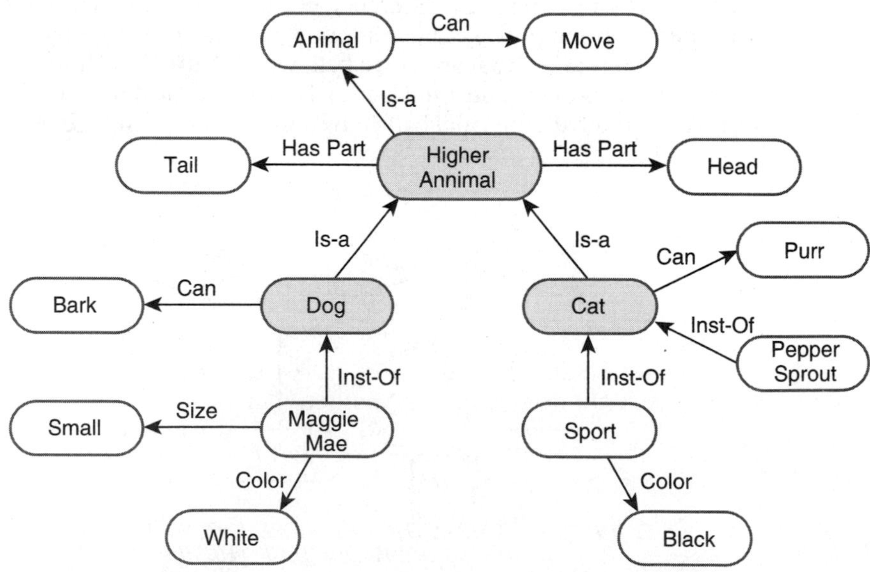

FIGURE 3.3 A SET OF HIERARCHICAL RELATIONSHIPS

This figure illustrates several things. First, there is inheritance of properties. We can infer that Maggie Mae, Sport, and Pepper Sprout can all move and that they each have a tail and a head. Since Maggie Mae is a dog, she has all of the characteristics (attributes) associated with a dog, including the ability to bark. The net also tells us that Maggie Mae is small and is white in color.

A new relationship, "is-a," is shown in this figure. The is-a relationship is similar to the inst-of relationship, but a distinction is made by AI workers. To understand the difference between the use of is-a and inst-of, make an analogy to the relationship of element and subset as they relate to the concept of set. Inst-of is a specific instance (member) of a class (as element is related to set). On the other hand, is-a represents a subclass of a larger class (as subset is related to set). That is why, in Figure 3.3, Maggie Mae is an instance of the class of dog and the class of dog is a subclass of the class of higher animal.

One of the main advantages of the use of semantic nets is that almost any object, attribute (characteristic), or concept can be defined and relationships created. But this flexibility carries a trade off. There are no standard guidelines for forming semantic nets, so their form can vary from system to system. In contrast, prediction rules have a standard format.

The amount of detail included in a semantic net depends on the type of problem to be solved. General problems require less detail than do more specific problems. For example, in Figure 3.3 even more descriptors about animals, dogs, or Maggie Mae could be added so that a user's very specific questions could be answered. In Figure 3.1 more information about Mary and Jack (age, physical condition, and so on) could be added so that desired questions could be answered. Clearly, semantic net representations can become very complex.

The graphic model is not how a semantic net is represented in a computer's memory. The objects and relationships are stored in a form dictated by the computer's language, using some data structure. A program for accessing the semantic net starts at some point (node) and uses various search and pattern-matching techniques to locate the answer to the question (problem).

Frames

A **frame** is a third technique for representing knowledge. It is a way of organizing knowledge into a collection or *bundle* where the knowledge contained in the frame is common to the concept, object, or situation. For example, in Figure 3.3, all of the knowledge about dogs might be collected in one frame, and all of the knowledge about cats might be collected in another frame.

Frames are often used in conjunction with other forms of knowledge representation, such as production rules and semantic nets. We might have one frame of rules about a particular concept and another frame of rules about a different concept.

Why is it necessary to use frames? Recall that an expert system must use search techniques on its knowledge base to obtain the necessary knowledge to solve a given problem. If the knowledge base is very large, the search may take a great deal of time. In the process, the search technique, because it is systematic in nature, may search through knowledge that is not relevant. If we are searching for knowledge about automobiles, then we do not want to search through knowledge about airplanes or trains. Human beings have the capability of knowing the difference between automobiles and airplanes, for example, but computers do not. Using frames makes expert systems capable of differentiation.

A frame consists of two parts: a name and a set of attributes associated with the name. Each attribute occupies a place in the frame called a slot; each slot has a slot name. Each attribute may have a value, known as a filler. As an example, consider Figure 3.4.

Slot name:	Filler
Frame name:	Immediate family
Father:	John
Mother:	Sue
Sister – 1:	Judy
Sister – 2:	Jill
Brother:	Nil

FIGURE 3.4 EXAMPLE OF SLOTS AND FILLERS

This example shows the names of the members of the immediate family for some person. In this case, the person has a father, mother, two sisters, and no brothers.

It is often the case that a frame may be associated with a class of objects or a category of situations. For example, a "trip" frame may contain slots for useful information about a trip: when, where, length of time, mode of transportation, and cost. Associated with this frame may be other frames that represent specific trips. Consider Figure 3.5.

Each slot can have any number of procedures attached to it. These procedures perform specified tasks for the slot. The most common types of procedures are "if-added," "if-deleted," and "if-needed." The if-added procedure executes when new information is to be added to the slot. In the trip I frame of Figure 3.5, the where slot may need a new value each time we take a vacation.

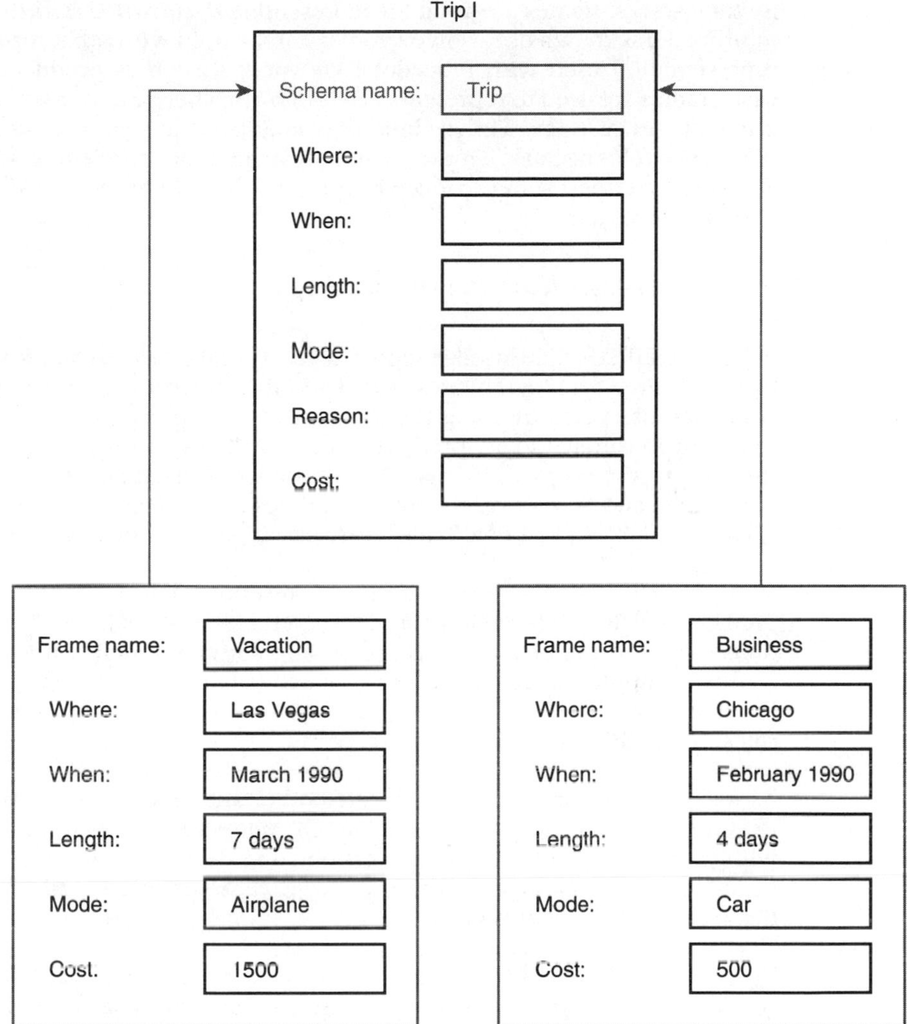

FIGURE 3.5 AN EXAMPLE OF A SCHEMA AND TWO INSTANCES OF TRIP

The if-deleted procedure executes when a value is to be deleted from a slot. Suppose that we change our mind about our vacation destination in Figure 3.5. The value Las Vegas is then deleted; of course, this change may necessitate a change in values of the other slots.

The if-needed procedure executes when a value is needed for an empty slot. In Figure 3.5, we might leave the value of the cost slot empty because this value depends on the values for the where, when, length, and mode slots. When we have made decisions about these values, the procedure can then calculate a value for the cost slot.

By themselves, frames are good for representing *declarative knowledge* as opposed to procedural knowledge. However, they can be used with other forms of knowledge representation used with procedural knowledge, such as production rules. In this case, frames are used to represent classes of rules where each class of rules represents a related set of rules. This technique is employed in expert systems with a large number of rules because it makes searching much more efficient. Instead of searching all rules, the expert system proceeds to a frame containing the rules needed to solve a particular problem.

Propositional and Predicate Logic

A fourth method of knowledge representation is called **predicate logic** or **predicate calculus**. Predicate logic allows us to determine the truth of assertions about objects, concepts, or events by using the techniques of propositional logic (calculus). But, unlike propositional logic, predicate calculus allows us to describe attributes about the objects, concepts, or events. We will begin our discussion by examining propositional logic and then cover the full capabilities of predicate calculus. Of course, both predicate and propositional calculus are based on mathematical logic.

Mathematical logic is appropriate for representing knowledge in many situations. Predicate calculus is sometimes used in expert systems and is highly regarded because of its structure and power relative to dealing with inferencing. It is, in fact, the basis for the computer language PROLOG, which has been prominent in expert systems development. However, as we shall see, it is not as easy to implement as production rules, semantic nets, or frames.

As we mentioned in Chapter 2, the method of knowledge representation used in an expert system is closely tied to the inferencing technique used. Because inferencing with predicate calculus is well understood and developed, some system developers believe that it should dictate the knowledge representation method used. However, this view is not widely held.

Propositional Calculus. A **proposition** is a statement that is either true (T) or false (F). As such, it is a premise that can be used to derive new propositions or inferences. Rules are then used to determine the truth value (T or F) of the new proposition.

In propositional logic, propositions are represented with letters. This is done to simplify the notation because connectives are used in the declaration of the propositions. So, we might have the following:

P Our newspaper is published Sunday through Friday.

Q Today is Saturday.

R Our newspaper is not published today.

By themselves, these propositions are not very useful. They can be combined using logical connectives to form new propositions that can be used in more complex problem-solving situations. The most commonly used connectives are AND, OR, NOT, and IMPLIES. We will illustrate these forms and then summarize the results in the form of a truth table.

NOT Q is the negation of Q.

From the example:

Q = Today is Saturday.

NOT Q = Today is not Saturday.

If Q is true, then NOT Q is false; if Q is false, then NOT Q is true.

When the connective AND is used, the new proposition formed is true when both original propositions are true. For example:

J = The ball is green

K = The ball has white stripes.

L = J AND K.

The ball is green and has white stripes.

If either J or K is false (or if J and K are both false), then L is false.

The connective OR forms a new proposition that is true when at least one of the two original propositions is true. Using the J and K example:

M = J OR K

= The ball is green or has white stripes.

The OR used in this case is the inclusive form of OR. The proposition formed by the exclusive OR (XOR) is true when either proposition is true but not both are true. The XOR connective is useful in digital circuit logic but is not often used in propositional logic. The use of the IMPLIES connective means that if proposition C is true, then proposition D is true where we know that C implies D. This also can be written as:

$C \rightarrow D$

Consider the truth values for E where E = C IMPLIES D. By definition, E is evaluated logically as NOT C OR D. Hence, the only time that E is false is when C is true and D is false; otherwise, E is true. Let's look at an example:

C = It is raining at noon.

D = The baseball game is canceled.

$E = C \rightarrow D$

The possible cases are the following:

1. C and D are both false.

   This can be stated as:

   If it is not raining at noon, then the baseball game is not canceled.

   E is true in this case.

2. C is false and D is true.

   This case can be stated as:

   If it is not raining at noon, then the baseball game is canceled.

   E is again true since the baseball game could be canceled for other reasons.

3. C is true and D is false.

   This can be stated as:

   If it is raining at noon, then the baseball game is not canceled.

   In this case, E is false since we have already said (from C → D) that if it is raining, then the game is canceled. This is the only time that E is false.

4. C is true and D is true.

   This case can be stated as:

   If it is raining at noon, then the baseball game is canceled. Clearly, E is true.

We can summarize all of the connectives and their results in Table 3.4 using statements A and B.

TABLE 3.4 TRUTH TABLE

A	B	NOT A	A AND B	A OR B	A → B
T	T	F	T	T	T
F	T	T	F	T	T
T	F	F	F	T	F
F	F	T	F	F	T

We can now extend our discussion to more complex situations in predicate calculus by using these ideas from propositional calculus.

Predicate Calculus. Propositional calculus is adequate for some problem-solving situations. However, it suffers because its propositions cannot be broken apart; that is, the proposition must be treated as a whole. This condition is not satisfactory for dealing with most real-world situations in which finer detail is needed.

Predicate calculus allows a proposition to be broken into two basic parts: arguments (objects) and predicates (assertions about or characteristics of objects). It also allows the use of variables and functions of variables. It makes use of symbolic logic statements and allows the use of all properties of propositional calculus. These characteristics of predicate calculus make it much more powerful than propositional calculus and, hence, a more desirable knowledge-representation method.

Here are some examples of how predicate calculus can be used with several propositions.

Proposition: Jack threw the ball.

Predicate calculus: THREW (Jack ball)

Proposition: The airplane is on the ground.

Predicate calculus: ON (airplane ground)

In the first example, "THREW" is the predicate and the arguments are "Jack" and "ball." In the second example, "ON" is the predicate and "airplane" and "ground" are the arguments. The order associated with the arguments is not coincidental.

Predicates can also have only one argument. Here is an example:

Proposition: The apple is red.

Predicate calculus: RED (apple)

An alternative form of predicate calculus is sometimes used. From the examples, this form may appear as the following:

(threw Jack ball)

(on airplane ground)

(red apple)

In predicate calculus, **variables** may be used. A variable, represented by a letter, can take on a number of specified values, all of which must make the proposition true. For example, if we encountered the expression:

red(X)

any name that can be substituted for X must have the color red as a property.

The power of variable usage is that of generalization. A proposition that can be stated in general terms is more useful than one that can be stated only for a specific case.

Predicate calculus also makes use of quantifiers, represented by symbols that state the range of a variable. A *universal quantifier* is one that states that a proposition containing a variable is true for all possible values of the variable. An *existential quantifier* states that a proposition is true for at least one value of the variable. For example:

All Hoosiers are U.S. citizens.

(A Hoosier is a (legal) resident of Indiana.)

Using the universal quantifier, we have:

(for all x)[Hoosier (x) → U.S. citizen (x)]

This last expression says that if x is a Hoosier, then x is a U.S. citizen.

The *existential qualifier* (there exists) is used with a proposition such as the following:

Some dogs like cats.

Using the existential qualifier, we have:

(there exists x)[dog (x) → likes (x, cats)]

This expression says that there exists at least one x such that if x is a dog, then x likes cats.

We can use the predicate calculus to deduce certain facts. Suppose we want to show that Tweetie is yellow, given that Tweetie is a canary and all canaries are yellow. This task can be done with the following expression:

(for all x) [(canary) (x) → yellow (x) AND canary

(Tweetie)] IMPLIES yellow (Tweetie)

We will discuss the deduction technique used here in greater detail in Chapters 4 and 6.

Case Study: Knowledge Representation in WPHELP

Now let's turn our attention to knowledge representation in our expert system WPHELP. As you recall, this expert system advises a small business on what word processing packages to buy to meet determined word processing needs.

An overview of the factors affecting the selection of a word processing package is timely. Because this is an illustrative expert system, we will simplify the number and complexity of factors involved. Let's assume that two factors are critical in the selection of a word processor for the office. First, the type of documents produced by the office must be identified. Typing a memo is quite different from producing a complete book on a word processor. Reports, letters, charts, and so on, all have different word processing needs. A second important factor is cost. Five dollars versus 500 dollars is definitely a consideration to any small business. Even though there are other factors, such as training of office workers, compatibility with other software packages, availability of telephone assistance, and more, we will consider the selection of a word processor to rest mainly on identifying the functionality required by the office and the price allocated by budgetary concerns.

The knowledge can be represented in four different ways: production rules, semantic nets, frames, and predicate calculus. We will attempt to sketch out the representation of the WPHELP knowledge in each of these four ways.

Production rules (or rules) express the knowledge in a series of IF-THEN or IF-THEN-ELSE rules. Our WPHELP knowledge base might consist of the following variables and rules, as shown in Table 3.5 and Figure 3.6.

TABLE 3.5 VARIABLES FOR WPHELP

VARIABLES	MEANING VALUES OF	VARIABLES
Functionality	Degree of complexity of desired document	Simple, Complex
Cost	Amount of money budgeted or available for purchasing a word processor	Low, High
Word Processor	Software package or program to produce documents	A, B, C

```
Rule 1:    IF (functionality is simple)
              AND (cost is low or high),
           THEN (select a word processor equal to A).

Rule 2:    IF (functionality is complex)
              AND (cost is low),
           THEN (selecting a word processor is not possible).

Rule 3:    IF (functionality is complex)
              AND (cost is high),
           THEN (select word processor B or C).
```

FIGURE 3.6 RULES FOR WPHELP

The rules for WPHELP are summarized in Table 3.6.

TABLE 3.6 A TABLE FOR WPHELP

FUNCTIONALITY	COST	WORD PROCESSOR
Simple	Low	A
Simple	High	A
Complex	Low	None possible
Complex	High	B or C

The rules and relationships can be expressed in a tree or branching structure as shown in Figure 3.7.

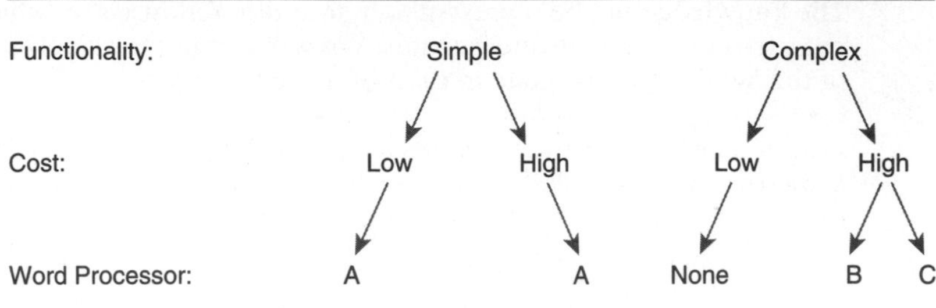

FIGURE 3.7 TREE FOR WPHELP

The values of the functionality and cost variables determine which rules are used to select a word processor. Again, for simplicity of understanding, only selected rules and variables are being considered at this time.

The second method of knowledge representation, the semantic net, uses nodes and arcs to display relationships and variables. A possible semantic net for WPHELP is given in Figure 3.8.

The variables are pictured by the shaded nodes (ovals) and the values (or instances) are indicated by the other nodes (ovals). The is-a relationship could be broadened from "is a" to other equivalent words. For example, "A is a Low" could be interpreted as "Word processor A is a low-priced word processor."

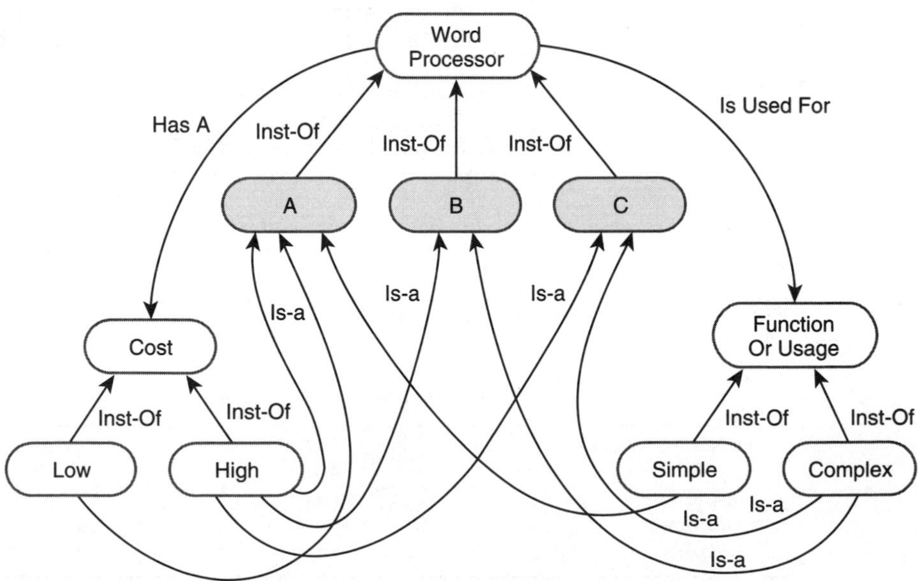

FIGURE 3.8 SEMANTIC NET FOR WPHELP

Using the third method of knowledge representation, frames, we could represent the knowledge base of WPHELP, as shown in Figure 3.9, as a schema with frames.

In this arrangement, information about each word processor is grouped in a separate frame. The schema is the outline, or template, for the information in each of the three word processor frames (A, B, C).

In the fourth method, predicate calculus, we could state the same information in the following mathematical structures:

F: Functionality

C: Cost

WP: Word Processor

Given: F and C

Find: WP

SELECTION (C, F) = WP

In this statement, SELECTION is a function or operation that depends on the variables C and F.

Thus, we have expressed the knowledge base of WPHELP with production rules, semantic nets, frames, and predicate calculus. This case study will use rules, the most common method, in subsequent chapters.

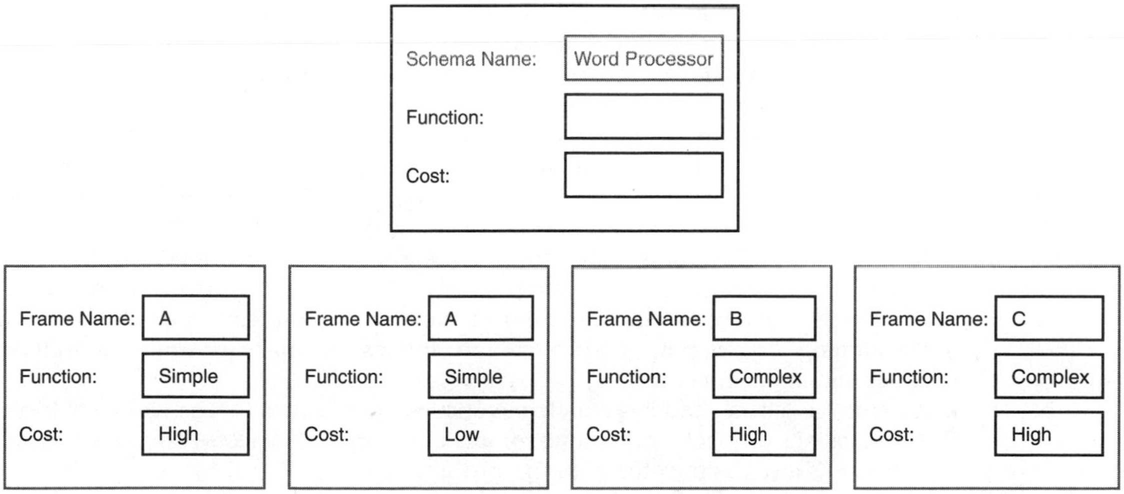

Schema Name: Word Processor
Function:
Cost:

Frame Name: A
Function: Simple
Cost: High

Frame Name: A
Function: Simple
Cost: Low

Frame Name: B
Function: Complex
Cost: High

Frame Name: C
Function: Complex
Cost: High

FIGURE 3.9 FRAME REPRESENTATION FOR WPHELP

Key Points

- Three terms—data, information, and knowledge—should not be used interchangeably.
  - Data is facts, words, numbers, or measurement.
  - Information is data organized to answer questions.
  - Knowledge is refined information.
- Models represent the structure or organization of knowledge.
  - Each model can be transferred into digital data stored in a computer and accessible through a language or expert system tool.
  - The contents of the model can be used in the reasoning process.
- Models of knowledge representation in an expert system may be either declarative (facts or assertions) or procedural (actions or procedures).
- There are four common models of knowledge representation:
  - Production rules (procedural model)
  - Semantic networks (declarative model)
  - Frames (declarative model)
  - Predicate calculus (declarative model)
- Production rules, the most popular type of model, infer a set of rules of IF-THEN relationships. The symbol → may be used instead of IF-THEN to represent the same rules. The parts of the rule are known as follows:
  - IF antecedent, THEN consequent.
  - IF situation, THEN action.
  - IF premise, THEN conclusion.
- The THEN part of a rule is considered valid (True) when the IF part is valid (True).
- The IF-THEN parts of a production rule may contain compound statements with AND or OR.
- The second model for knowledge representation, semantic networks (nets), is a graphic scheme where points (nodes) are connected by directed links (arcs or arrows) that show hierarchic (tree) relations between points.
- Common relationships in semantic nets are is a (is-a) and has part (has-part). Other relationships indicate the characteristics of a node (thing).
- Semantic nets are a very flexible tool for knowledge representation, but there are currently no standards for the symbols used in the nets. The net can vary depending on the type of knowledge and the degree of detail desired.
- A third method for knowledge representation is frames. This method organizes, bundles, or clusters knowledge in frames or partitions. For example, all knowledge about dogs could be in one frame and all knowledge about cats in another frame.
- Partitioning, or grouping, knowledge into frames can make computer searching easier and more efficient.
- A frame has a name and a set of characteristics or attributes (of the named object). The divisions or places in a frame for attributes are called slots. The value of the attribute is given as the "filler" to the attribute.
- The fourth model for knowledge representation is predicate calculus or predicate logic.
- Predicate calculus includes propositional calculus, which maintains that all statements (A, B, C, and so on) are either true or false, but not both.

- Predicate calculus allows a statement (proposition) to be broken into two parts: arguments (objects) and predicates (assertions about or characteristics of objects). It also allows for the use of variables and functions of variables.
- Predicate calculus uses quantifiers, such as:
  - for all
  - there exists at least one
- How these models for knowledge representation—production rules, semantic nets, frames, and predicate calculus—are converted into usable computer code has not been explained here. The logical representation of the knowledge, not the physical implementation, has been the theme of this chapter.
- The knowledge base of WPHELP, the expert system on selecting a word processor, can be expressed in any of the four given methods; however, the production rule method will be emphasized in this text.

Exercises

1. Organize the following character string into information:
   therearefourmodelsforknowledgerepresentation
   productionrulessemanticnetsframespredicatecalculus
2. Name the four common models for knowledge representation.
3. Given the statement:

   If today is Tuesday and the temperature is above 60 degrees F, then I will go fishing.

   What is the antecedent? What is the consequent? Using the form: IF A and B, THEN C, what is A? B? C?

   At this time, what is the truth value of A? B? A AND B? C? IF (A AND B), THEN C? Would this production rule be a good rule in an expert system concerning your hobbies? Why?
4. State four production rules that apply to your business or educational institution.
5. Given the semantic net, fill in the names or values of the nodes in Figure 3.10 (select your favorite car and write it in the car node).

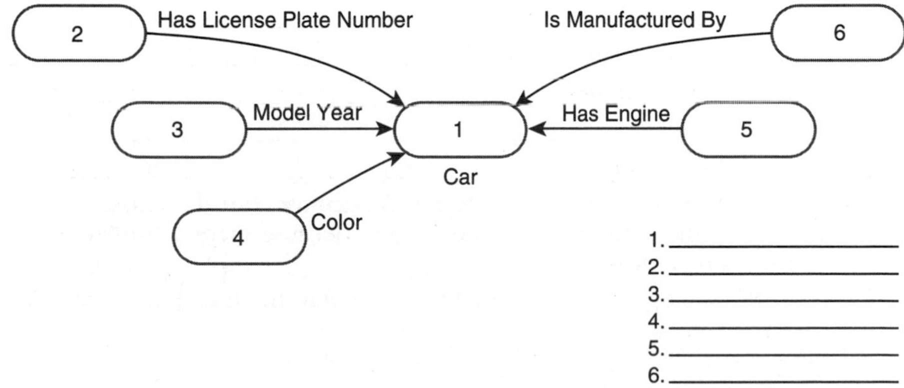

FIGURE 3.10 SEMANTIC NET FOR CAR

6. Given the attribute values in Figure 3.11, fill in the possible relationships of the six nodes.

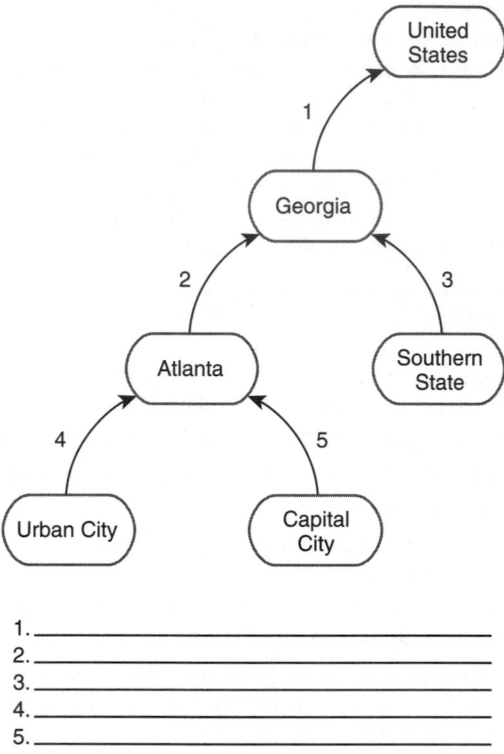

FIGURE 3.11 SEMANTIC NET FOR ATLANTA

7. Convert the information in Exercise 5 to the following frame:

   Frame name ____

   License plate number ____

   Year of model ____

   Color ____

   Type of engine ____

   The manufacturer could be an attribute, but the chosen representation has the manufacturer in a higher frame (because many manufacturers make more than one type of car).

8. What procedures would be relevant to the frame in Exercise 7?

9. Fill in the following table:

A	B	NOT A	NOT A AND B	A OR NOT B	B → A
T	T	F			
T	F	F			
F	T	T			
F	F	T			

10. Convert the following predicate calculus propositions to English sentences:
    a. DROVE (Mary car)
    b. ON (back, floor)
    c. BANANA (yellow)
    d. BLUE (x)
    e. (for all y) [(Buick) (y) → CAR (y)]
11. Using the rules from WPHELP, rewrite rule 1 by splitting it into two separate rules.

    IF functionality is simple and cost is low,
    THEN ______________________________

    IF functionality is simple and cost is high,
    THEN ______________________________
12. Using the definitions of variables given in the case study WPHELP, circle the correct classification, simple or complex.

DOCUMENT	SIMPLE	COMPLEX
a. Memo	S	C
b. Book of 1,000 pages	S	C
c. One-page business letter	S	C
d. Two-column newsletter with pictures	S	C
e. German to French dictionary	S	C

13. What is the difference between the procedural model of knowledge representation and the declarative model of knowledge representation?
14. What model of knowledge representation is used by VP-Expert?
15. Contrast the terms *data*, *information*, and *knowledge*.
16. Classify each of the following as data, information, or knowledge:
    a. brown ______________________________
    b. The special costs $2.49. ______________________________
    c. 47 degrees Fahrenheit ______________________________
    d. The outside temperature is 47 degrees Fahrenheit. ______________
    e. Because the temperature is 47, I will need a light jacket. ______________

Assignments

1. Using the WPHELP expert system, increase the possible values of functionality to:

   Simple

   Standard

   Complex

   Using the following chart, rewrite the production rules for the expert system:

FUNCTIONALITY	COST	WORD PROCESSOR
Simple	Low	A
Standard	Low	B
Complex	High	C

2. a. List five factors other than cost and functionality that could affect the selection of a word processor.

   b. Now transform these factors into heuristic (rule of thumb) IF-THEN rules.
3. Give five heuristic IF-THEN rules that save you time (in your job or at school).
4. List five heuristic IF-THEN rules for finding a job.
5. Itemize five heuristic rules for purchasing a used car.
6. Give five heuristic IF-THEN rules for excuses that a college student might use for late assignments.
7. Another good source of understanding knowledge representation is an article by B. Buchanan and E. Shortliffe, "Production Rules as a Representation in a Knowledge-Based Consultation System," *Artificial Intelligence*, February 1977. Read and report on this rule-based system.
8. Frame-based systems are used in MIS and finance. For a detailed description of a frame-based system, INVEST, for advising financial investments, read and report on the article "INVEST—An Expert System for Financial Investments" in the Summer 1988 issue of *IEEE Expert*.
9. Consult your periodical resources for articles on the four models of knowledge representation. Read and summarize the articles and share your findings with a classmate or friend.

   You will no doubt find articles about other models of knowledge representation. After you read and summarize each article, determine if the model described is different from each of the four models or simply an extension of one of them. If an extension, which model(s) is the extension based on?

Reasoning

As we have seen in Chapter 3, the knowledge possessed by an expert system is essential to the performance of the system in some given problem-solving situation. The knowledge must be organized and represented in such a way that it can be stored in the computer's memory and later retrieved as needed. The knowledge base, by itself, is not sufficient to solve problems. However, it is in a form that the expert system can use in the problem-solving process.

To use the knowledge it possesses, an expert system uses an inferencing mechanism to access the knowledge, reason about it, and draw conclusions or generate solutions. This can be done by using search and pattern-matching techniques. The inference mechanism searches the knowledge base, looking for appropriate knowledge to deal with the problem at hand. Pattern-matching techniques are used to match a desired goal or to determine if the criteria associated with a particular hypothesis can be satisfied.

In this chapter, we will discuss several important concepts related to reasoning. We will examine methods of reasoning, inferencing strategies, and the relationship between uncertainty and reasoning. Finally, we will discuss two reasoning control strategies, forward and backward chaining, which are the primary control strategies used in expert systems.

Methods of Reasoning

Search

We are all familiar with the concept of search. We search for answers to questions using books or databases and for objects that we need or have misplaced. Computers use search to locate data. Search implies moving around, looking at or listening to things, and making judgments about the pertinence of our findings to what we are trying to locate.

The search techniques we use as human beings are often systematic and predetermined. Often, however, our search techniques are rather haphazard, with the result that they are not very efficient.

Computers can be programmed with search techniques or algorithms that are quite efficient. In expert systems, one or more search algorithms are used on the knowledge base to locate knowledge to *potentially* solve the problem. The search algorithms do not solve the problem but provide the means for the expert system to find the solution.

Search is essentially a trial-and-error technique. Unlike an algorithm or formula, which solves a problem directly, it is more random and less organized in finding answers or solutions. This should not be surprising or troublesome because we, as human beings, use search techniques when we are solving problems. We search our memory for relevant knowledge that we need in a particular situation. Our search techniques are not always successful, and sometimes we must turn to other sources (people, books, databases, and so on) for help.

Like knowledge-representation techniques, search has been a major focus of work in the fields of AI and expert systems. A number of search methods have been used with knowledge bases in expert systems. Before we discuss some of these methods, let's examine some fundamental elements of problem solving involving the use of *states*.

The Essence of Problem Solving

The general purpose of problem solving is to move from a problem (initial) state to a goal (solution) state. This movement is done through the use of procedures or operations.

Problem states define the problem situation and the existing conditions. The initial problem state is a statement of the problem and preexisting conditions. Beginning with the initial state, a problem solver devises a plan for reaching the **goal state**. The plan may be a single procedure that allows direct movement from the initial state to the goal state. However, this is not usually the case, particularly in complex problems. In these situations the expert system will move through a hierarchy of **intermediate problem states** to the goal state; each problem state along the way represents a partial solution or a part of the goal state.

Let us illustrate these points using the game of tic-tac-toe. The initial state is the empty 3 × 3 grid. The preexisting conditions include two players wishing to play the game and some way (pen, pencil) for recording player moves on the grid. A goal state is represented by the grid form when one player has three Xs or three Os in a row; each player has a different goal state.

Because each player can make only one mark at a time, it is not possible to go directly from the initial state to a goal state. In this case, a series of intermediate problem states will occur after each player makes a mark. Note that the set of intermediate problem states will vary from game to game. The procedure for moving from one state to the next is that each player makes a mark.

The goal state possibly may not be reached with the devised plan. In the case of the tic-tac-toe game, the intermediate problem state can occur in which neither player can make a mark that will allow movement to the desired goal state. The alert player will alter the plan if the same problem situation occurs in the next game.

Revising a plan as movement is made from one intermediate problem state to another is often necessary. In our game example, each player may have devised an initial plan or strategy. However, after each turn, each player must examine the consequences of the last action and possibly revise the original plan. If a revision is made, it is usually based on several anticipated intermediate problem states closer to the desired goal state. The player may reason "If I make this move now and these moves the next two times, I can get into a position where my opponent cannot win."

Our discussion can be summarized with Figures 4.1 and 4.2. In Figure 4.1, there is movement from the initial state to the goal state through a series of intermediate states. The control strategy defines procedures (operations) that determine how the movement is made from state to state. From our tic-tac-toe example, the initial state was the empty grid, the goal state was a grid of either three Xs or three Os in a row, and the intermediate states were the grid form after each player had a turn. The **control strategy** defines how the game moves from one state to the next.

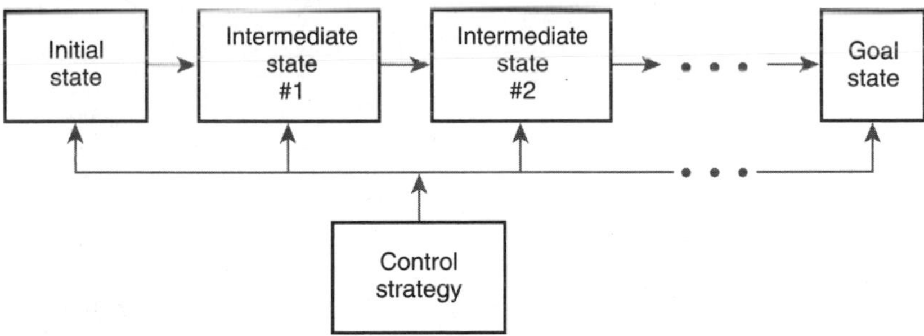

FIGURE 4.1 RELATIONSHIP AMONG INITIAL STATE, INTERMEDIATE STATES, GOAL STATE, AND CONTROL STRATEGY

Figure 4.2 illustrates the relationship between the control strategy and the operators. The control strategy calls on an operator (or operators) to move from one state to another. Information about the states is stored in the database; this information includes the initial state, goal, current state, and previous state transitions.

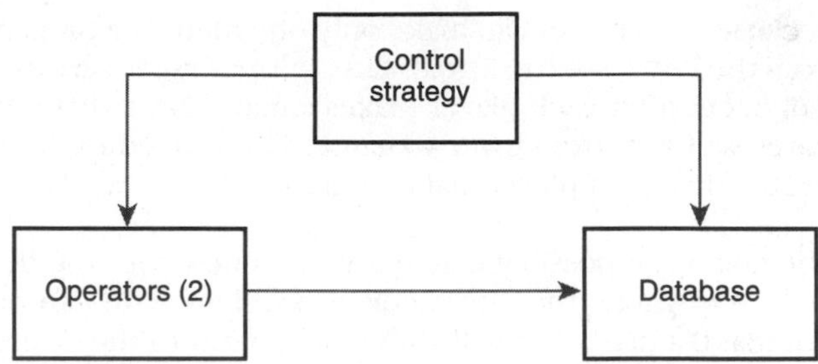

FIGURE 4.2 THE ESSENCE OF CONTROL STRATEGY

State and Search Space

To extend our discussion of problem solving, it is often helpful to use a graphical method to visualize the problem and possible solutions. This graph is sometimes known as a **state graph**.

Think of the state graph as a map of how to go from a starting point to a destination (goal). This map is illustrated in Figure 4.3.

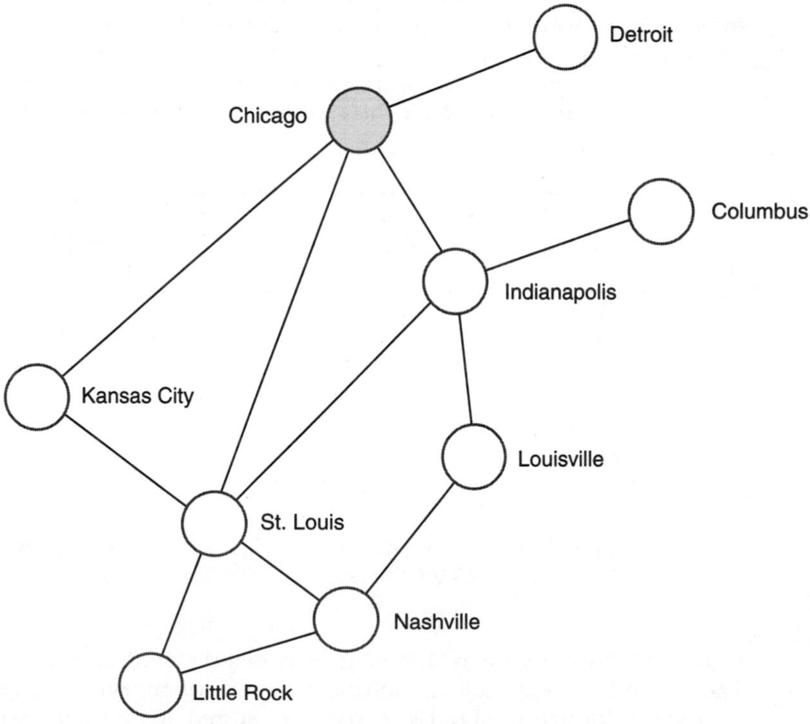

FIGURE 4.3 A STATE GRAPH

The cities in the graph are nodes that must be traveled through on the way from a starting point, say Chicago, to a destination point, say Nashville. There are several ways to travel between Chicago and Nashville, and the route will depend on several factors. If the lowest mileage is the criterion, then we would take one route. If stopping to see friends or relatives along the way is important, then we might choose another route. Of course, we may change our mind part way through a particular route; this change could be caused by road conditions or other constraints.

Relating this example to problem solving, the nodes represent the problem states (initial, goal, intermediate). The links between the nodes represent the transition (operator or procedure) for moving from one state to the next. Distance between states may be a constraint; the distance can be thought of as a measure of the complexity of interstate movement.

As we have pictured it, representing a state graph in computer memory is difficult. For this reason, we can think of this representation as conceptual or implicit, with its actual representation embedded in the computer language we are using.

The state space representation can be altered using a data structure known as a *tree* in general and as a *search tree* in particular. (There are other kinds of trees, but they are beyond the scope of this text.)

A **search tree** shows all possible paths from the initial state node to other nodes. Some of the paths may not lead to the goal node. A search tree for our travel example is shown in Figure 4.4. This tree illustrates several things of note; we will also use it to introduce some terminology about tree structures.

First, several nodes are repeated in the search tree because there may be more than one path through a particular node (city). For example, St. Louis appears on three different paths.

Note that not all paths lead to the goal (Nashville). One path leads to Detroit, one leads to Columbus, and several lead to Little Rock. Although these paths are not successful (they do not lead to the goal), they are nonetheless paths that have been defined and must be considered in a search process.

A third point to note is that the search tree is divided into levels, with one or more nodes on the same level. The reason for the use of levels will become clear in our later discussion of search methods. A final point to note is that the lengths of the links between the nodes is a unit distance, not actual distance. The choice of the cities (nodes) for each path is arbitrary; choosing additional cities along any path would change the appearance of the graph. The choice of cities is also arbitrary in that there are other possible nodes (cities) and paths from Chicago to Nashville. Figure 4.4 does illustrate the concept of a search tree. In an actual problem-solving situation, making an exhaustive list of possible intermediate states based on the given conditions would be important.

A number of terms are associated with trees. Many of the terms are familiar; we know them in the context of a family tree.

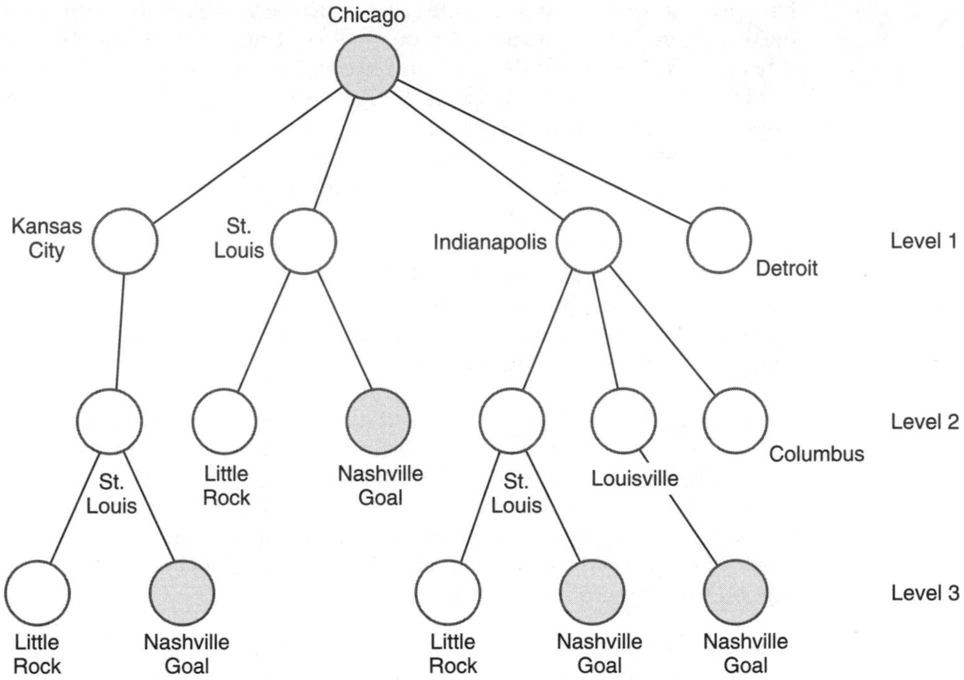

FIGURE 4.4 EXAMPLE OF A SEARCH TREE

The entry node for a tree is called the *root node*. In Figure 4.4, Chicago is the name of the root node. *Branches* (links) run from the root node to nodes on the next level; these nodes are known by several descriptors—child, successor, or descendant. For example, Indianapolis is a child of Chicago, Louisville is a child of Indianapolis, and Nashville is a child of Louisville. In the other direction, Louisville is a parent or antecedent of Nashville, Indianapolis is a parent of Louisville, and so on. The end of any path is called a *leaf node*. In our example, Detroit, Columbus, Nashville, and Little Rock are all leaf nodes.

It may seem strange that St. Louis is a child of both Chicago and Indianapolis. This occurs because of the nature of the search tree; it is not characteristic of trees in general.

Search Methods

There are basically two methods of search: *systematic* or *blind search* and *heuristic search*. Both methods are used in AI and expert systems.

Strictly speaking, there is another search method, known as *random search*. As human beings, we all have used this method, usually with less than spectacular results. Because of its inherent likelihood of failure and inefficiency, it is not a viable method for AI and expert systems work.

As we shall see, heuristic search is, in a sense, a specialized form of systematic search. It is systematic search that has the benefit of some sort of guidance.

Systematic search is a set of procedures used to search a state space. The procedures vary but they all have the common characteristic of searching the entire state space, or searching until the first occurrence of the goal node is found. These procedures are complete; that is, they do not miss a single node. At the same time, these procedures have inherent inefficiency. A great deal of time can be wasted before the goal node is found. However, even with these limitations, systematic search may be the only tool we have to use—and it is certainly better than a random search.

A number of systematic search algorithms have been developed for AI and expert systems work. We will examine four of them: depth-first, breadth-first, forward, and backward. The underlying approaches of these four are the basis for most of the others.

The first systematic search method is **depth-first search**. This method begins at the root node and proceeds down each path from the root to a leaf node, searching for the goal. When it reaches the end of the path with no success, the algorithm backtracks to an unsearched node and starts down the next path. This algorithm is illustrated by Figure 4.5.

The search begins at A and continues until it reaches the goal node I. At each node, the algorithm checks to see if it is the goal node. If the goal node is found, the search terminates. If the goal is not found, the algorithm proceeds along the same path or goes to the next path.

In Figure 4.5, the algorithm would move from A to B, B to C, over to D (B and A have already been determined to be unsuccessful), and then to E and F. Next G is searched and then H. Finally, I is found to be the goal and the search terminates. The algorithm does not research nodes previously found to be unsuccessful.

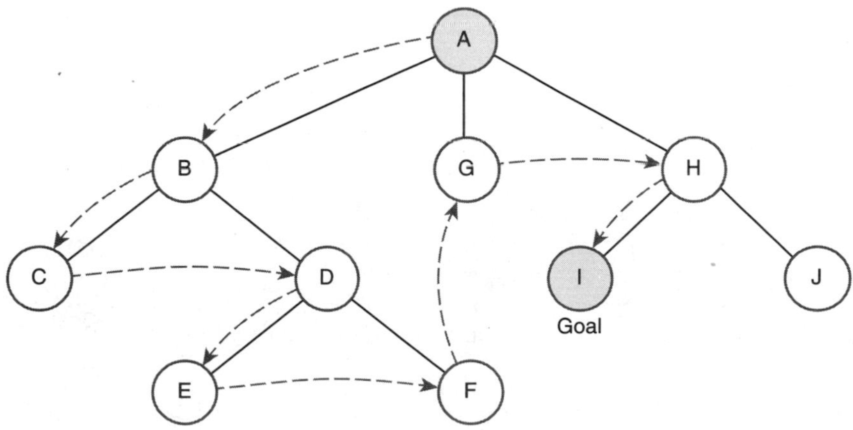

FIGURE 4.5 DEPTH-FIRST SEARCH

The amount of search needed directly depends on where the goal node is located. In general, the goal node location cannot be determined in advance, and so the algorithm cannot be made more efficient.

The second systematic search method is **breadth-first search**. As with depth-first search, this method also begins at the root node. However, instead of proceeding down to the end of the leftmost path, the algorithm proceeds to the first level of nodes below the root node and then searches all nodes at that level. It then proceeds to the next level and searches all nodes on that level. The process continues until the first occurrence of a goal node or until every node has been found to be unsuccessful.

The breadth-first search algorithm is illustrated in Figure 4.6. The search begins at A and then proceeds to B as before. At this point, the search proceeds to G and then to H. Having had no success (finding the goal node), the algorithm moves to the leftmost node on the next level, which is C. It then proceeds to D and then reaches I, which is the goal state.

As with depth-first search, breadth-first search is complete but relatively inefficient. The amount of search needed again depends on the location of the goal node, which is usually not predetermined.

There is really no good way to compare the performance of breadth-first search to that of depth-first search. Both methods work but suffer from such limitations as inefficiency and time.

Recall our earlier discussion of state space and how we moved from the initial state to the goal state. As we searched for a path to the goal state, we used a third systematic search technique known as **forward search** or **forward chaining**. We can think of it as moving forward from the initial state to the goal state; we could use the same search tree as in Figure 4.4.

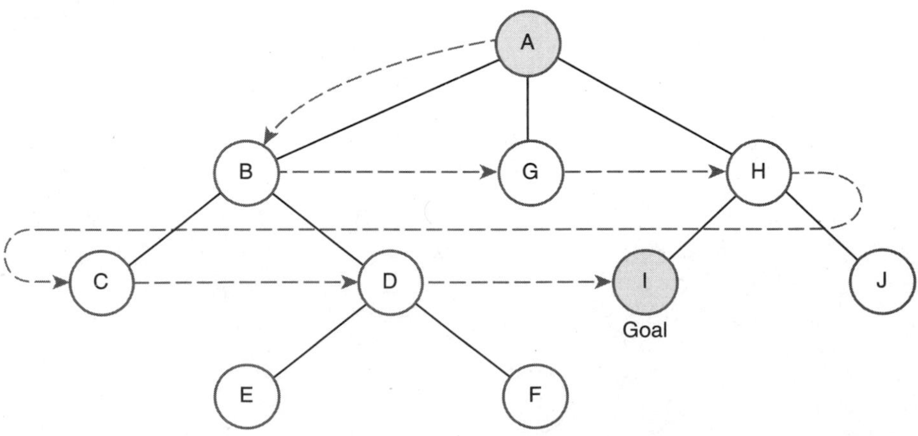

FIGURE 4.6 BREADTH-FIRST SEARCH

We also can search for the initial state by starting at the goal state. This technique is the fourth form of systematic search, called **backward search**. To illustrate this, we can redraw Figure 4.4 to reflect the movement from goal state to the initial state. **Backward chaining** is illustrated in Figure 4.7.

The search tree in Figure 4.7 also could be used with either the depth-first or breadth-first search methods. In this case, we are again searching from the root to a leaf. The difference is that the positions of the goal and initial states have been interchanged.

Note that a search can begin from any node in a search tree. It can then proceed forward (toward the goal state) or backward (toward the initial state).

We have seen that systematic search will locate the goal node if it exists. This type of search is sometimes referred to as *blind search* because the search process has no idea where the goal node is until it is located. As we saw before, this type of search technique can be very inefficient in terms of time.

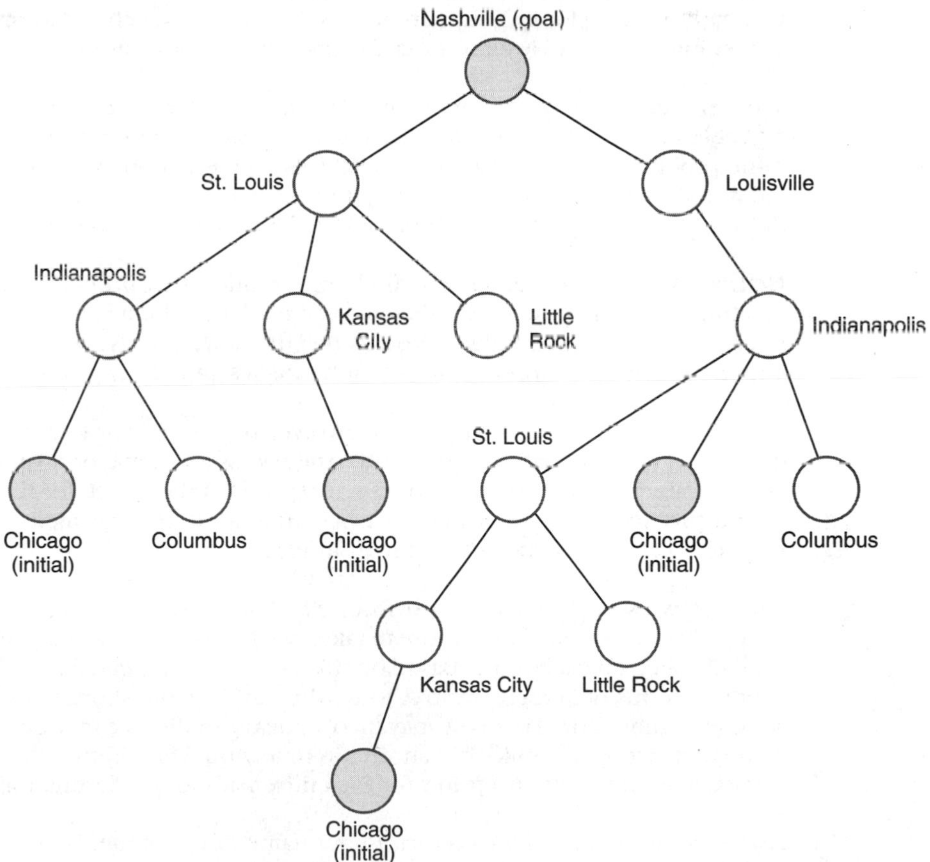

FIGURE 4.7 BACKWARD SEARCH FROM GOAL STATE TO INITIAL STATE

The problem of inefficiency is often referred to as "cost." We are all familiar with the notion of cost although we usually think of it in terms of money. However, cost can be measured in other ways, in terms of time or distance units, for example. The cost will be less if the solution is found in less time, or it will be less if we don't have to travel so far to reach the destination (goal).

Fortunately, there are ways to reduce the cost of search; we can classify these techniques as heuristics. Heuristics may be general in nature or associated with a specific problem domain.

Heuristics are guides, hints, or rules of thumb. That is, if we have several alternative paths to choose from in a search tree, a heuristic gives us a hint about which path may be most likely to lead to the goal node. It thereby makes our searching easier by taking less time (or whatever criteria) and causing perhaps even less frustration.

To illustrate the concept of heuristic, let us return to our childhood for a moment. Think of the times your parents would hide a little gift or surprise in the house and you had to search for it. Your parents would give hints such as "warmer" or "colder" as you moved either closer to or farther away from the treasure.

We can think of many times when we have used a heuristic to search for a solution to a problem. Often, the heuristic that we used was based on our experience with the same problem or one closely related. We also remember that the heuristic we chose did not always work. What we thought would work to solve the problem sometimes didn't, and we lost time and effort backtracking to try other techniques.

Heuristics are not a guarantee of finding a solution to a problem on the first try or locating the goal in the most efficient manner. In the long run, however, the use of heuristics is likely to produce desired results more quickly. And, as we learn from experience, we will improve our set of heuristics and add new ones.

The use of heuristics, in many cases, involves the use of some type of mathematical evaluation. The evaluation may involve the use of a formula or a value to determine the desirability of moving to the next state (node) in the search for the solution (goal) to the problem. This mathematical evaluation is often performed by an evaluation function. Let us illustrate with some examples.

Suppose we wish to travel by car from city A to city B. We have several alternative routes. In deciding on which route to take, we may study a road map and calculate the total distance for each route. Based on this evaluation, we can determine the shortest route, in terms of mileage, from A to B. Alternatively, the shortest route may require the use of substandard roads or may involve heavy traffic. We may decide that another route that is longer in miles but shorter in time would be a better choice. In this case, our evaluation was done a priori for the entire path to our destination.

As a second example, let us return to our game of tic-tac-toe. Is it possible to use an evaluation technique to determine our next move?

The answer is "yes." To do this, we can assign values to particular moves. We are interested in maximizing our chances of winning and minimizing our opponent's chances of winning. This technique is known as Minimax.

Suppose we are in the middle of a game and the current board configuration appears as shown in Figure 4.8; we can build a game tree of possible moves using this configuration as the root node for the tree. A partial game tree is shown in Figure 4.9. For purposes of our discussion, player A will represent our interests in the game and player B will represent our opponent's interests. Of course, when the game first began, player A could have plotted a strategy that would have won the game (three Xs in a row). But player B likely would not have cooperated; after two or three moves, player A might use a "wait and see" strategy. Then a reevaluation of the original strategy would have been necessary.

In Figure 4.9, node 1 represents A's move. Then the next level of nodes (2, 6) would represent B's possible move. The bottom level (nodes 3, 4, 5, 7, 8, 9) represent A's possible moves, and so on.

Suppose player A decides to use the following evaluation function.

$$100A + 10B + C - (100D + 10E + F)$$

Where

A = number of current rows with 3 Xs

B = number of current rows with 2 Xs

C = number of current rows with 1 X

D = number of current rows with 3 Os

E = number of current rows with 2 Os

F = number of current rows with 1 O

Note that the use of "row" may represent a row, column, or diagonal and that, to be counted, the rest of the row, column, or diagonal must be blank.

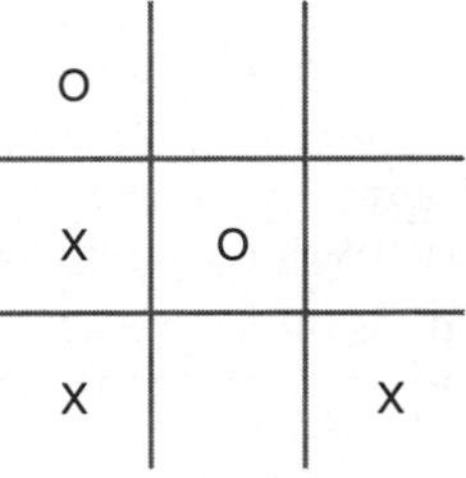

FIGURE 4.8 CURRENT TIC-TAC-TOE BOARD CONFIGURATION

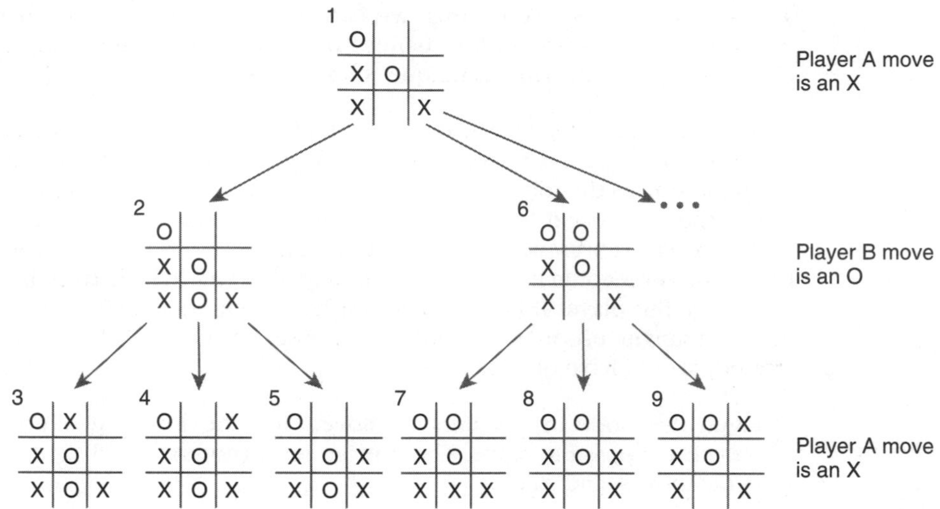

FIGURE 4.9 PARTIAL GAME TREE FOR TIC-TAC-TOE

Using the evaluation function, player A calculates that the value of nodes 3 through 5 are 1, 0, and -1, respectively, and the values of nodes 7 through 9 are 91, 0, and 10, respectively.

In analyzing the results, player A recognizes that the consequence of his or her move of node 1 has a maximum value of 1 considering node 3. Nodes 7 and 9 have higher values, but player B probably will never choose node 6; hence, the path to nodes 7 and 9 is blocked.

Reasoning Strategies

Now that we have examined the notion of search, we are ready to look at **reasoning strategies**. Whenever we think of reasoning, we think of logic.

Logic can be described as the scientific study of the process of reasoning and the system of rules and procedures used in the reasoning process. As such, it is of interest to us in our study of expert systems as they are used to solve problems. We now will examine the two basic types of reasoning strategies: deduction and induction.

Deduction

Deduction is the process of taking known general principles and applying them to specific instances to infer a conclusion. For example, we know that the interest on a loan is given by the product of the principal, interest rate, and duration of the loan.

We can deduce what our interest would be for a given principal, interest rate, and duration.

Not all general principles can be quantified or are widely accepted. We may hold an assertion or belief to be true that others might contradict. For example, we may assert that if our (body) temperature is over 100 degrees, then we will not go to work today. Someone else might not use this parameter.

To use deduction, we must be able to state the situation using *major* and *minor premises*. For example, in the situation just described, we have

> Major premise: IF my temperature is over 100 degrees, THEN I will not go to work today.
>
> Minor premise: My temperature is 100.6 degrees.
>
> Deduction: Therefore, I will not go to work today.

As a second example, suppose that our company will not ship parts unless they have been quality tested. Our process of deduction would appear as the following:

> Major premise: All parts must be quality tested.
>
> Minor premise: At least one part has not been tested.
>
> Deduction: The lot (of parts) cannot be shipped.

In both cases, the deduction (conclusion) is valid if both the major and minor premises are true.

A major premise can have several minor premises. In this case, all of the minor premises must be true to arrive at a valid deduction. This requirement is, unfortunately, a limitation of deductive reasoning. In real life, in a complex situation, not all minor premises may be known to be true with 100% accuracy. A valid deduction then cannot be made.

Induction

Like deduction, **induction** uses facts to infer conclusions. However, unlike deduction, the conclusions are not based on a general principle or assertion.

Conclusions are inferred only from known facts. For example, suppose we have the following facts (premises):

> Premise: Defective spark plugs cause poor engine performance.
>
> Premise: Defective alternators cause poor engine performance.
>
> Premise: Dirty crankcase oil causes poor engine performance.
>
> Conclusion: Defective spark plugs, defective alternators, and dirty crankcase oil are the causes of poor engine performance.

This example represents some interesting facts. First, induced conclusions may not be entirely accurate. We know that there are certainly other causes of poor engine performance. As more facts are added to the inferencing process, the conclusion will increase in accuracy. The conclusion will be most accurate if all possible premises are known to be true, which is usually impossible. Hence, inductive reasoning must deal with uncertainty in obtaining conclusions.

Second, the inferred conclusion is changeable. As new facts are known, the conclusion will change.

We human beings use inductive reasoning on a daily basis. From a few examples or facts, we infer generalizations about things and people. For example, we may think that a particular spreadsheet software package is the best one to use. This package was the first one we learned and we have used it for several years. We have tried other packages but none seems as good. Similar new packages may appear on the market, but we are convinced that none is better than ours; we won't even consider anything new.

As another example, when we were growing up, our parents always owned a certain brand of appliance or car. When it came time to buy our first appliance or car, we believed the brand our parents bought must have been the best so we purchased that brand.

Sometimes we are correct, but often we are not. There will, in general, always be a certain degree of uncertainty.

Abduction

A third type of inferencing strategy is **abduction**. Like deduction, it also makes use of major and minor premises. Recall from the prior example:

> Major premise: IF my temperature is over 100 degrees, THEN I will not go to work today.
>
> Minor premise: I did not go to work today.
>
> Conclusion: My temperature was over 100 degrees.

Unlike deduction, abduction is not a legal form of inference. In our example, there may have been other reasons why we did not go to work today (the car did not start; we needed a mental health day).

Even though abduction is not a valid inference technique, it is very useful and sometimes necessary. It is useful in diagnostic types of expert systems because it can generate possible explanations. It must be carefully used.

For example, suppose we know the following rule:

> IF the car is out of gas,
>
> THEN it will not start.

We find that the car will not start. We cannot definitely conclude that the car is out of gas. There may be some other explanation, but our abduction, that the car is out of gas, gives us a plausible explanation that we can easily check. Because there are a number of possible explanations, we have to start somewhere. Our abduction acts as a kind of heuristic; it gives us a place to start.

Uncertainty and Reasoning

Uncertainty is very much a part of our reasoning process whether we use deduction or abduction. In the case of deduction, it is often impossible to know all of the possible facts needed to reach a desired conclusion. We must decide at what point we are willing to make a deduction, short of complete information. As we have pointed out, a similar problem exists in the use of inductive and abductive reasoning.

Expert systems are designed to deal with incomplete or uncertain knowledge. The system may ask the user for input and receive the reply that he or she does not know the answer. The expert system must then take this lack of knowledge into account in its inferencing process. The system will still come up with an answer to the problem, but it may be qualified or may be different than anticipated.

Such is not the case with algorithmic systems. If data is incomplete or missing, the algorithmic system will come up with inaccurate results or no results at all. It cannot deal with incomplete or uncertain data.

Certainty Factors

Some expert systems handle uncertainty by using **certainty factors**, also known as **confidence factors**. These factors are a measure of the degree of uncertainty about the knowledge involved.

Certainty factors should not be confused with probability numbers. We will discuss this point in greater detail in Chapter 6.

Certainty factors reflect confidence or belief, not a probability. For example, what does it mean when the weatherperson says that the chance of snow is 50%? It does not mean that a probability of 0.5 has been assigned to the chance of snow. It does mean that, on days with similar weather conditions, it snowed 50% of the time. The weatherperson is 50% confident that it will snow, but this confidence is based on historical statistics.

To further illustrate this point, a set of certainty factors does not have to contain numbers between 0 and 1, although these numbers are certainly valid. Certainty factors could come from a set of, say, -100 to +100, where +100 refers to absolute belief and -100 refers to absolute disbelief. Numbers in between -100 and +100 refer to the relative degrees of believability.

Certainty factors are almost always assigned by experts, based on their intelligent judgment. Certainty factors can change as new knowledge is gained. But, at any point in time, they are static values.

Certainty factors can be associated with premises (IF) or conclusions (THEN). Here are some examples:

IF the car has a defective battery (CF = 0.6),

THEN the car will not start.

IF it rains today,

THEN the game will be canceled (CF = 0.8).

These rules establish what is known as a *threshold level*. That is, it is decided beforehand what CF value is needed by a rule to be fired (executed). The knowledge engineer (system developer) establishes this level in consultation with the expert(s). The threshold value will, of course, vary from rule to rule.

We also must consider the effect of threshold values when several rules are combined (chained) together. Techniques such as averaging are often used, but there is no general agreement among workers in the field as to which technique is best. It is also obvious that the validity of any threshold value will be tested during system usage.

On occasion, a rule will have multiple premises. The aggregate effect of the certainty factors assigned to each premise is not standard among expert system developers. For example, consider the rule:

IF A (CF = 0.6)

AND B (CF = 0.3)

AND C (CF = 0.8),

THEN D

What certainty factor should be assigned to D? Some would argue that the confidence is as good as the highest certainty factor of the premises, which is 0.8. Others would argue that D should not have a certainty factor greater than the lowest of the premises' certainty factors, which is 0.3. Still others would say that the certainty factor of D should be the average (0.57) or the median (0.6). In any case, the knowledge engineer and the expert(s) must make the decision.

A similar problem occurs when there is a chain of rules, as mentioned earlier. Again, there are no generally accepted guidelines, but note that if an error (relative inaccuracy) exists in one or more rules in the chain, the inaccuracy will be compounded as the various rules move through the chain.

Clearly, while the ability to deal with uncertainty is a major advantage of expert systems, it can also be a major disadvantage if not thoroughly considered by the knowledge engineer and the expert(s). For the system to maintain its credibility, the knowledge engineer and the expert(s) must study the alternatives and reach a defensible conclusion when dealing with uncertainty.

Fuzzy Logic

One of the more interesting techniques for dealing with the question of uncertainty is **fuzzy logic** (Zadeh 1988). Fuzzy logic is a way of dealing with ambiguous criteria or values.

For example, we may say, "the girl is tall." What do we mean by tall? Is 5 feet 6 inches tall? Is 6 feet 2 inches tall? A particular height is tall to one person but not to another. It depends on one's relative definition of tall.

Fuzzy logic represents an attempt to define such ambiguous terms as tall, short, heavy, and so on. It defines such a term as a range, with a probability associated with that range. We can then assign a probability that indicates whether tall is in the specified range. If ambiguous terms such as tall can be defined quantitatively, then they can be used more readily and reliably.

Representing ambiguous terms is very complex and difficult. For this reason, fuzzy logic currently is not widely used in AI and expert systems. This may or may not change as work in fuzzy logic continues.

Reasoning Control Strategies

Earlier in this chapter, we discussed the notion of a control strategy relative to the movement from one state to another in a problem-solving situation. This control strategy determines how a search is conducted as it attempts to find an occurrence of the goal state.

Expert systems make use of control strategies in their reasoning process. These strategies are used to search the knowledge base for the occurrence of facts and/or rules that can be used to obtain a problem solution.

The two most popular control strategies used in expert system development work are *forward chaining* and *backward chaining.* As we shall see in this section and in later chapters, both control strategies are very powerful and relatively easy to implement.

Forward Chaining

From our earlier discussions in this chapter, we learned two ways to deal with a search space for a given problem. One way is to consider given facts and move forward to reach a solution to the problem. Rules are used in this forward movement; the rules generate new facts that can then be used with other rules. This process is known as **forward chaining**.

To illustrate forward chaining, consider the following example of initially known facts and rules.

Initial Facts: A, B, C, D, E

Rules

1. IF A AND B THEN F
2. IF C AND D THEN G
3. IF F AND G THEN H
4. IF E AND H THEN I

Note several things about this example, as pictured in Figure 4.10. Rules 1 and 2 contain information about given facts (initial facts A, B, C, D, and E). Rule 3 deals with facts F and G, which are not initially known, while Rule 4 is concerned with one initially known fact (E) and one unknown fact (H).

Forward chaining works like this. Since we know A and B, Rule 1 allows us to know F, making F a known fact. Since we know C and D, Rule 2 allows us now to know G. Since we now know F and G, Rule 3 is used; now we know H. Since we now know H and we already know E (initial fact), we now know I from Rule 4. Let us suppose that I is the solution/conclusion to the problem; we need go no further. We have used a chain of rules and facts to find out what we wanted to know.

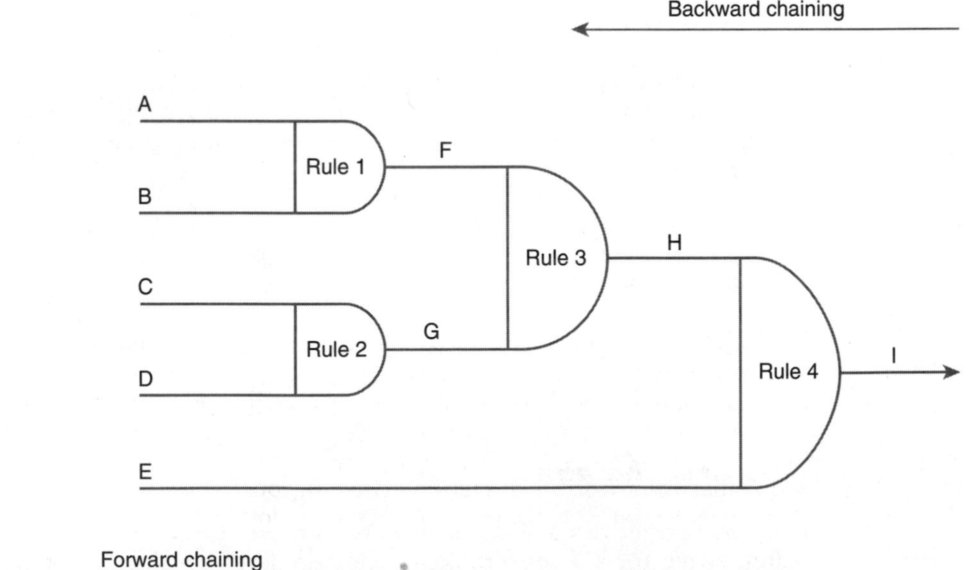

FIGURE 4.10 FORWARD AND BACKWARD CHAINING

In our example, the chain could be completed; this will not always be the case. If we had not known E initially, the chain would have been broken and we could not have known I because E could not be derived from any of the given rules. Just because we have a mechanism (chain) for determining I, we cannot guarantee that we can determine I, unless we somehow manage to find fact E or generate another rule that will have I as a conclusion but with different premises than Rule 4.

Backward Chaining

The other way we use facts and rules to obtain a conclusion (solution) is known as **backward chaining**. Here we move backward from what we finally want to know to see if we have the necessary initial facts.

Let's use our previous example of initially known facts and rules in Figure 4.10. We begin by supposing that we know I, which is our conclusion/solution. To know I, we need to know E and H through Rule 4. We initially know E. To know H, we need to know F and G by Rule 3. To know F, we need to know A and B by Rule 1, and to know G, we need to know C and D by Rule 2. As it happens, we know A, B, C, D, and E (initial facts). Hence, we can verify that we know I; we worked backward through the rules and facts.

Of course, the chaining could have broken down. Again, suppose we did not know E. Our chain would have stopped with Rule 4, and we could have gone no further. Given this set of rules, there was no way to know I.

Case Study: Reasoning Strategies for WPHELP

Let's incorporate some of the concepts from this chapter into our case study of WPHELP, the expert system for selecting a word processor. We will refine the fuzzy concept of functionality, use forward and backward chaining for deductive reasoning, and explore certainty factors for our rules.

The concept of simple or complex functionality is indeed a little nebulous; it needs to be refined. Just how do we know if a particular word processing document is simple or complex? To clarify the fuzzy notion of functionality, let's use a more precise and descriptive characterization of the types of documents that could be produced with a word processor.

We will assume that there are four types of basic classifications of word processing documents: Type 1, Type 2, Type 3, and Type 4. A Type 1 document is a simple sequential list of words like a letter, memo, or rough draft of a book chapter. A Type 2 document has all the features of a Type 1 document with the addition of page numbering, titles, headers, or footers. An example of a Type 2 document would be a report, term paper, or book chapter of text only. If a Type 2 document requires a structured layout, then we will classify it as a Type 3 document. Examples of Type 3 documents would include resumes, charts, and reports with multiple margins. Centering, italics, and form letters with insertion capability from a database would be in this category. A Type 3 document is labeled a Type 4 document whenever a typographical layout is needed. This implies the need for different-sized characters and different fonts (styles of lettering). Type 4 documents include newsletters, advertisements, and brochures. If we mix text and illustrations (graphics) in the same document, we might be using desktop publishing. Examples of each type of document and the characteristics of each type are summarized in Tables 4.1 and 4.2, respectively.

TABLE 4.1 EXAMPLES OF DOCUMENT TYPES

DOCUMENT	EXAMPLES
Type 1	Sequential text Letters, memos, rough drafts
Type 2	Paginated text, reports, term papers, book chapters
Type 3	Structured text, resumes, charts, documents with multiple margins, form letters with database insertion capabilities
Type 4	Typographical layout, newsletters, advertisements, brochures, reports with different text sizes and character types

TABLE 4.2 CHARACTERISTICS OF DOCUMENT TYPES

DOCUMENT	CHARACTERISTICS
Type 1	Simple, sequential text
Type 2	Page numbering, headers, footers, titles
Type 3	Multiple margins, centering, italics
Type 4	Different fonts, superscripts, integrated graphics, desktop publishing

Let's now expand the rules for WPHELP to include two new rules that link type to functionality. If the needed document is Type 1 or Type 2, then the functionality is simple. If the needed document is Type 3 or Type 4, then the functionality is complex. The rules for our expanded expert system with certainty factors between 0 and 1 (in parentheses) are given in Figure 4.11.

These five rules can be expressed as search trees, as shown in Figure 4.12, with the goal states (word processor A, B, or C) as the root and the initial states Type 1, Type 2, Type 3, Type 4, Low, and High as the leaves of the trees. Intermediate states are simple and complex.

If two child nodes from a parent node have an arc connecting the links, then the diagram is interpreted as an AND rule such as

IF child1 AND child2, THEN parent node.

If there is no arc connecting the links, then an OR is inferred as

IF child1 OR child2, THEN parent node.

For example, Rule 3 in Figure 4.12 is

IF complex AND high, THEN B or C.

Rule 1:	IF functionality is simple AND (cost is high or low) (0.90), THEN select word processor A.
Rule 2:	IF functionality is complex AND cost is high (0.90), THEN selecting a word processor is not possible.
Rule 3:	IF functionality is complex AND cost is high (0.95), THEN select word processor B or C.
Rule 4:	IF the document is Type 1 (0.85) OR Type 2 (0.90), THEN the functionality is simple.
Rule 5:	IF the document is Type 3 (0.90) OR Type 4 (0.95), THEN the functionality is complex.

FIGURE 4.11 EXPANDED RULES FOR WPHELP

An OR condition is expressed in Rule 5 with

IF Type 3 OR Type 4, THEN complex.

Even though "Not possible" is not a desired goal state, it is shown so that each rule is included in the tree.

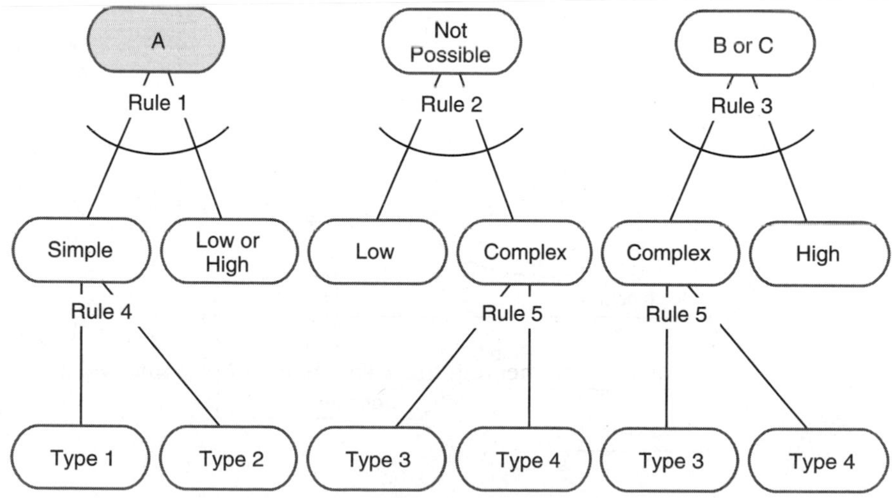

FIGURE 4.12 SEARCH TREE FOR WPHELP

In Figure 4.12, we can visualize forward and backward chaining as control strategies. If we start with facts (initial states) at the bottom of any of the trees and "climb" upward to its root, we use forward chaining. For example, if we know that the document is Type 2 and the cost is low, then Rule 4 would be true because of the OR and Rule 1 would be true because both premises (simple and low) are true. Hence, the goal, A, is reached. Forward chaining is *traveling toward a goal state*.

The opposite approach, backward chaining, can also be illustrated with Figure 4.12. Assume the goal state as A. For A to be true, both simple and low must be known to be true. Simple is true if the document is known to be Type 1 or Type 2. In this diagram with the goal at the top (root), backward chaining is searching down the tree, whereas forward chaining is searching up the tree.

These concepts also can be illustrated in the left-to-right format shown in Figure 4.13.

An AND is shown as a half-circle; OR is shown as a triangle. The inputs are to the left of the rule, and the output(s) are to the right. For example, Rule 4 would be interpreted as:

IF TYPE 1 OR TYPE 2, THEN the task is simple.

Initial states are on the far left, and goal states are on the far right. Hence, searching from left (initial state) to right (goal state) is reaching conclusions by forward chaining. Moving from the right (goal state) to the left (initial state) is backward chaining. Sometimes we are not able to move across the diagram; hence, a goal state

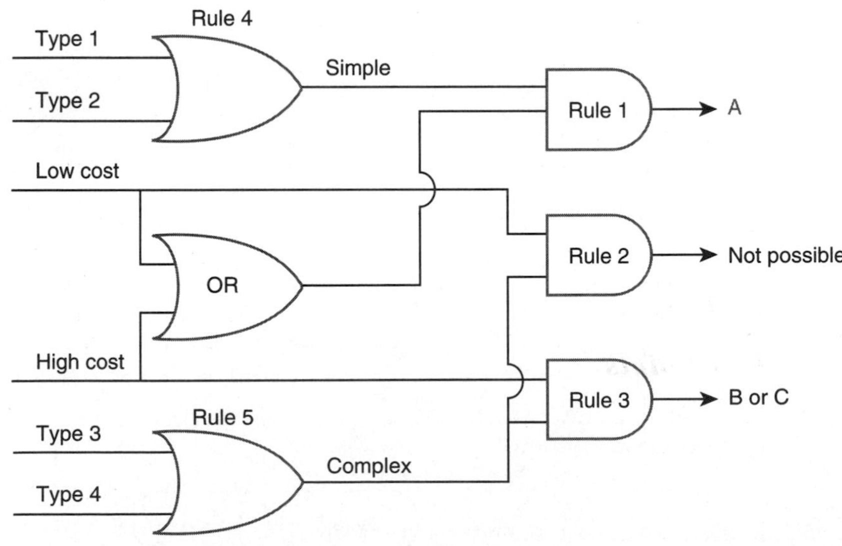

FIGURE 4.13 LEFT-TO-RIGHT DIAGRAM OF RULES FOR WPHELP

cannot be reached. Regardless of the visual orientation of the rules, always determine the goal state (THEN) from the initial state (IF).

Moving on to certainty factors (CF) with our expanded rules, Figure 4.14 shows the search trees for WPHELP with the certainty factors from the rules in Figure 4.11. The initial states have been assigned certainty factors between 0 and 1 (in parentheses). For example, Type 1 (0.85) means that we are 85% certain that the classification scheme for Type 1 documents is correct. The question now is, how do we determine certainty factors for the THEN part of rules? We will discuss this question in the case study in Chapter 6.

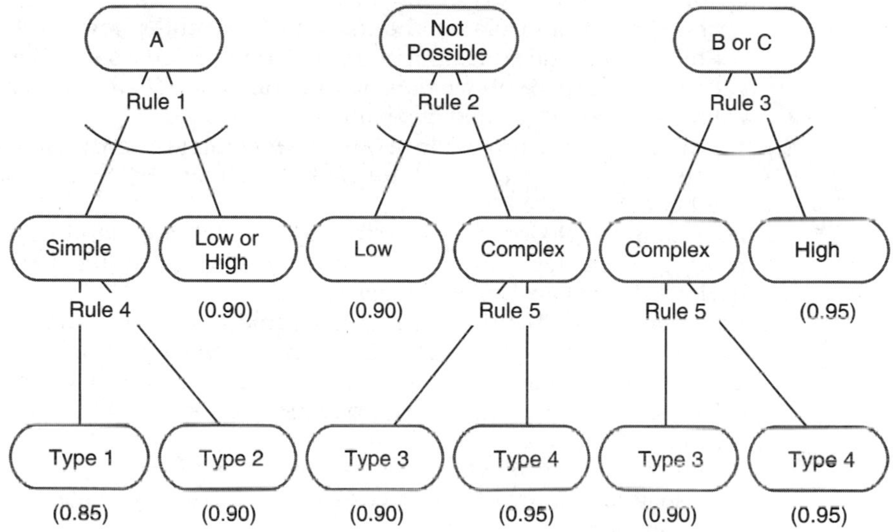

FIGURE 4.14 CERTAINTY FACTORS FOR WPHELP

In this section, we have refined the fuzzy concept of functionality, depicted forward/backward chaining in search trees and gate diagrams, and explored certainty factors for the WPHELP expert system. In Chapter 5, we will concentrate on the knowledge base instead of the reasoning mechanisms.

Key Points

- Search and pattern-matching techniques are needed to search the knowledge base of an expert system.
- The general purpose of problem solving is to move from a problem (initial) state to a goal (solution) state. This is accomplished by procedures or operations.
- In moving from the initial state to the goal state, an intermediate state can occur.
- The control strategy of an expert system determines how a move is made from one state to another state.

- Information about states (initial, goal, current, and previous) is stored in a database.
- A state graph is a visualization of the alternative moves from state (node) to state.
- The links in a state graph represent the transition (operator or procedures) for moving from one state to another.
- One way to represent a state graph (in a computer's memory) is to use a search tree structure.
  - Nodes may be repeated in a search tree.
  - Not all paths lead to the goal (node).
  - The search tree is divided into levels.
  - The length of the links between nodes does not reflect actual distances.
- Search-tree terminology includes:
  - Root node—The entry or top node
  - Branches—Links from the root node to nodes at the next level
  - Child—The node at the end of one link (also called a successor or descendant)
  - Parent—The node above the link to a child node (also called an antecedent)
  - Leaf—A node at the end of a path
- In search trees, the root node may represent either the goal state or the initial state.
- There are two types of search methods used in expert systems: systematic (blind) and heuristic (guided by hints).
- Four common algorithms for systematic searches are: depth-first (down from the root); breadth-first (across each level); forward (from initial state to goal state); backward (from goal state to initial state).
- An example of a heuristic search is a game tree for tic-tac-toe. Mathematical or evaluation functions can guide players in sequential moves.
- Three basic types of reasoning are deduction, induction, and abduction.
  - Deduction is applying known principles (major and minor premises) to specific instances to reach a conclusion.
  - Induction is inferring conclusions from known facts.
  - Abduction is a form of inference that is not legally valid but may generate possible explanations.
- Uncertainty of either premises or conclusions can be expressed as certainty or confidence factors.
- The threshold level of a rule is the predetermined certainty value assigned to the premise or conclusion in a rule.
- There is no general agreement on how to assign a certainty factor to a conclusion derived from multiple premises with certainty factors. Methods include:
  - Highest
  - Lowest
  - Average
  - Median
- Fuzzy logic is a method for dealing with ambiguous criteria or values.
- Knowledge engineers determine the search strategies, reasoning methods, and certainty factors in an expert system.
- The case study WPHELP was expanded to define functionality more precisely. The five rules were expressed in a search tree and a gate diagram. Reasoning strategies and certainty factors were discussed.

Exercises

1. What is the control strategy for moving from an initial start state (off) to the goal state (on) for turning on a light in your home?
2. Using Figure 4.7, name the root node, the level one nodes, the level two nodes, the level three nodes, the leaves of the tree, and the children of the Nashville node.
3. Using the tree in Figure 4.15, name the root node, the level one nodes, the level three nodes, the leaves of the tree, the children of node B, the parent of node I, and the successor(s) of node D.

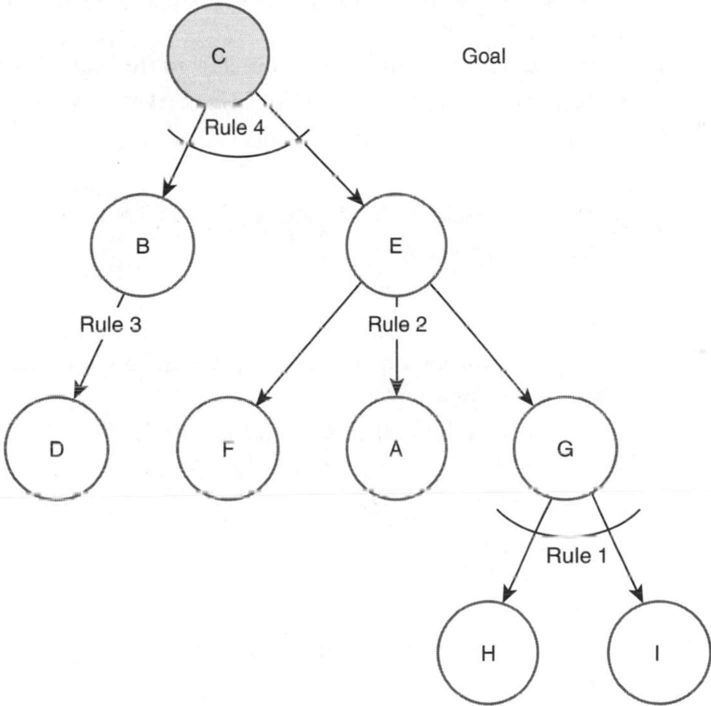

FIGURE 4.15 SEARCH TREE

From the same tree, list all the nodes after first a depth-first search, then a breadth-first search. If the goal state were C, which search method, depth-first or breadth-first, would be more efficient?

4. Given the search tree in Figure 4.15, write the IF-THEN rules using AND/OR. Assume that the goal state is the root node and that initial states are the leaves.

5. Using Figure 4.10, write the four IF-THEN rules expressed in the diagram.

   Rule #1: ______________________________

   Rule #2: ______________________________

   Rule #3: ______________________________

   Rule #4: ______________________________

6. What certainty factor (between 0 and 1) would you give if:
   a. Your car didn't start on a morning when the temperature is -40 degrees F?
   b. You call a friend on the telephone; it rings ten times and no one answers; hence your friend is not at home?
   c. You want to buy a microcomputer on a budget of $100?
7. Verify the computation on nodes 1 through 9 for the partial game tree for tic-tac-toe in Figure 4.9.
8. What are the deductive conclusions from the stated premises?
   a. Major premise: IF today is Tuesday, THEN tomorrow is Wednesday.

      Minor premise: Today is Tuesday.

      Conclusion: ______________________________
   b. Major premise: IF A = B AND C = D, THEN F < G.

      Minor premise: A = B

      C = D

      Conclusion: ______________________________
9. Given the information below, what would be your inductive or inferred guess for the next element in each list?
   a. Red, Green, Blue, Red, Green, Blue, Red, Green, ______________
   b. 1, 2, 3, 4, 5, 6, 7, 8, 9, 10, 11, 12, 13, ______________
   c. 1, 1, 2, 3, 5, 8, 13, 21, 34, ______________
10. What certainty factors between 0 (not possible) and 1 (certain) would you assign to the following:
    a. IF my car is orange,

       THEN the car will be hard to resell (CF = _____).

       Why? ______________________________
    b. IF Miami beats Notre Dame in next year's football game

       (CF= _____),

       THEN Miami will go to a bowl game.

       Why? ______________________________
    c. IF inflation increases next year,

       THEN interest rates on mortgages will rise (CF = _____).

       Why? ______________________________
11. What is the difference between systematic reasoning and hueristics?
12. What are the four methods of systematic reasoning? Briefly describe each.

13. What is the difference between a search tree and a game tree?
14. Contrast deduction, induction, and abduction.

Assignments

1. Complete the following search-tree problem. A microcomputer program controls a robotic arm that can reposition blocks X, Y, and Z one at a time. See Figure 4.16.
   a. Complete the search tree in Figure 4.16 for the robotic arm to move the blocks from the original state (root) to the goal state.
   b. Change the goal state to ZYX and redraw the search tree.
2. Convert the rules in Figure 4.17 to a search tree and a left-to-right diagram. The following summary may be helpful: The seven rules given in Figure 4.17 are for

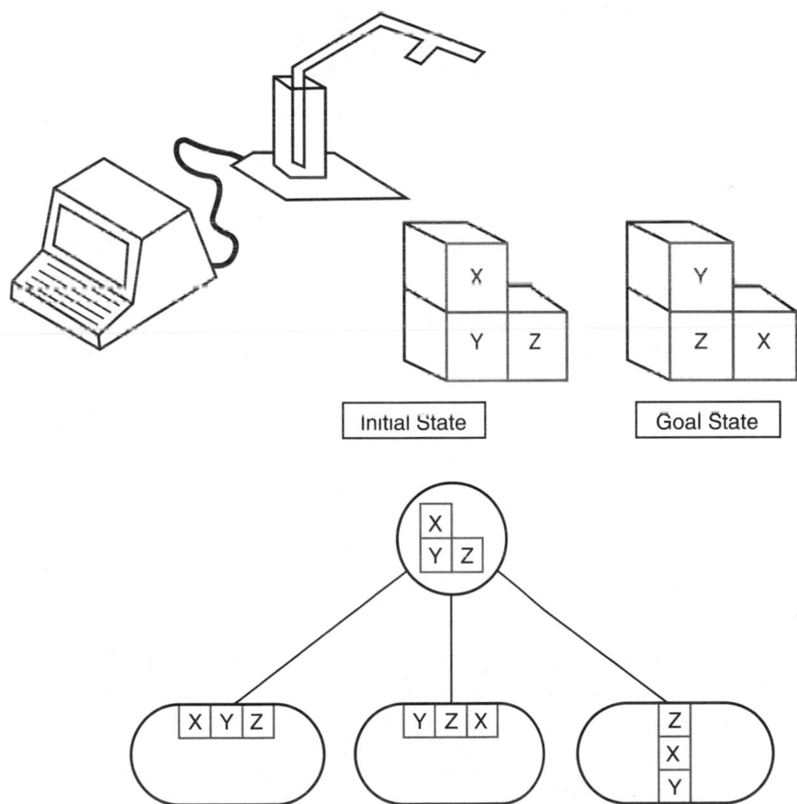

FIGURE 4.16 ROBOTIC-ARM PROBLEM

Rule 1:	IF status is married AND savings is satisfactory AND insurance is adequate, THEN action is to invest.
Rule 2:	IF status is single AND savings is satisfactory AND insurance is adequate, THEN action is to invest.
Rule 3:	IF savings is not satisfactory OR insurance is not adequate, THEN action is not to invest.
Rule 4:	IF status is married AND savings is six times monthly take-home wages or more, THEN savings is satisfactory.
Rule 5:	IF status is married AND savings is less than six times monthly take-home wages, THEN savings is not adequate.
Rule 6:	IF status is single AND savings is four times monthly take-home wages, THEN savings is satisfactory.
Rule 7:	IF status is single AND savings is less than four times monthly take-home wages, THEN savings is not adequate.

FIGURE 4.17 RULES FOR INVESTMENT EXPERT SYSTEM

advising a person to invest or not to invest; only three factors—insurance, marital status, and amount of savings—are considered. The rules utilize the following variables:

Status: Single or married

Savings: Satisfactory or unsatisfactory

Insurance: Adequate or inadequate

Action: Invest or not invest

The expert system must have the value of the saving accounts and the monthly wages to advise on whether to invest.

3. What is the action (goal state) from the expert system in Assignment 2 for each of the following (initial states)?

INITIAL STATE		GOAL STATE
a. Married Adequate insurance Monthly wages: $1,700	$10,000 in savings	__________

INITIAL STATE		GOAL STATE
b. Single Adequate insurance Monthly wages $1,700	$10,000 in savings	__________
c. Single Adequate insurance Monthly wages $4,000	$20,000 in savings	__________

4. Expand the rules in the investment system in Assignment 2 to be more precise about insurance. Assume that a single person needs insurance of at least five times his or her annual salary and a married person needs at least eight times his or her annual salary.
5. Using the investment expert system, do you think monthly wages (gross) or take-home wages should be used as predictors? Why?
6. List several additional factors that should be included in a comprehensive expert system advising a person to invest or not invest.
7. To obtain a more in-depth understanding of inductive reasoning, read Chapter 4, "A Theory and Methodology of Inductive Learning" from *Machine Learning* by R. S. Michalski (and others), Los Altos, CA: Morgan Kaufmann Publisher, Inc.
8. Using your library resources, look for books/articles on fuzzy logic as a way of dealing with uncertainty. Summarize your findings into a short paper (six to eight pages) that gives an overview of fuzzy logic and how it can be applied to expert systems technology.

5

The Knowledge Base

We have examined a number of general concepts associated with artificial intelligence and expert systems so far; we are now ready to discuss the components of an expert system in greater detail. They are the knowledge base, the inference engine, and the user interface.

In this chapter we will examine the creation and maintenance of the knowledge base. Whether a given problem is a suitable candidate for an expert system, the choice of an appropriate development tool, the role of the human expert and how his or her knowledge is to be acquired, and finally, the need for organizing the knowledge base will be explored.

Is the Problem Appropriate?

The question of whether a problem is **appropriate** for expert system development is a difficult one at best. No specific criteria can be applied to a problem that will answer the question. We discussed some generic types of problems that are probable candidates for expert systems development in Chapter 2. These classifications, however, are general; they give us a place to start but greater specificity is needed.

Basic questions must be answered before using an expert system in an organization. Waterman (1986) suggests that three questions need to be answered. Their role is illustrated in Figure 5.1.

1. Is expert systems development possible for the problem solution?
2. Is expert systems development justified for the problem solution?
3. Is expert systems development appropriate for the problem solution?

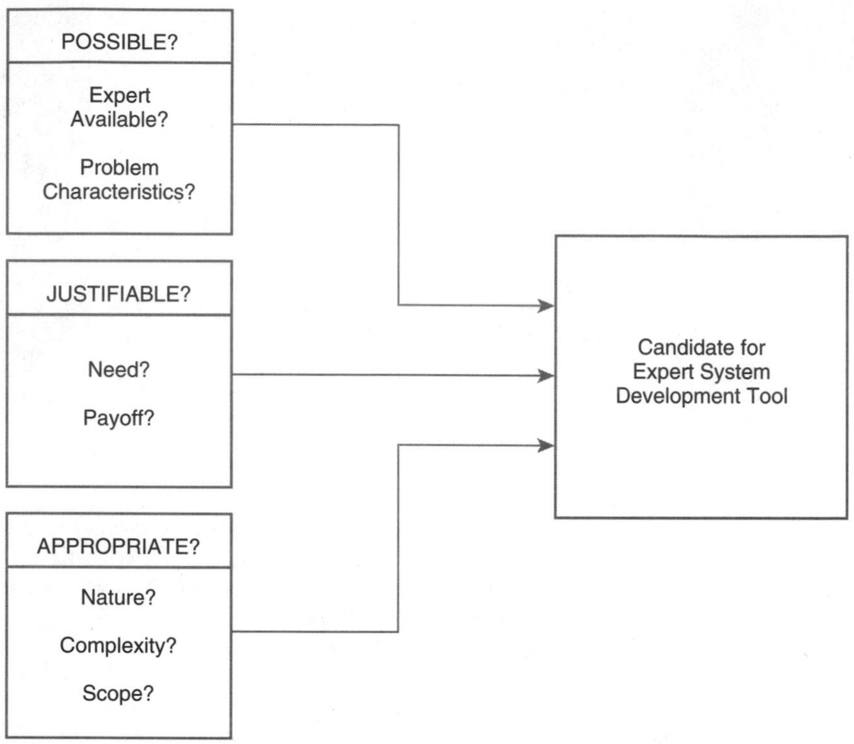

FIGURE 5.1 QUESTIONS RELATIVE TO APPROPRIATENESS OF THE PROBLEM

To determine whether a given problem is appropriate for expert system development, all three questions must be answered affirmatively. Each question can be answered by considering a series of subquestions.

Relative to the questions of *possible*, we need to consider both the expert who will serve as the model for the expert system and certain characteristics of the problem that make expert system development possible.

The *availability* of a true expert is an initial key consideration. If one does not exist, the question of the use of an expert system is a moot one. If such an expert does exist, does he or she have an interest in developing an expert system? Finally, do other true experts exist who disagree with the initially designated expert? If so, then the appeal of the given problem as a candidate for expert system development diminishes considerably, if not completely.

What about the *characteristics* of the problem? It must deal with the cognitive skills of the expert, not physical skills. If a problem deals with both cognitive and physical skills, then the cognitive part alone might be used in the expert system development. For example, the XCON/XSEL expert system configures VAX computer systems to meet specific customer orders. XCON/XSEL can advise how the computer system will be arranged, but it cannot go to the customer site and install the cables and wiring; these tasks must be completed by the human expert.

Some problems may be too difficult to consider. If an expert takes days, weeks, or even months to solve the problem, then developing an expert system for that problem is not possible.

Finally, problems that require on-the-job experience or common sense are not likely candidates for expert system development. However, this state of affairs may change in the future.

The second question deals with the problem of *justification* of using such resources as time, personnel, and money. Justification is always a major hurdle in any organizational structure for any type of project.

Justification usually involves two questions: Is there a need? and What is the payoff? *Need* is determined by how the expert system will be used or implemented. If the knowledge possessed by the expert exists in no other individual (or in very few individuals), then a need can be demonstrated on several grounds. First, retirement or even death could cause the loss of the expert's knowledge. If the expert's knowledge is needed in several places at the same time or in a physically dangerous environment (such as a fire or nuclear explosion), the development of an expert system can be justified. Multiple copies of a completed expert system can be made and distributed in these cases.

Two schools of thought address the *payoff* question. Some would argue that the largest possible payoff to the organization, such as saving the organization from demise, should be the determining factor. This is a noble goal but is also a big gamble. If the expert system development project succeeds, all is well; however, if it fails, the results can be very damaging, with considerable loss of time, personnel, and money.

The other school of thought takes a more conservative approach: build one or more expert systems that solve smaller problems. Although the total payoff (or individual system payoff) may be somewhat smaller, the resources committed by the organization will probably cost much less. A failure at this point will not cause as much damage. The choice of a large or small payoff depends on the attitudes of the organization's management.

We wish to note that the problem of justification is not unique to expert systems. The methods used in the justification process also apply to other proposed projects. For more details on methods of justification, we suggest that you consult a systems analysis text such as *Systems Analysis, Design, and Implementation* by John Burch (boyd & fraser publishing company, 1992).

The final question, related to the *appropriateness* of the problem, deals with the complexity and scope of the problem. Some aspects of the nature, complexity, and scope were considered in the basic question of whether using expert system development is possible.

The *nature* of an appropriate problem requires the use of symbolic manipulation and heuristics. If these are not required, the problem likely could be solved with a conventional algorithmic program; there would be no need to develop an expert system. Symbolic manipulation and heuristics must not always be used, but their use should be the primary focus needed to solve the problem.

Complexity addresses the issue of problem difficulty. A problem should not be too trivial. Even if a problem takes only a few seconds to solve, it still can be considered for expert system development. Finding an accurate way to make this judgment is difficult because equating complexity with the time it takes to reach a solution could be misleading. A possible heuristic (not to be taken too literally) might be, "If the problem appears too easy, then it may be about right."

To summarize, for a problem to be chosen for expert systems development, it must be possible, justifiable, and appropriate. If the determination process is hurried or cursory, the result is likely to be less than desirable.

Choosing the Development Tool

Once a given problem is deemed acceptable for expert system development, a **development tool** must be chosen. Again, a systematic approach is useful.

In Chapter 2, we classified expert system development tools as either high-level languages or shells. Following are some of the features or characteristics of each class.

High-Level Languages

With **high-level languages**, the knowledge engineer has several options. A number of languages, such as LISP, PROLOG, FORTRAN, and Pascal, are available. Most of these high-level languages are mature, and some of them have a number of support facilities, such as debuggers, graphics interfaces, and full-screen editors. These facilities are helpful to the programmer.

High-level programming languages have the advantage of flexibility. They are not constrained by the type of problem or by the desires of the knowledge engineer. They will support different forms of knowledge representation, although some languages are better suited to particular representations than others. For example, PROLOG is best suited to a logic form of knowledge representation.

Using high-level programming languages also has some disadvantages. For one, these languages require sufficient skill and understanding. The structures of both the knowledge base and the inference engine must be created from scratch, utilizing personnel, time, and money.

Shells

The class of tools known as **shells** has become increasingly popular recently. The shells provide an environment for creating an expert system that can be used by an inexperienced knowledge engineer. The shell requires only that the knowledge base be created; it provides the structure of the inference engine. A number of shells such as VP-Expert have been created with support facilities that make using the shell much easier.

Even with all of their popularity, shells have some disadvantages. The shell dictates the form of knowledge representation and the type of inferencing to be used. The human/machine interface is physically dictated, even though the knowledge engineer usually can control the content of the interface through available programming tools.

Shells are sometimes more expensive initially than high-level programming languages; however, expert systems developed with shells usually take less time to complete so that total development cost may be less. A particular shell usually can deal with only one type of problem, whereas high-level languages deal with a broader range of problems.

Choosing High-Level Languages or Shells

Which is better? There is no definitive answer. The choice depends on the application, the time allotted to the development of the expert system, and the skill level of the knowledge engineer. We will focus our attention on the actual choice of development tools in the remainder of this section.

Choosing a development tool is not always easy. The initial choice of a particular development tool may need to be changed because the tool does not do what the knowledge engineer anticipated. This may be a greater problem for an inexperienced knowledge engineer. Keep in mind that expert systems design often involves scrapping the project and starting over.

Even the experienced knowledge engineer faces problems in tool choice. Often the knowledge engineer develops a preference for one particular tool or set of tools, which may not be a good choice for the problem to be solved. The engineer may be tempted to alter the problem to fit the tool. This approach will often result in less-than-satisfactory performance of the finished expert system.

What are the appropriate considerations in the choice of an expert system development tool? Waterman (1986) suggests that there are six considerations; they are summarized in Figure 5.2.

The first of the considerations is *developmental constraints*. As we already pointed out, these constraints involve time, personnel, and money.

The constraint of development time can be a major one. If the knowledge engineer has a specified time period in which to complete the work, then a tool must be

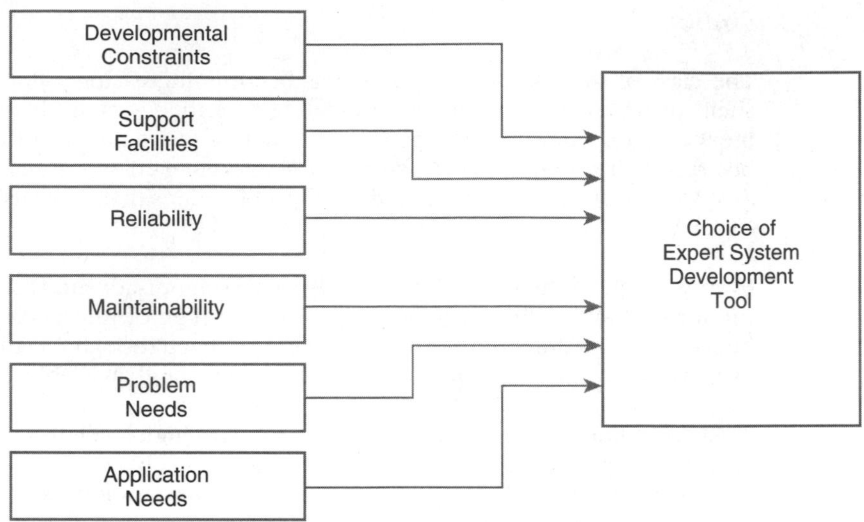

FIGURE 5.2 DEVELOPMENT TOOLS CONSIDERATIONS IN THE CHOICE OF AN EXPERT SYSTEM

chosen that can be easily learned and used (if not previously known to the system developer). The time constraint can lead the experienced engineer to choose the "preferred" tool—a potential problem.

Personnel can also be a constraint. Often a team approach is used in expert systems development because of the complexity of the problem. However, no linear relationship exists between the size of the team and the amount of time used to develop the expert system; for example, doubling the size of the team does not cut development time in half. The knowledge engineer/team leader must be wary of any offer by management to increase the number of team members (Waterman 1986).

Money is the third development constraint. The development costs for an expert system are high. The development of an expert system cannot be hurried; simply increasing the number of people in the development effort will not alleviate the time factor.

The cost of personnel is high. If an outside knowledge engineer is used, his or her services will be expensive. Good and experienced knowledge engineers are in short supply. If internal personnel are used, they often require training. If they are assigned to the expert system development team, they must be replaced in their usual positions within the organization.

Finally, the preliminary cost of development may be high. A high-level language may cost less initially, but it will require a skilled programmer whose services could be expensive. An expert system shell also is likely to have a high initial cost.

A second consideration is *support facilities*, which were described earlier in this section. An adequate set of support facilities will be a major asset to the development process.

Reliability is a third consideration. Beware of choosing a tool that is still in the experimental stage or is being revised, such as a Beta release. Companies that develop these tools may hurry to get the tool to the marketplace; testing and other tool construction work may sometimes be slighted. If possible, talk to other people who have used the tool; they can give you a great deal of information about the performance of the tool.

Maintainability is a fourth consideration. Be sure the manufacturer of the tool will be available to deal with technical questions and problems with the use of the tool after you have purchased it. If possible, deal only with companies that have an established record of product service.

A fifth consideration is that of *problem needs*, characteristics of the problem input and output. For example, what kind of data will be given to the expert system? Is the data easily accessible? The data might vary over time or be very costly to acquire. The output may be in the form of a summary report. Perhaps the problem must be broken into a series of subproblems whose solutions may or may not have an effect on the solution of the other subproblems. A clear understanding of the problem to be solved obviously is a consideration in the choice of a development tool.

The final consideration is that of *application needs*. The type of application and the end users the expert system is being designed for determine the application needs. We discussed a generic classification system for applications in Chapter 2; each application type has a set of needs, although these classes of problems are not mutually exclusive. Will the expert system be an assistant to the user, or will it perform in an autonomous fashion? Again, a clear understanding of the application is an important consideration in the choice of the expert system development tool.

The weight given to these various considerations will vary depending on the person(s) making the choice. This set of considerations, although not exhaustive, probably represents the majority of factors to be considered. Initial care in the choice of an expert system development tool for the given problem will influence the eventual success (or failure) of the development efforts. The choice process should not be taken lightly.

Acquiring and Organizing the Knowledge

The knowledge engineer's key tasks are the *acquisition* and *organization* of the knowledge. The knowledge engineer must aim to make the knowledge base as effective as possible.

In this section, we will examine some of the desirable characteristics of the knowledge engineer and several techniques used to acquire the knowledge. Finally, we will discuss how the knowledge engineer can organize the acquired knowledge for efficient use.

Acquiring the Knowledge

What kind of person makes a good knowledge engineer? Because most expert systems are built in cooperation with one or more human experts, the knowledge engineer must have good communication skills. He or she must be skilled in the arts of interviewing and listening and must be able to translate the knowledge gained into a representation that the computer can understand. In addition, the person must be thorough and well organized so that sessions with the experts are as productive as possible.

The Interaction Process. The key to the knowledge-acquisition process is the **interaction** between the knowledge engineer and the human expert. Figure 5.3 represents this interaction. The knowledge engineer queries the expert to try to capture his or her knowledge and to determine what strategies he or she uses to solve a given problem. The query/response process should be, ideally, a flow of the expert's knowledge and strategies with minimal or nonexistent gaps.

Prior to the initial session with the expert, the knowledge engineer must gain a basic understanding of both the expert's domain and the problem(s) to be solved. Experts often find expressing thoughts pertinent to subject domain difficult. Because an expert has a deep knowledge about the subject, he or she will make assumptions about the qualifications of the listener (knowledge engineer). An unprepared listener may find the expert's information meaningless, thus causing frustration during the session and wasting time as each tries to understand the other.

The knowledge-acquisition process usually involves more than just consulting an expert; it also may involve posing sample problems for solution and discussing their solutions. The knowledge engineer may alter the problem situations and pose them again to the expert. A representative sample of problems and the expert's solutions and analysis may aid the knowledge engineer.

The Interview Process. The **interview** is probably the technique most often used to acquire knowledge from the human expert. Good interviewing is an art, but the

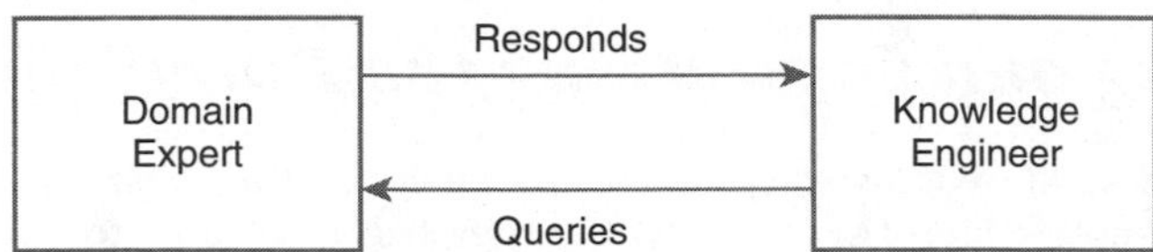

FIGURE 5.3 THE DOMAIN EXPERT/KNOWLEDGE ENGINEER CONNECTION

knowledge engineer can use some tested techniques to make the process more effective.

The first goal of the knowledge engineer is to understand the structure of the expert's thinking. Understanding this structure will allow the knowledge engineer to maximize the transfer of knowledge from the expert to the knowledge base.

The key to understanding the expert's thinking structure is to use *introspective modeling*, which involves examining portions of the expert's previous experience. These portions are made up of such things as sensations and sensuous memories (by sight, sound, and so on), thoughts (such as visual images or auditory dialogue), and feelings (emotions such as fear, anger, happiness). It is not enough to simply transfer knowledge. At any given moment, the expert's thinking can be considered a structure that can be explored. For example, we might ask the question: "What direction is Los Angeles from Chicago?" The response might come slowly or automatically. If the response was not immediate, the expert might have erected a visual image of a United States map, marked the approximate locations of both cities, and then answered the question. Knowing how the expert proceeded to answer the question may give the knowledge engineer insight.

To this end, the knowledge engineer should frame questions such as, "What was your thinking as you saw the accident?" or "Did you use past experiences to solve the problem?". If confused, the knowledge engineer must review the expert's responses until they become clear; if the knowledge engineer has prepared prior to meeting the expert and understands the expert's thinking process, confusion is less likely to occur. The knowledge engineer must also be objective as he or she elicits knowledge from the expert; he or she must not let personal biases affect what the expert has said.

The interview process is based on verbal communication between the expert and the knowledge engineer. With verbal communication comes a constant risk that words do not communicate the message. To minimize this danger, we suggest that the knowledge engineer gain a thorough understanding of the expert's thought processes and an initial understanding of the problem content.

The knowledge engineer also must be alert to the ambiguities of language. A statement by the expert, such as, "This is the key to the problem," may indicate gaps in the knowledge that the expert can bridge but that the knowledge engineer may not see; these gaps must be filled. The expert may use comparative words; for example, the expert may proclaim, "This is a better system." The knowledge engineer should then respond, "Better than what?"

Some words incorrectly imply universality. A statement such as, "This technique will never work" is an example. The engineer may want to know, "Could a major modification be made so that it will work?"

Some words have unspecified references. An example of this difficulty is a statement such as, "This component forms a protective shield." Again, the knowledge engineer's response is, "Which component is this?"

Words can also imply causal relationships that do not exist. One example is this statement: "If we continue to run this much water through the system we could have some expensive repair bills."

To summarize, the knowledge engineer must gain an understanding of the expert's thinking process. This is not an easy task, but careful questioning techniques to elicit maximum knowledge and introspective modeling can be very useful. Finally, the knowledge engineer must be continually aware of the imprecise nature of language in verbal communications.

Organizing the Knowledge

Periodically throughout the knowledge-acquisition process, the knowledge engineer must be concerned with how to organize the knowledge in a form suitable for the knowledge base. The process of **organizing the knowledge** involves deciding how to represent the knowledge, verifying the knowledge, and validating the inference chain. We will examine each of these areas in this section.

Representing the Knowledge. The acquired knowledge must be converted into a form that the expert system can understand. In Chapter 3, we discussed several methods for **representing knowledge**.

The choice of the method of representation depends on the application for which the expert system is designed and the choice of the system development tool. Recall that procedural knowledge is best represented by production rules and that declarative knowledge is best represented by semantic nets, frames, or predicate calculus. Many applications have a mixture of these two types of knowledge; consequently, a choice of representation is not hard and fast. Many applications tend to be procedural in nature, and so the use of rules has been predominant. Rules have also been favored because they are easy to create and interpret.

The particular development tool to be used also has a pronounced effect on the elected knowledge representation model. Ideally, the representation model should be chosen first, then the tool. For a variety of reasons, this sequence is not always the case.

If a high-level programming language is going to be used, any of the four models of knowledge representation discussed in Chapter 3 could be used. Some languages lend themselves to specific representation models; for example, PROLOG, works well with predicate calculus.

Most expert system shells use rules as knowledge representation. Some people view this as a limiting factor in the use of a shell, but there is little question about shells' value and impact. A number of significant expert systems, such as COMPASS (Goxal et al. 1985) and LITHO (Bonnet and Dahen 1986), have been built with shells; this will continue to be the case. At this point in the evolution of expert system research and development, it is not generally feasible to mix different knowledge representation models in the same expert system. The only exception to this statement has been the organization of the rule-based system into a frame representation to make search techniques more efficient.

Verification. The importance of **verification** cannot be overemphasized. It is the process of ensuring the accuracy of the acquired knowledge; it gives credibility to the expert system's knowledge.

Verification must be performed in a systematic and meticulous fashion; it is a determining factor in communication between the knowledge engineer and the human expert.

As the knowledge is acquired, the knowledge engineer must organize it so that meaning can be derived from it. The expert must verify the accuracy of the derived meaning.

Verification is best achieved periodically. If inaccuracies occur, they can be corrected before they become a major problem.

The verification process also can assist in locating gaps in the knowledge base. It is relatively easy for these gaps to occur, so the knowledge engineer must be sensitive to this potential problem.

Validation. Equally important to the verification process is the **validation** process. Validation is used to check the accuracy of inference chains that may be used on the knowledge base; the question is, "Does the expert system give appropriate and correct results?"

Validation also is best accomplished at the beginning of the development process and on a periodic basis. The validation process is most effective when test problems are submitted to the expert for solution. The results should be valid and consistent. Validation also can assist in locating gaps in the knowledge base. If the inference chain produces invalid results (or no results at all), the validation process can assist the knowledge engineer in locating and repairing troublesome spots.

Using the validation process when creating the knowledge base will keep small difficulties from becoming large ones. The result should be an expert system that is much more likely to perform to predetermined specifications.

Updating and Maintaining the Knowledge Base

Once created, no knowledge base should be considered static. As new knowledge becomes available, the knowledge base will need to be updated. Further, new problems may be posed to the knowledge base that use the new knowledge, existing knowledge, or a combination of the two.

The process of adding new knowledge or correcting existing knowledge in the expert system must be carefully monitored to maintain verification and validation standards. Failure to do so may result in a performance level that is less than acceptable.

If a problem is encountered because of the addition to or a change in the knowledge base, the "quick-fix" method, although tempting, may create an even larger problem. The system must be thoroughly tested during the updating process to ensure that changes are implemented without degradation of system performance.

Maintenance of the knowledge base is also an important consideration. No matter how thoroughly the knowledge base has been tested, errors will occur. As new knowledge is added or changed, the system's performance may degrade because of a lack of computer memory or resulting inefficiency in the search techniques used on the knowledge base. In a rare situation, the knowledge base may need to be redesigned to correct these complex problems. However, the point is that the knowledge base will always need some maintenance services.

The knowledge engineer is the one who creates the knowledge base, but the responsibility for updating and maintaining the knowledge base is often delegated to an individual or group of individuals within the organization. As the reins of control pass from the knowledge engineer to others, appropriate and adequate documentation also passes. We will say more about documentation in Chapter 8.

Case Study: WPHELP

To this point, we have concentrated on details and the development of rules for our example expert system. The user will input the data the WPHELP system will use in choosing an appropriate word processing package.

Data to be used by the knowledge base can be input from a database or a spreadsheet. Large volumes of data can be input efficiently this way without requiring the user's time. We will discuss these possibilities, but we will not attempt to write any rules for WPHELP.

There are several advantages of using a database in conjunction with an expert system. The database can provide a great deal of data in a very short period of time. Consequently, an expert system with many rules can perform its function in a relatively short period of time. In addition, the user is freed from the tedious and time-consuming task of entering large amounts of data. Input from the database guarantees the accuracy of data, whereas the user may make errors when entering the data manually.

Using the database does not necessarily eliminate the problem of dealing with uncertainty (incomplete or missing data) for the expert system. The database itself may have incomplete or missing data. Updating the database, which is especially important when dealing with fast-changing data, is also a problem.

WPHELP might use a database in several ways. First, considerable word processing product information might be contained in the database. The database would allow the use of a larger number of rules to better assist in the final product choice. For example, the cost variable could have a number of values other than low and high.

Spreadsheets can also be used effectively with an expert system. The accounting information in a spreadsheet can be used to make financial decisions in which a number of calculations are needed as new data becomes available. As with databases, spreadsheets can provide large amounts of data in an efficient manner and free the user from data entry. They also provide the opportunity to use the what-if capability that exists in most expert systems. The what-if facility allows the user to see what would happen if the figures in the spreadsheet were changed. As with databases, spreadsheets also may have incomplete or missing data, but the expert system can deal with this problem.

WPHELP could use a spreadsheet in its choice of a word processor. Budget information about the cost of the package and its effect on the business could be stored in the spreadsheet. Also, if the word processing applications for the organization need special equipment such as monitors, screens, or printers, the financial effect of obtaining the equipment could be analyzed; amortization and depreciation of the equipment as well as employee training costs could be determining factors in the final product choice.

Key Points

- Determining whether a given problem is appropriate for expert system development is difficult because there are no established criteria.
- Problems are unique and often complex so that only one or more parts of the problem may be candidates for expert system development.
- Although not all expert systems are developed in an organizational setting, those that are must have access to organizational resources.
- For a problem to be a candidate for expert systems development, its solution must be possible, justifiable, and appropriate.
- For a problem to be considered for expert system development:
  - Genuine expertise must exist.
  - The problem must involve only cognitive skills.
  - The problem cannot be too difficult to solve in terms of time.
  - The problem cannot require on-the-job training or common sense to obtain a solution.
- To justify expert system development:
  - There must be a need within the organization.
  - There must be a recognized and appropriate payoff to the organization.
- For expert system development to be appropriate:
  - The problem should require the use of symbolic manipulation and heuristics.
  - The problem must not be too difficult to solve.
- High-level programming languages are good expert system development tools because they are flexible. However, they require a great deal of programming skill

and are generally difficult for an inexperienced knowledge engineer to use. Updating and maintaining WPHELP is further complicated by new software developments. The emergence and acceptance of the Microsoft Windows environment complicates choosing a word processor.

- Shells as expert systems development tools are easier to use than high-level languages and are appropriate for inexperienced knowledge engineers. However, the shells are less flexible and usually dictate the knowledge representation, the inferencing strategies, and the format of the user interface.
- The choice of an expert system development tool should be based on the following:
  - Developmental constraints
  - Support facilities
  - Reliability
  - Maintainability
  - Problem needs
  - Application needs
- The key tasks for the knowledge engineer in creating a knowledge base are the acquisition and organization of the knowledge. A good knowledge engineer must:
  - Have good communication skills.
  - Be skilled in the arts of interviewing and listening.
  - Be thorough and organized.
  - Be able to interpret the acquired knowledge and convert it into the appropriate knowledge representation.
  - Acquire adequate background knowledge before consulting with the expert.
- The key to the knowledge-acquisition process is the interaction between the knowledge engineer and the human expert.
- The first goal for the knowledge engineer in the interview process is to understand the structure of the expert's thinking.
- Introspective modeling may be very helpful to the knowledge engineer in understanding the expert's thinking process.
- Verbal communication between the human expert and the knowledge engineer must be thorough and meaningful.
- The knowledge engineer must be continually aware of the imprecise nature of language during verbal communication in the interview process.
- Organizing the knowledge involves decisions about how the knowledge is to be represented, verification of the knowledge, and validation of the inference chain.
- Updating and maintaining the knowledge base are very important to the continued successful performance of the expert system. These tasks are often performed by someone other than the knowledge engineer; this means that appropriate and adequate documentation about the expert system must be provided by the knowledge engineer.

Exercises

1. What factors are involved in determining whether a problem is appropriate for expert systems development?

2. What factors must be considered in the choice of an expert system development tool?
3. What difficulties can occur during the interaction between the human expert and the knowledge engineer?
4. What are the different techniques for acquiring knowledge? Give one advantage of and one potential problem with each technique. How could the potential problem be solved or avoided?
5. What is meant by introspective modeling? How does its use aid in the interview process?
6. What is meant by verification of the knowledge base? What is meant by validation of the inference chain? Why are these two tasks important to the expert system?
7. What is meant by the art of listening? Why is listening important to a knowledge engineer?
8. What potential problems are caused by the processes of updating and maintaining the knowledge base? How can these problems be solved?
9. Contrast the use of high-level programming languages and shells as expert system development tools.
10. What factors must be considered in the choice of a development tool? Explain what each factor means.
11. Describe the interaction process between the knowledge engineer and the human expert.
12. What are the characteristics of a good knowledge engineer?
13. Why is documentation of the expert system important?

Assignments

1. We have said that the knowledge engineer must have good communication skills. It also has been said that college graduates entering the business world must have good communication skills. Are these two sets of communication skills the same? How are they different?

   To assist you in developing your answer, ask three people in your local business community who are responsible for hiring college graduates. Use their opinions to form a list of good communication skills. What skills seem to be most important? Why?

   After you have developed the list of communication skills, look at the list of courses in your department's curriculum. Where is emphasis placed on the development of each of the skills? Are some skills neglected? If so, how might they be included?
2. Talk to a top-level management person at one or more of the local businesses about problems the business is trying to solve. Make a list of the problems and determine if each problem is a candidate for expert systems development. If a problem is not a candidate, state why.

3. Identify at least one problem on your campus that could be solved with the aid of an expert system.
4. Find product descriptions for three rule-based expert system development tools. Good sources of information for these descriptions are magazines such as *Byte*, *PC AI*, and *AI Expert*. How do the products seem to compare (from the descriptions) to the criteria for selecting a tool? What appear to be the strengths and limitations of each shell? You may choose your desired application. If necessary, write to the company for further information using your school or company letterhead. See also Assignment 4 in Chapter 6 and Assignment 1 in Chapter 8.
5. The purpose of this assignment is to build a small knowledge base using rules. In addition to writing the rules, show the knowledge base as either a search tree or a left-to-right diagram; the format for both of these illustrations was discussed in Chapter 4. Do not assign certainty factors. Use the criteria specified in the description of each problem. What is your source of expertise for each problem?
   a. The problem is to find student or personal housing for the next term. Factors to be considered include:
      1. distance from the main classroom building or from your job
      2. type of housing
      3. number of people

      Add any other factors you would like to consider.
   b. The problem is to determine what you are going to do for your next vacation. Factors to be considered include:
      1. when
      2. number of days
      3. cost
      4. main activity during the vacation
      5. number of people

      Add any other factors you would like to consider.
   c. The problem is to pick an ideal partner for a date. Factors to be considered include:
      1. type of activity
      2. cost
      3. hobbies
      4. personality
      5. likes and dislikes

      Add any other factors you wish to consider.
   d. The problem is to decide what kind of car to purchase. Factors to be considered include:
      1. gas mileage
      2. mileage from school or work to home

3. maximum number of passengers
4. cost

Add any other factors you wish to consider.

6. A number of approaches to the interviewing process have been developed and discussed in books on systems analysis and design. What are the major techniques? What are the characteristics of each? Could these techniques be used by a knowledge engineer in expert systems development? Why?
7. Find and summarize articles about the knowledge acquisition process. What techniques are being used? What are the characteristics of the automated methods? Do the automated methods show promise of acceptance and use?
8. What are the characteristics/features of the VP-Expert knowledge base? How is acquired knowledge represented?

6

The Inference Engine

In Chapter 5, we examined in some detail the process of creating the knowledge base. We now will turn our attention to issues involved in creating the inference engine. Our discussion here expands on the topics presented in Chapter 4. We will deal with the two areas with which most, if not all, expert systems must grapple: uncertainty and inexact reasoning because of the nature of the knowledge in the knowledge base.

Uncertainty

Uncertainty results from a lack of adequate information needed to make a decision. Uncertainty is a problem because it may prevent us from making the best decision; at worst, it can cause us to make a bad decision. In the business world, uncertainty may mean financial loss or less-than-appropriate profit.

As human beings, we have become experts, to some extent, in dealing with uncertainty. If we could not do so, we would not survive. For example, we are confronted with uncertainty about job, school, weather, traffic, and so on. We become experts at maximizing our performance by learning how to select easy classes, what to do if the weather is inclement, and how to drive under adverse traffic conditions. To deal with uncertainty, we must reason under uncertainty.

Recall our discussion in Chapter 4 of methods of reasoning. The **deductive** method of reasoning is also referred to as **exact reasoning** because it deals with exact facts and exact conclusions. Some expert system applications permit the use of exact reasoning, making it relatively easy to find the best conclusion.

Unfortunately, when some of the data is not exact, the problem of finding the best conclusion becomes more complex. The number of possible conclusions increases and the additional task of determining which conclusion is best arises. Several methods for dealing with this task have been developed, and it is up to the expert system designer to determine which method is appropriate for the given application.

Sources of Uncertainty

The knowledge we have may seem uncertain for a number of reasons. Sometimes the sources of uncertainty can be easily eliminated, but many times they cannot be resolved; then we must live with them. Clearly, we want our expert system to be able to deal as effectively as possible with uncertainty; this feature must be taken into account in the design of the inference engine.

Sources of errors that create uncertainty in knowledge can be classified as follows (Giarratano and Riley 1989):

1. Ambiguity
2. Incompleteness
3. Incorrectness
4. Measurement
5. Systematic errors
6. Random errors
7. Reasoning process

Ambiguity errors occur when a statement can be interpreted in more than one way. For example, the statement *Sell the bad stock* is ambiguous. Which company's stock should be sold?

Incompleteness errors occur when a piece of knowledge is missing. The statement *Turn the valve* is incomplete. Which way should the valve be turned?

Errors of **incorrectness** occur when a piece of knowledge is wrong due to human error or equipment malfunction. It also could occur in the case of a hypothesis that is an assertion or the consequent of a rule. An example of a hypothesis is *This key will open the door*. One type of incorrect knowledge is *false positive*, which means accepting a hypothesis that is not true. A *false negative* means rejecting a hypothesis that is true. From our example, if the key will not really open the door, then accepting the hypothesis gives a false positive error (known to statisticians as a *Type I error*). On the other hand, if the key will really open the door, then rejecting the hypothesis gives a false negative error (known to statisticians as a *Type II error*).

Errors of **measurement** occur when a numerical fact is reported incorrectly about a person or object. These errors are caused by mistakes in precision and accuracy.

Although the terms *precision* and *accuracy* sometimes are carelessly used interchangeably, they are quite different. For example, suppose we wish to measure a person's height. We could do it with a centimeter rule or a meter ruler. Because it is unlikely that the person's actual height falls exactly in a gradation mark on either ruler, we would have to estimate the height based on visual inspection of the two rulers. Our estimate would be more precise with the centimeter ruler because the distance between gradations is less, making it easier to approximate. Accuracy would come into play if we used two centimeter rulers but the gradations on one ruler were marked incorrectly. Hence, one ruler (correctly marked) would be more accurate than the other ruler (incorrectly marked).

Systematic errors are errors that are made consistently. For example, if we use an incorrectly marked centimeter ruler, we will always get an inaccurate measurement.

Random errors sometimes occur when electrical equipment is used to give a measurement, and the equipment is not reliable. Sometimes the equipment gives the correct reading and sometimes it doesn't; we never really know.

Finally, errors can occur in the **reasoning process**. These errors can be inductive or abductive; these two forms of reasoning were introduced in Chapter 4.

Recall that induction reasons from the specific to the general case. For example,

> We have never had a customer complaint about our service.
>
> Therefore, we will never have a customer complaint about our service.

We would not be very confident that there will never be a complaint about our service since eventually someone likely will complain. There is a good likelihood of error if we accept our conclusion.

Consider another example.

> The car will not start.
>
> Therefore, the battery is dead.

We would have little confidence in this conclusion. Suppose we had more information.

> The car will not start.
>
> The headlights will not work.
>
> Therefore, the battery is dead.

Now we would have more confidence in our conclusion, but we still could not be completely sure that the battery is dead. Note that the addition of more specifics raises our level of confidence in the conclusion.

Human experts tend to make use of inductive processes in making their judgments. They are able to alter their conclusions as they acquire new knowledge. Human experts are usually quite good at this process. However, human experts are not infallible, and they can and do make errors.

Programming a computer to use inductive processes is extremely difficult. Hence, most current expert systems are not able to generate new conclusions on their own; new conclusions must still be preprogrammed.

Errors also can occur in the abductive process. Recall that abduction uses this form of reasoning:

> *p* implies *q*
>
> *q*
>
> therefore *p*

This example illustrates abductive reasoning:

> If the battery is dead, then the car will not start.
>
> The car will not start.
>
> Therefore, the battery is dead.

We know that there may be many reasons for the car not starting, so our confidence in the conclusion would not be very great.

Even though abductive reasoning has a good possibility of error, it is nevertheless useful in diagnostic expert systems. It must be used carefully, however.

Uncertainty and Rules

Because most expert systems are rule-based, we will examine in greater detail the effect of **uncertainty** on the rules in the expert system. Giarratano and Riley (1989) note that there are three basic areas of uncertainty pertaining to rules; these are illustrated in Figure 6.1.

Uncertainty can occur in individual rules. Recall that a rule is made up of an *antecedent* (IF part) and a *consequent* (THEN part). As we have discussed, various errors can occur in either the antecedent or the consequent. Reasoning errors can occur in the rule as a whole. Uncertainty factors can affect the evidence in both the antecedent and the consequent. Finally, uncertainty is involved in determining how the knowledge should be combined in the antecedent. Should AND be used exclusively? Should OR be used exclusively? Should a combination of AND and OR be used? The knowledge engineer must make these decisions as the rules are formulated.

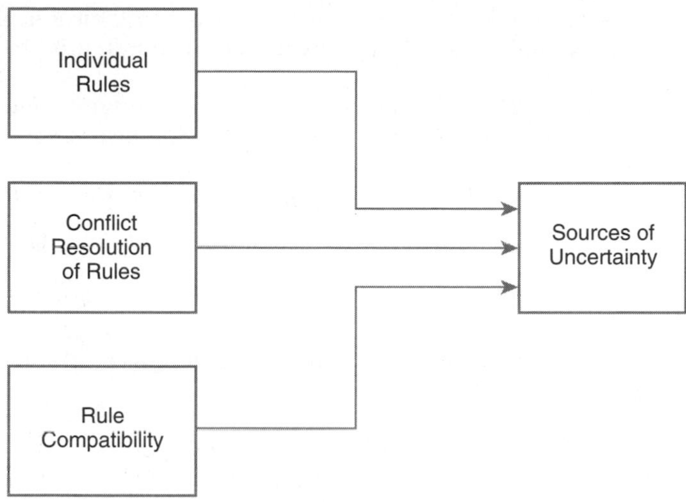

FIGURE 6.1 SOURCES OF UNCERTAINTY IN RULES

As the knowledge engineer formulates the individual rules, each rule must be verified for correctness. At this point, the knowledge engineer tries to minimize the inherent uncertainties described above.

Even if each individual rule can be verified, there is no guarantee that the inference chain will give valid results. Uncertainties exist as inference chains are formulated. These uncertainties involve situations in which there is conflict between rules or when rules are found to be incompatible. *Conflict resolution* is used to deal with conflicts between rules; it is also a source of uncertainty.

The knowledge engineer may establish an *explicit* priority that determines the order in which rules are executed or fired. A problem can occur if the established priorities are not optimal or even correct.

Implicit rule priority can also be a problem. If the antecedents of two rules are both satisfied simultaneously, the inference engine must decide which rule is to be fired first; strategies such as specificity (based on number of conditions—more conditions imply more specificity) of the antecedents or order in which the rules were entered are possible resolutions of the conflict. In the case of antecedent specificity, consider an example:

Rule 1: IF there is a fire, THEN use a fire extinguisher.

Rule 2: IF there is a fire and it is burning grease, THEN use the fire extinguisher.

In this case, Rule 2's antecedent is more specific than Rule 1's; Rule 2 would fire first. This order assumes that it is known that there is a fire and that it is from burning grease. If it is known only that there is a fire, then Rule 1 would be fired. We will discuss conflict resolution in greater detail later in this chapter.

Incompatibility of rules is also a source of uncertainty. Incompatibility can occur, for example, if two contradictory rules exist. Suppose we have these rules:

Rule 3: IF there is a fire on the assembly line, THEN throw water on it.

Rule 4: IF there is a fire on the assembly line, THEN don't throw water on it.

These two rules are obviously contradictory. Closer inspection indicates that the action depends on the type of fire. Water is appropriate for extinguishing one type of fire but not for other types. If the antecedent were more specific, as in Rule 2 above, the contradiction would not exist.

Rule incompatibility and rule conflict resolution are much more of a problem with large knowledge bases that have a great number of rules. The knowledge engineer must test carefully to validate the inference chains and minimize uncertainty as much as possible.

Dealing with Uncertainty

Over the years, a number of methods have been developed for dealing with uncertainty in a quantifiable way. AI researchers do not agree on which method is best because the choice of method depends on the class of applications being developed. All do agree that uncertainty cannot be ignored.

It appears that *classical probability theory* is the best way to deal with uncertainty; a group of researchers subscribe to its use. When classical probability is used, a formula known as *Baye's Theorem* is often used. If it were applied to a medical diagnosis situation, it would be done as follows:

$$P(D \text{ given } S) = [P(S \text{ given } D)P(D)]/P(S)$$

where

D is the disease.

S is the set of symptoms exhibited by the patient.

$P(D$ given $S)$ is the probability that the patient has the disease D, given that the patient has symptoms S.

$P(S$ given $D)$ is the probability that the patient will exhibit symptoms S, given that disease D is present.

$P(D)$ is the prior probability that the patient has the disease D before any symptoms S are known.

$P(S)$ is the prior probability that the patient has the symptoms S before the disease D is known.

If all of these probabilities on the right side of the equal sign were known, it would be quite easy to calculate $P(D$ given $S)$. However, it usually would be impossible to obtain the needed probabilities in this medical situation, so Baye's Theorem is of little or no use. Although classical probability theory is not useful in a medical diagnosis expert system, for example, it has been successfully used in a geological expert system called PROSPECTOR, which we discussed in Chapter 2.

Another major problem with the use of classical probability in the medical diagnosis expert system was discovered by the developers of MYCIN. They found that physicians were quite willing to assign a probability to a patient's having a disease, but they were not willing to assign a probability to the patient's *not* having the disease.

Let's examine this point in more detail. In classical probability,

$$P(\text{event will occur}) +$$

$$P(\text{event will not occur}) = 1$$

For example, if we roll a single die,

$$P(\text{rolling a five}) = 1/6$$

$$P(\text{not rolling a five}) = 5/6.$$

(We could get a one, two, three, four, or six.) The sum of the two probabilities is 1.

$$P(\text{five}) + P(\text{not a five}) = 1$$

It then follows that

$$P(\text{five}) = 1 - P(\text{not a five}).$$

In the MYCIN situation, the physicians (experts) were willing to say that given a particular set of symptoms, the probability that the patient had a particular disease was, for example, 0.8. They were not willing to say that given the same symptoms, the probability that the patient did not have the disease was 0.2.

The MYCIN knowledge engineers found that the judgments of the experts were based on belief and disbelief. From these findings, the MYCIN group developed the notion of certainty factors and made use of them in MYCIN (Giarratano and Riley 1989, Buchanan and Shortliffe 1985). A number of expert systems and expert system shells have made use of MYCIN's certainty factor structure; others have used variations on the MYCIN model.

Certainty Factors

MYCIN knowledge engineers developed the **certainty factor** as a way of combining a measure of belief with a measure of disbelief relative to a given hypothesis H and observable evidence E. The certainty factor (CF) was originally defined as

$$\mathrm{CF}(H,E) = mB(H,E) - mD(H,E)$$

where

$\mathrm{CF}(H,E)$ is the certainty factor for the hypothesis H with evidence E.

$mB(H,E)$ is the measure of belief in H given E.

$mD(H,E)$ is the measure of disbelief in H given E.

In the MYCIN model, mB and mD were defined as probabilities assigned by the physicians (experts); the values of both mB and mD could range between 0 and 1. Hence, the value for the certainty factor could range between 1 ($mB = 1$ and $mD = 0$) and -1 ($mB = 0$ and $mD = -1$).

A positive certainty factor ($mB > mD$) means that the given evidence supports the hypothesis. On the other hand, a negative certainty factor ($mB < mD$) indicates that the evidence does not support the hypothesis. If the certainty factor is 0, then one of two situations has arisen. Either both mB and mD are 0 (there is no evidence), or mB and mD have the same nonzero value (the belief is canceled out by the disbelief).

The original MYCIN definition of the certainty factor presented some interesting difficulties. A piece of disconfirming evidence (mD) could have a drastic effect on the certainty factor for a particular hypothesis. For example, suppose that $mB = 0.85$ and $mD = 0.60$. Part of the MYCIN method involved the setting of a *threshold (minimum) value* for the firing of a rule. For a rule to be fired, the certainty factor of the hypothesis had to exceed the threshold value. Suppose that the threshold value was 0.25. The rule in our example would not be fired since the certainty factor was not greater than the threshold value.

To lessen the possible effect of such a situation, the MYCIN definition was changed in 1977 to

$$\mathrm{CF}(H,E) = (mB - mD)/(1 - \min[mB,mD]).$$

In our previous example, the effect of the change in the definition can readily be seen. Under the original definition, the certainty factor was calculated to be $0.85 - 0.60 = 0.25$. Under the new definition, the certainty factor would be as follows.

$$\mathrm{CF} = (0.85 - 0.60)/(1 - \min[0.85,0.60])$$

$$= 0.25/0.40$$

$$= 0.625$$

The new definition thus softened the effect of a piece of disconfirming evidence.

Since the certainty factor concept was developed, a number of expert systems and expert system shells have adopted a variety of conventions for representing certainty factors as numbers. For example, VP-Expert has confidence (certainty) factors (CNF) that range from 0 to 100. These numbers represent percentages as a measure of confidence (belief) with 100 representing 100% confidence (total belief), equivalent to the MYCIN certainty factor of 1. The confidence factor of 0 represents 0% confidence (total disbelief), equivalent to the MYCIN certainty factor of –1. Thus, a CNF of 65 means that a confidence of 65% has been assigned. Another expert system shell, EXSYS, has three different schemes for representing certainty factors: –100 to 100, 0 to 10, and 0 to 1. In both VP-Expert and EXSYS, the initial values are assigned subjectively by the expert or the knowledge engineer.

The MYCIN knowledge engineers also noted an extension to the definition. Given the rule

IF E THEN H

then

$$CF(H,E) = CF(E)CF(H)$$

where

$CF(H,E)$ is the certainty factor of the hypothesis H given the evidence E.

$CF(E)$ is the certainty factor of the evidence E.

$CF(H)$ is the certainty factor of the hypothesis H.

If the evidence is certain (CF = 1), then $CF(H,E) = CF(H)$, which is intuitively what we think should be the case. For example, if we have the rule

IF E THEN H (CF = 0.75)

then

$$CF(H,E) = 0.75.$$

Note that if no certainty factor is specified for E, it is assumed that $CF(E) = 1$. The same assumption is also made for H; in our example, $CF(H)$ was specified as 0.75.

As we mentioned in Chapter 4, both the antecedent and the consequent of a rule have certainty factors associated with them. The certainty factors may or may not be specified; if they are not specified, they are assumed to be 1.

So far, we have looked only at certainty factors as they apply to the consequent (hypothesis) of a rule. The MYCIN model also considers certainty factors applied to antecedents. We have two possible situations: a single piece of evidence in the antecedent or multiple pieces of evidence in the antecedent whose certainty factors must be combined.

Consider, first, the situation in which there is a single piece of evidence in the antecedent. What is the effect if we have uncertain evidence (CF < 1)? What could be said about the certainty of *H*?

The MYCIN model solved this problem by using the relation

$$CF(H,e) = CF(H,E)\ CF(E,e)$$

where

CF(H,e) is the certainty factor of the hypothesis based on uncertain evidence e.

CF(H,E) is the certainty factor of the hypothesis based on the evidence E.

CF(E,e) is the certainty factor of the evidence E.

For example,

IF CF(H,E) = 0.8 AND CF(E,e) = 0.5, we have

CF(H,e) = 0.8 * 0.5

= 0.4

Intuitively, the result should not be surprising. If we are less than certain about the evidence, then we should be less confident about the certainty of the hypothesis, which was originally based on certain evidence.

Now consider the situation in which there are several pieces of evidence in the antecedent. These pieces of evidence are connected by ANDs, ORs, NOTs, or some combination thereof.

In the MYCIN model, when several pieces of evidence are connected by the logical operator AND, the final certainty factor of the antecedent is the minimum of the set of certainty factors contained in the antecedent. For example, the rule

IF *E*1 AND *E*2 AND *E*3 THEN *H*

has the pieces of evidence in the antecedent: *E*1, *E*2, and *E*3. If CF(*E*1) = 0.6, CF(*E*2) = 0.7, and CF(*E*3) = 0.5, then the certainty factor of the antecedent is

min[CF(*E*1),CF(*E*2),CF(*E*3)]

or 0.5. If CF(*H*) = 0.9 then the adjusted certainty factor for *H* is

0.5 * 0.9 = 0.45,

which follows from our earlier discussion.

The logical operator OR works in a similar fashion but uses the maximum value of the set of certainty factors in the antecedent. Consider the rule

IF *E*1 OR *E*2 OR *E*3 THEN *H*.

The certainty factor of the antecedent is given by

max[CF(*E*1),CF(*E*2),CF(*E*3)].

If CF(*E*1) = 0.3, CF(*E*2) = 0.6, and CF(*E*3) = 0.5, then the certainty factor of the antecedent is 0.6. As before, if CF(*H*) = 0.7, then the adjusted factor for *H* is given by

0.6 * 0.7 = 0.42.

In the case of the NOT logical operator, the certainty factor of the negative of a piece of evidence is the negative of the certainty factor. For example, if CF(*E*) = 0.6, then CF(not *E*) = -0.6.

The logical operators can be combined in the MYCIN model. Suppose we have the rule

IF (*E*1 AND *E*2) OR (*E*3 AND NOT *E*4) THEN *H*.

The certainty factor of the antecedent would be given by

max[min[CF(*E*1)],CF(*E*2)],[min[CF(*E*3),−CF(*E*4)]].

If we have

CF(*E*1) = 0.4

CF(*E*3) = 0.5

CF(*E*2) = 0.7

CF(*E*4) = 0.6

then the certainty factor for the antecedent is given by

max[min[0.4,0.7],min[0.5,−0.6]]

= 0.4.

If CF(*H*) = 0.8 then the adjusted factor for *H* is given by

0.4 * 0.8 = 0.32.

As with the range of certainty factor values, not all expert systems and expert system shells use the same techniques for calculating certainty factors as does the MYCIN model. The techniques used by other systems are similar but have some variances.

For example, VP-Expert, an expert system shell, assigns a confidence factor (CNF) to a rule in the same way as MYCIN. That is, for the rule

IF E THEN H

the confidence factor is given by

$$\text{CNF}(H,E) = \text{CNF}(E)\text{CNF}(H).$$

Confidence factors are entered in VP-Expert as whole numbers, but the system converts them to decimal form for purposes of calculation.

VP-Expert also uses the MYCIN technique for dealing with multiple pieces of evidence in the antecedent where AND is used. For example, the confidence factor for the antecedent in the rule

IF $E1$ AND $E2$ AND $E3$ THEN H

would be calculated as

$$\min[\text{CNF}(E1),\text{CNF}(E2),\text{CNF}(E3)].$$

VP-Expert uses a different technique than MYCIN for calculating the confidence factor of an antecedent that makes use of the logical operator OR. For the rule

IF $E1$ OR $E2$ THEN H

VP-Expert uses the formula

$$\text{CNF}(E1) + \text{CNF}(E2) - [\text{CNF}(E1) * \text{CNF}(E2)].$$

For example, if $\text{CNF}(E1) = 70$ and $\text{CNF}(E2) = 60$, a confidence factor of 88 would be assigned to the antecedent.

$$0.70 + 0.60 - [(0.70)(0.60)]$$

$$= 1.30 - 0.42$$

$$= 0.88$$

Recall that the CNF of 88 means 88% confidence or 0.88 confidence.

VP-Expert also treats the combination of the operators AND and OR in the antecedent. For example, in the rule

IF $E1$ OR $E2$ AND $E3$ THEN H

VP-Expert separates the antecedent into two statements to calculate the CNF. The antecedent

$E1$ OR $E2$ AND $E3$

is treated as

$E1$ AND $E3$ OR $E2$ AND $E3$.

This expression would be translated to

min[CNF($E1$),CNF($E3$)]

OR

min[CNF($E2$),CNF($E3$)].

If CNF1 = min[CNF($E1$),CNF($E3$)] and CNF2 = min[CNF($E2$),CNF($E3$)], then the expression becomes

CNF1 OR CNF2,

which becomes

CNF1 + CNF2 – [CNF1 * CNF2].

For example, suppose

CNF($E1$) = 40, CNF($E2$) = 50, and CNF($E3$) = 60.

Then the antecedent

$E1$ OR $E2$ AND $E3$

becomes

min[0.40,0.60]

OR

min[0.50,0.60]

= 0.40 OR 0.50

= 0.40 + 0.50 – [(0.40)(0.50)]

= 0.90 – 0.20

= 0.70

Recall that the confidence factors for VP-Expert range from 0 to 100. Hence, VP-Expert does not support the NOT logical operator because it would require the use of a negative number as a confidence factor.

Resolving Rule Conflicts

Earlier in the chapter, we briefly mentioned that rule conflict had a certain amount of uncertainty associated with it. We now want to examine this problem in greater detail.

One of the responsibilities of the inference engine is to resolve conflicts between rules. A priority system must be present so that rules can fire in some order for purposes of execution; a conflict resolution scheme is critical because only one rule can fire at a time.

A number of **conflict resolvent techniques** have been developed. All expert systems and expert systems shells use one or more such techniques in the inference engine. Conflict resolution can be classified as shown in Figure 6.2.

Rules may be fired in the order set by **explicit** priorities, which are attached to the rule and assigned by the expert or knowledge engineer. This priority technique works best for smaller expert systems. It is relatively inflexible to changes in the knowledge base because the priorities are static.

Implicit priority techniques are more flexible, especially when more than one technique is employed in the expert system. If the first technique used does not resolve the conflict, then a second technique is employed, and so on.

A number of implicit techniques have been used. Among others are the following:

1. The use of certainty factors associated with the rules
2. The specificity of the rules' antecedents
3. The order in which rules are entered into the knowledge base

PRIORITY	DESCRIPTION
Explicit	Priority Assigned by the Expert or Knowledge Engineer
Implicit	Priority Assigned by the Inference Engine

FIGURE 6.2 CLASSIFICATION OF PRIORITIES FOR RULES

4. The order of rules in the knowledge base
5. The recency of fact entry into the knowledge base

Using certainty factors means that the rule with the higher certainty factor should be fired first. Recall that the certainty factor of the rule is calculated by multiplying the certainty factors of the antecedent (evidence) and the consequent (hypothesis). For example, suppose we have the following rules:

Rule 1: IF $E1$(CF = 0.8) THEN $H1$(CF = 0.9).

Rule 2: IF $E2$(CF = 0.95) THEN $H2$(CF = 0.75).

The certainty factor for Rule 1 is

$$(0.8)(0.9) = 0.72.$$

The certainty factor for Rule 2 is

$$(0.95)(0.75) = 0.71.$$

Rule 1 would be fired first.

A second technique considers the specificity of the rules' antecedents. For instance, recall an earlier example:

Rule 3: IF the car will not start,
THEN the battery is dead.

Rule 4: IF the car will not start
AND the headlights will not work,
THEN the battery is dead.

If we know that the car will not start and the headlights will not work, then Rule 4 would be fired because its antecedent is more specific than that of Rule 3. If it is known only that the car will not start, then Rule 3 would be fired first.

The order in which the rules were entered into the knowledge base is a third conflict resolution technique. When the rule is entered into the knowledge base, it is marked with a time-stamp or another indicator. If there is a conflict, the rule entered first would be given priority.

A fourth technique is to use the current order of the rules in the knowledge base. This technique differs from the previous one in that a new rule may be inserted between two other existing rules. This technique is the simplest one to implement because it requires no additional overhead, such as calculation or reading an indicator of entry order, for the inference engine. With some of the previous techniques, a second conflict resolution technique may be needed, as in the case where two rules in conflict have the same certainty factors.

A final technique for our discussion is to use the recency of data entry into the knowledge base to resolve rule conflicts. For example, if data used in the antecedent of Rule 5 was entered at 12:30 p.m. today and Rule 6's antecedent data was entered yesterday, then Rule 5 would fire before Rule 6. This technique is most often used in real-time applications where facts in the knowledge base are updated continually.

Case Study: WPHELP

One of the features of an expert system is its capability to deal with uncertainty. Many expert systems make use of certainty factors to perform this task. The certainty factors, initially assigned by the knowledge engineer or a human expert, represent a quantitative degree of belief (confidence) in a rule. How can we incorporate certainty factors into WPHELP? In the choice of a word processor, we must consider our confidence about the vendor that has produced the package we are considering. Some of the questions that should be considered might be:

1. Does the vendor have a record of longevity in software development?
2. What kind of support does the vendor provide after the package has been purchased?
3. What is the quality of the documentation provided with the package?
4. What is the reaction of users of the package? Are they satisfied?

If we are very confident that the response to each question will be positive, then we might assign a high certainty factor (such as 95% or even 100%) to the choice of that package. If we are very confident in the positive response to three of the questions but somewhat uneasy about the other question, we might assign a certainty factor of 80% to the choice of the package. This certainty factor could be lower if we are uneasy about the response to a question that we consider to be more important than the others. For example, the opinions of others who have used the package might be much more important to us than the other considerations.

If you think a lot of subjectivity is involved in the assignment of certainty factors, you have a good perspective of how difficult the process can be. The use of subjectivity is common among human experts, but their subjectivity has been tempered by experience—that is why they are called "experts."

Let's consider how certainty factors can be assigned to the antecedents of the rule. Recall from the Case Study in Chapter 3 that functionality was described as simple or complex and that cost was described as low or high. Clearly, these descriptors can have different meanings. At what point does functionality change from simple to complex or cost from low to high? Also, a complex function to one person may be simple to someone else and vice versa. These perceptions can be reflected with certainty factors. If the functionality level of a task is somewhere between simple and complex, this can be reflected in the certainty factor. If we feel that the tasks will be more simple than complex, then we might assign a certainty factor of 40% to complex. It would not necessarily follow that we should assign a certainty factor of 60% to simple.

We might establish a minimum level (threshold) of acceptability for the certainty factor. That is, we may say that if we are less than 50% confident in the choice of a particular vendor, then the vendor will not be considered. This could also be true for our declaration of cost: if we are less than 50% confident that the cost is high, then we will use the value of low. This choice can affect the rules that we will finally consider because only rules with low as a value for cost will be considered.

In Module 6, we will look at some examples of the effects of certainty factors on rules, both individually and collectively. Mathematical calculations for the combining of certainty factors will also be explored.

Key Points

- Uncertainty can be considered as the lack of adequate information needed to make a decision.
- Deductive reasoning is referred to as exact reasoning.
- When facts are not exact, the number of possible conclusions increases, causing a new problem — that of determining which conclusion is best.
- There are several sources of errors that cause uncertainty, including:
  - Ambiguity
  - Incompleteness
  - Incorrectness
  - Measurement
  - Systematic errors
  - Random errors
  - Reasoning process
- Uncertainty can occur in individual rules, the conflict resolution techniques used, or incompatible rules.
- In individual rules, errors can occur in the antecedent, the consequent, or both, as well as in the rule as a whole.
  - The likelihood of both the antecedent and the consequent must be considered.
  - Techniques for combining the evidence (where more than one piece of evidence is available) can have an effect on the rule.
- Conflict resolution is used to determine the order in which rules are fired (executed).
- Conflict resolution strategies may be explicit (priorities assigned by the knowledge engineer) or implicit (priorities assigned by the inference engine using an algorithm).
- Rule incompatibility occurs when two rules have the same antecedent but have conflicting consequents.
- Rule incompatibility and rule conflict resolution are much more serious problems with large knowledge bases that have a great number of rules.
- A number of methods have been developed over the years to deal with uncertainty in a quantifiable way, but researchers do not agree on which method is best.
- Classical probability theory has been used successfully in the PROSPECTOR expert system, but it could not be used in the MYCIN system.
- The MYCIN knowledge engineers developed certainty factors to deal with the notions of belief and disbelief.

- The original MYCIN definition of the certainty factor for a given hypothesis *H* and given evidence *E* used the difference of the measure of belief (*mB*) and the measure of disbelief (*mD*).
- The MYCIN measures *mB* and *mD* were probability numbers assigned by the physicians (experts). Hence, the certainty factor value could range between −1 and 1.
- A single piece of disconfirming evidence could have a significant effect on the certainty factor value. The MYCIN definition was changed in 1977 to soften the effect of the disconfirming evidence. The difference (*mB* - *mD*) was then divided by 1 - min[*mB*,*mD*].
- A number of expert systems and expert system shells have adapted the MYCIN model for the certainty factor or have made variations in the definition.
- In the MYCIN model:
  - The certainty factor of the hypothesis *H* given the evidence *E* is the same as the product of the certainty factors of *E* and *H*.
  - If the evidence is uncertain (CF < 1), then the certainty factor of the hypothesis is affected; the certainty factor of the hypothesis is now the product of the certainty factor of the antecedent and the original certainty factor of the hypothesis (conclusion). If no certainty factor is specified, it is assumed to be 1.
  - The value of the certainty factor of an antecedent containing more than one piece of evidence connected by the AND logical operator is the minimum value of the set of the certainty factors of the pieces of evidence. If the pieces of evidence are connected by the OR logical operator, the certainty factor of the antecedent is the maximum value of the set of the certainty factors of the pieces of evidence.
  - The certainty factor of the negation of a hypothesis is the negative of the certainty factor of the hypothesis.
- VP-Expert is an example of an expert system shell that uses a different range of possible certainty factors than MYCIN. The calculation techniques for VP-Expert differ from those of MYCIN with respect to the OR and NOT logical operators and the combining of evidence in the antecedent where both the AND and OR logical operators are used.
- Rule conflict resolution techniques may be classified either as explicit or implicit.
- Explicit conflict resolution techniques use priorities that are established by the expert or the knowledge engineer; the priorities in implicit conflict resolution techniques are established by the inference engine.
- Among others, implicit conflict resolution techniques include:
  - The use of certainty factors associated with the rules
  - The specificity of the rules' antecedents
  - The order in which rules were entered into the knowledge base
  - The order of the rules in the knowledge base
  - The recency of fact entry into the knowledge base

Exercises

1. What is uncertainty?
2. What are the major sources of errors that cause uncertainty? Give an example of each.

3. What are the sources of uncertainty in individual rules?
4. What is the role of verification and validation pertaining to individual rules?
5. Why is there a need for conflict resolution between rules?
6. Contrast explicit and implicit rule conflict resolution.
7. Why is rule compatibility a source of uncertainty?
8. What were the major problems in the use of classical probability in the MYCIN expert system?
9. How were certainty factors originally defined in the MYCIN model?
10. How and why was the MYCIN definition of certainty factor changed in 1977?
11. How does VP-Expert define and use the certainty factor?
12. Using the MYCIN method, calculate the certainty factor for the hypothesis in each of the following rules.
    a. IF E (CF = 0.7) THEN H
    b. IF $E1$ (CF = 0.4)
       AND $E2$ (CF = 0.75)
       AND $E3$ (CF = 0.70)
       THEN H
    c. IF $E1$ (CF = 0.95)
       OR $E2$ (CF = 0.68)
       THEN H (CF = 0.75)
    d. IF E (CF = 0.75)
       THEN H (CF = 0.9)
    e. IF $E1$ (CF = 0.85)
       AND $E2$ (CF =0.65)
       OR $E3$ (CF = 0.75)
       THEN H (CF = 0.85)
    f. IF $E1$ (CF = 0.65)
       AND NOT $E2$ (CF = 0.8)
       THEN H (CF = 0.40)
13. Repeat Exercise 12 using the VP-Expert method, if possible.
14. Name and describe five implicit rule conflict resolution techniques.
15. What is the effect of n exact facts on an individual rule?
16. What is the role of hueristics in dealing with the problem described in Exercise 15?
17. How do the measures of belief and disbelief differ from probability numbers?

Assignments

1. Consider the small investment expert system from Assignment 2 in Chapter 4.
    a. Describe, by tracing the execution of the rules, how forward and backward chaining would be used on the knowledge base. Assume that the order of

the rules in the knowledge base is the order in which the rules would be checked.

b. Repeat part (a) but with the order of the rules reversed. What differences, if any, do you find in the chaining processes?

2. In the investment expert system of Assignment 2 in Chapter 4, the measure of what is adequate insurance is rather vague. We can use certainty factors to help better define the levels of adequate. Consider the figures below.

INSURANCE AMOUNT	*CF*
10,000–20,000	0.30
20,000–30,000	0.60
over 30,000	0.90

We can also apply certainty factors to the ratio of take-home pay and savings. Consider the figures below.

STATUS	*SAVINGS-TO-PAY RATIO*	*CF*
Single	4.00 to 5.00	0.30
	5.00 to 6.00	0.60
	over 6.00	0.90
Married	6.00 to 7.00	0.30
	7.00 to 8.00	0.60
	over 8.00	0.90

a. Assume the certainty factors for both married and single are 1.0. For each of the following sets of initial conditions, calculate the final certainty factor to invest using the inference chain for the knowledge base.

Set 1

Married
$25,000 insurance
$10,000 savings
$1,500 monthly take home pay

Set 2

Single
$15,000 insurance
$20,000 savings
$4,000 monthly take home pay

Set 3

Married
$100,000 insurance
$50,000 savings
$5,000 monthly take-home pay

b. Repeat part (a) using the VP-Expert method.

c. Compare the results of your work in parts (a) and (b).

3. In Assignment 4 in Chapter 5, you were asked to investigate the characteristics of three expert system shells. Let's look in greater detail at the operation of the inference engine.
   a. How does the inference engine in each shell deal with uncertainty?
   b. How does each shell deal with rule conflicts?
4. In Assignment 5 in Chapter 5, you were asked to develop a small knowledge base to solve the given problem in each part. You were also asked to identify the source(s) of expertise that you used. Using your source(s), assign certainty factors to the rules in each knowledge base. (See also Assignment 1 in Chapter 8.)
5. Contact vendors that have advertised expert system development tools to get details of their products' features. Make comparisons of the following attributes:
   a. structure of the knowledge base (rules, induction from examples, etc.)
   b. method(s) of reasoning used
   c. method(s) of handling uncertainty

   Ask if you can obtain a demonstration sample/evaluation copy of their product(s). What appear to be the strengths and weaknesses of each product?

The User Interface

As we have mentioned, an expert system is composed of three basic parts: a knowledge base (rules and facts), an inference engine (reasoning mechanism), and a user interface. The user interface is the total communications medium between the participant and the expert system. This user interface could involve typing answers to questions displayed on a computer screen or selecting icons (symbols) by touching a touch-sensitive screen with your finger. The user interface includes the entire communications scheme between the user and the expert system; it includes the design of the computer screen, the use of external input/output devices, and the schema by which information flows between the user and the expert system. In this chapter we will present an overview explaining how useful interfaces are formed, exploring various types, and providing guidelines for effective user interfaces. Finally, we will consider a user interface for the WPHELP expert system.

Overview of the User Interface

The **user interface** is the lifeline of an expert system. Just as a well-manufactured product is not useful or profitable without good marketing, a good expert system is not effective without an appropriate user interface. The user interface can determine whether an expert system is accepted and utilized by a given population. A microcomputer system that no one uses is worthless. Hence, to reap the benefits from a well-designed and informative expert system, the user interface must make using the expert system both easy and timely.

The user interface is the total communications channel between the participant and the expert system. If an expert system is perceived as a circle, then the *atmosphere* surrounding the circle is the user interface, as shown in Figure 7.1.

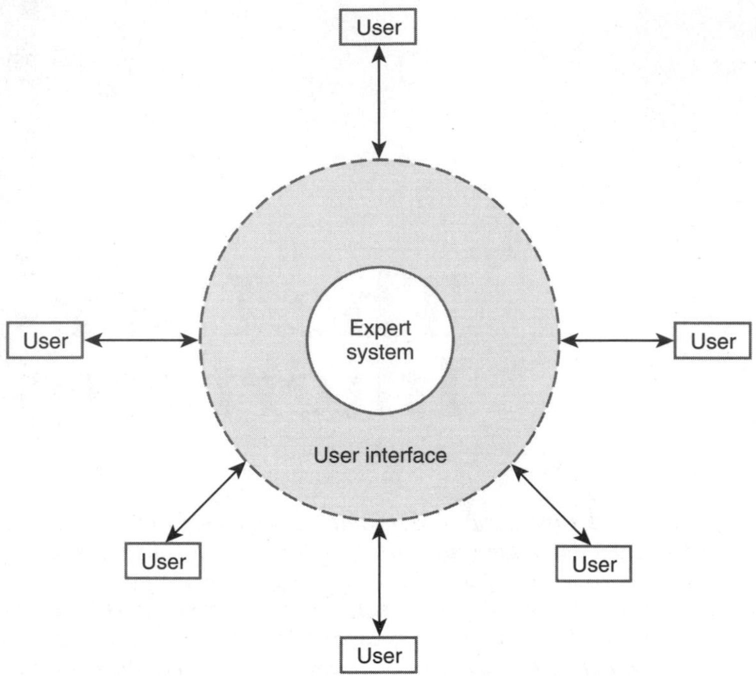

FIGURE 7.1 USER INTERFACE AS AN ATMOSPHERE

If the atmosphere is appropriate for the intended audience, the users will be able to communicate effectively with the expert system. If the information doesn't flow through the atmosphere between the user and the expert system, the system fails. The user interface is considered to be not only the physical method of sending and receiving information from the expert system, but also the perceptual and psychological media. Just as water is composed of hydrogen and oxygen, the user interface is composed of physical, perceptual, and psychological components. The user's perception of the system is critical to its success. If the tasks, functionality, and benefits are not readily understood by the user, the system will not be optimal. Moreover, from a psychological point of view, if the user perceives the system as confusing because of lack of proper instructions or thinks the system is hostile because of terse error messages or feedback, the user interface is inadequate. A poorly designed user interface can cause an expensive and well-developed expert system to fail.

Formation of the User Interface: Screen Design and Input Devices

User interfaces are made from three sources: the designer model, the user model, and the system model. A **model** is another name for image, schema, or configuration. You can think of a model as a set of eyeglasses through which an image or schema of an object, event, or person is formed, as illustrated in Figure 7.2.

Hence, the three models that help build a user interface are like three sets of eyeglasses, one from the system designer, one from an individual user, and one from all users of an expert system. The **designer model** includes the view of the person(s) designing, writing, and implementing the system. If the designers are a team of content experts, knowledge engineers, programmers, and users, then the designer model is their composite view and understanding of the system. As the system is being formulated, this model can vary from team member to team member. If the designer is one person, the designer model is that one person's mental configuration of the expert system.

The second building block, the **user model**, is a single individual's view of the system. This model can change from user to user.

The third building block, or **system model**, is the composite view of all users of the entire system. These three components are the pillars upon which the user interface is formulated.

For example, a manufacturer was interested in improving maintenance and trouble-shooting of shop floor equipment. An expert system designer and programmers were hired to create an expert system. The expert system on the shop floor was not successful, however, because the mechanics (the users of the system) had to communicate in terms that were not familiar to them. The vocabulary used by master mechanics was not incorporated in the user interface to the expert system. Hence, the user model and the system model were not identified correctly. The expert system interface was later rewritten to create an atmosphere more like the working vocabulary of the mechanics. This change made the expert system successful.

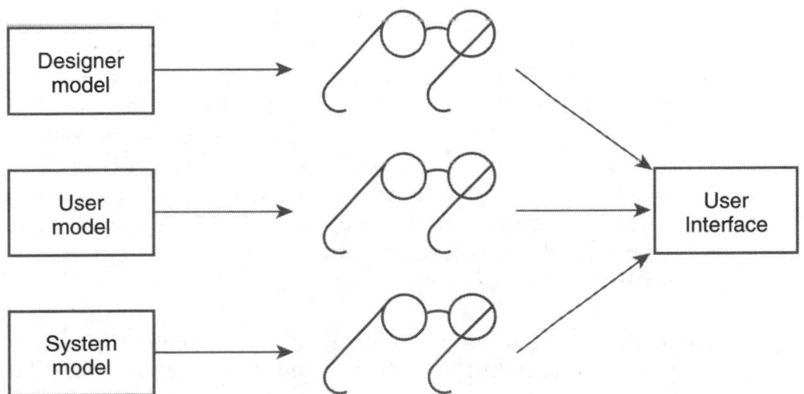

FIGURE 7.2 THREE MODELS (EYEGLASSES) FOR BUILDING A USER INTERFACE

Another example of an inadequate user interface occurred in a mortgage advisement system. A mortgage firm, eager to help its advisors become more proficient, hired a team of designers to formulate a mortgage advisement system for its employees. After many months of work and considerable expense, the firm was confused as to why the mortgage advisors were not using the system. After a thorough investigation, the designers of the system found that the users viewed the new expert system as an eventual threat to their positions. Hence, the view of the designers did not match the system model or composite image from the mortgage consultants. After considerable training and reworking of the user interface, the expert system was viewed as an assistant rather than a threat.

Underlying all three models that form a user interface are **tasks**. No user interface can be properly designed without understanding the overall, specific tasks to be accomplished. All three models must incorporate the same threads of task definition. If the expert system is used only by domain-specific experts (doctors, lawyers, airplane mechanics, and so on), then the tasks must be specific to that domain. If, however, the expert system is to be used by more general audiences (those seeking advice on investments, for example), the tasks must be appropriate and defined for more general-purpose users. The nature, background, and knowledge of the intended user are crucial to understanding the tasks to be incorporated in the user interface.

Screen Design

Today's expert systems employ many different types or styles of interface design. Even though the user interface is not exactly the same as the design and type of interaction between a computer keyboard and display screen, we will discuss some common screen formats and types of interaction. The more common screen layouts are: question and answer (teletype), menus (hierarchical, pull-down), icon-driven, and windows. Different types of interaction for any of these layouts are achieved with keyboard, mouse, touch-sensitive screen, light pen, or voice.

Many of the currently used screen layouts have been ported from the trends in microcomputer software applications development. The past decade has seen a decrease in hardware costs, increase in memory size, development of graphics tools, and increased concern for user communications. This trend has centered on microcomputer software development but has migrated to the mainframe software environment. The new range of applications with word processing, database, spreadsheet, and desktop publishing have offered rich and varied user interfaces that system designers are incorporating into expert systems. Even though video displays are not always required, we will restrict our comments to them because of their widespread use and availability in computer systems.

Questions and Answers (Teletype). Let's turn our attention to the design of a video display terminal (computer screen) and look at each of the different types of screen layout. The simplest and earliest is called the **teletype** or **question-and-answer** layout. The expert system, through the software, displays a question on the screen. The user then types a reply, using the keyboard or other device. The computer software of the expert system takes the reply and incorporates the information in the knowledge base. This method is quite linear; it is shown in Figure 7.3.

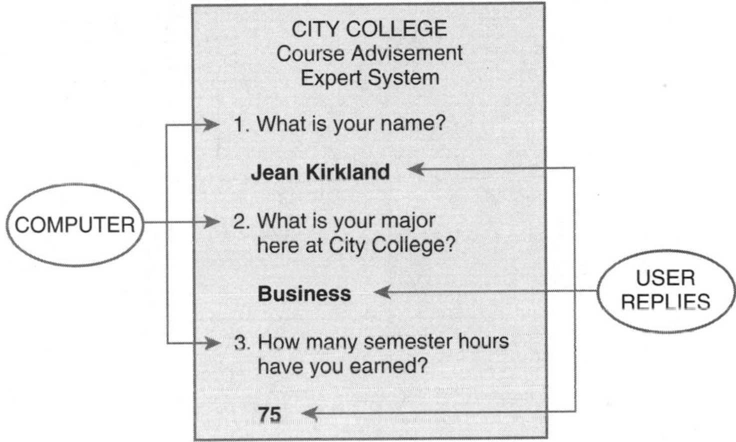

FIGURE 7.3 QUESTION AND ANSWER TYPE SCREEN DESIGN

This type of screen organization in a user interface can be more than adequate for some expert systems. Its advantage is that it is easy to design and implement in software; its disadvantage is that it allows the user to retype or re-enter incorrect or misspelled answers.

Menus. A **menu-driven** interface displays a list of choices, as shown in Figure 7.4. The user's selection could lead to another menu of choices. Users of information and database systems are quite familiar with menu-driven interfaces.

An issue with menu-driven user interfaces is the determination of the optimal number of displayed items on one menu screen (breadth of a menu) versus the optimal number of levels of submenus (depth of a menu). See Figure 7.5.

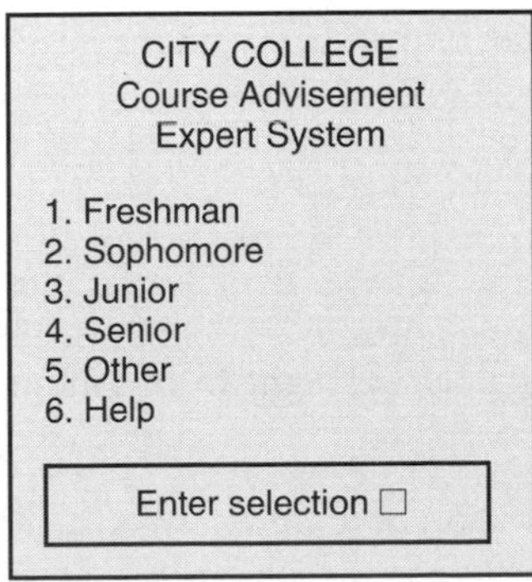

FIGURE 7.4 MENU TYPE SCREEN DESIGN

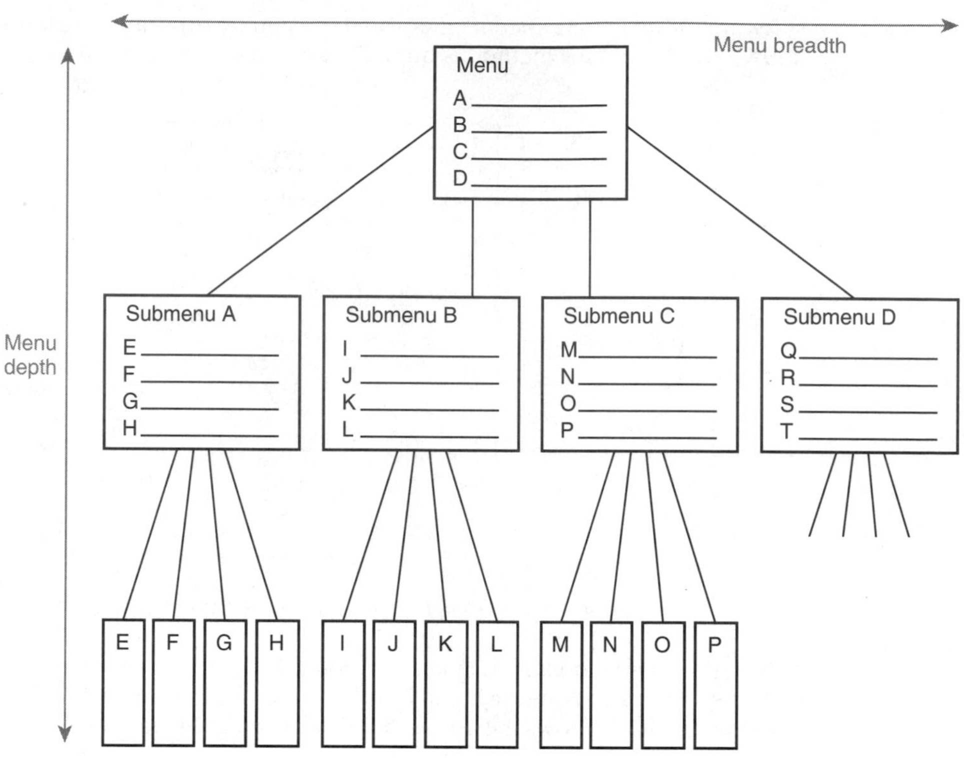

FIGURE 7.5 MENU BREADTH AND MENU DEPTH

Users like to get to specific information as soon as possible, yet too many choices in one menu (breadth) leads to a crowded and confusing organization of information. This *breadth-depth* menu structure can be designed in hierarchical and pull-down menu formats. Both formats display the same logical menu structure information as shown in Figure 7.5, but the arrangement of the items on the screen is different.

The hierarchical format for a menu structure displays the menu choices arranged horizontally, as shown in Figure 7.6, in a band called a *menu bar*. The user moves across the menu bar (A to B to C to D) by moving a cursor or highlighted box around the item. When the locator box is on A, the submenu for A is displayed for possible selection of E, F, G, or H. As the locator box is moved from A to B, submenu A disappears and submenu B becomes visible. This interface is common to many spreadsheet users; Lotus 1-2-3 employs this hierarchical type of menu interface. It is possible to move left or right in the menu bar and to have submenus in horizontal format under submenus.

The pull-down (pop-up) submenu format is shown in Figure 7.7. It has the same logical construction as the hierarchial arrangement except that the submenu choices are arranged in a vertical format. Only one submenu from the menu bar is visible at a time. This menu interface style (pull-down menu) is incorporated in dBASE III PLUS, a

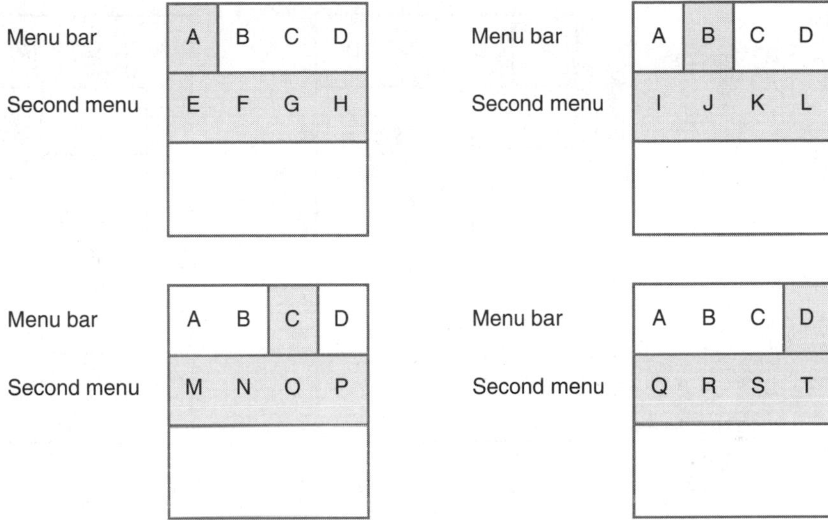

FIGURE 7.6 HIERARCHICAL ARRANGEMENT OF SUBMENUS FOR MENU-DRIVEN SCREEN DESIGN

popular microcomputer database software package. Both of these menu formats, hierarchical and pull-down, have the advantage of requiring less area on the screen, but both require considerably more screen programming tools and expertise. However, both formats are familiar to millions of office workers and professionals.

Icon. Another popular type of interface is the icon environment. In this style of interface, icons (symbols) represent the choices on a menu instead of words. Users select a noun symbol such as a letter, report, or drawing, as shown in Figure 7.8, and then select an action to be performed on the noun. Possible verb icons from Figure 7.8 are writing to disk, sending to the printer, or deleting (throwing in the trash can). A noun icon may be dragged and placed on top of a verb icon to tell the computer to perform that action on the selected noun. The icon environment relies heavily on visual and graphical communications and is used extensively in Macintosh software products. The icon environment is not restricted to graphical symbols, but uses text along with symbols to facilitate the user interface.

Windows. The last generic type of screen design is called the windowing environment, illustrated in Figure 7.9. Windows are rectangles or frames of information that appear on the computer screen. Users can view, edit, rearrange, enlarge, and perform similar manipulations on multiple windows and their contents. Windows may overlap each other. One window may contain the text of a memo and another window might contain a business graph. The user has considerable control over the windows and their appearance on the screen. Special windows, called *dialog boxes*, are windows where users enter information about a task to be performed. Another special window, called an *alert box*, may appear when the system wants to alert the user to unusual situations (printer is out of paper, disk is full, no file is available, and so on). The advantage of this interface is that multiple information boxes (windows) are available

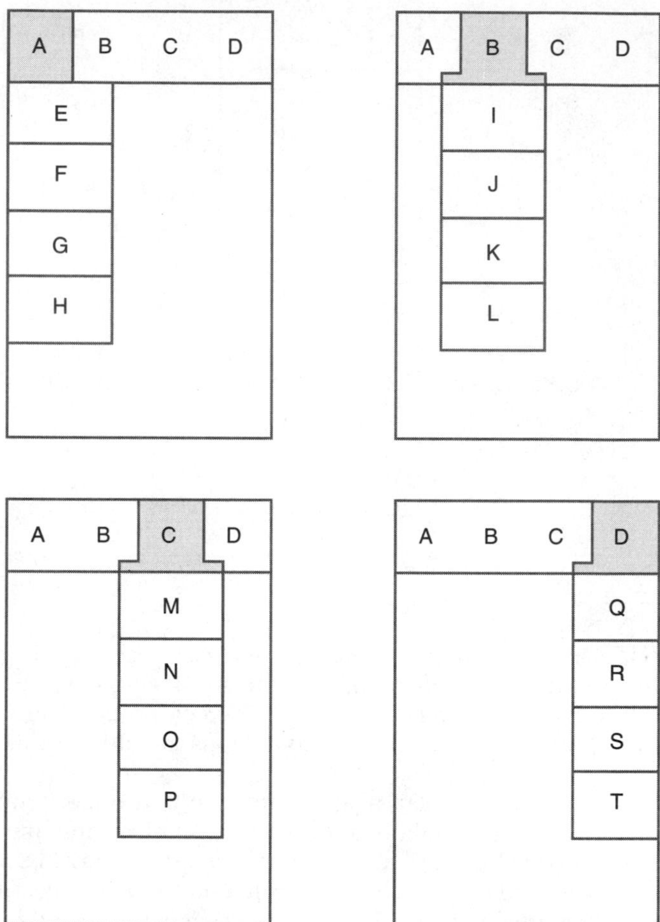

FIGURE 7.7 PULL-DOWN MENU SCREEN DESIGN

simultaneously; the disadvantages are that extensive tools and expertise are needed to create the windows on the screen.

These screen configurations—question and answer, standard menu, hierarchical menu, pull-down menu, icons, and windows—can be mixed. The Macintosh environment, for example, uses a standard interface of pull-down menus, icons, and windows. This interface is known to the user as the *desktop* because of its symbolic representations of the common tasks done at the desk. Each style has its own special advantages and disadvantages. Users, however, are getting more sophisticated in their use and expectations of user interfaces. An interesting psychological factor, the "homesick syndrome," is evolving from users of computer information systems. When asked what user interface design participants prefer, many prefer the interface design of the first computer application they learned. Very few users in the business environment are exposed to all interfaces in their daily routine.

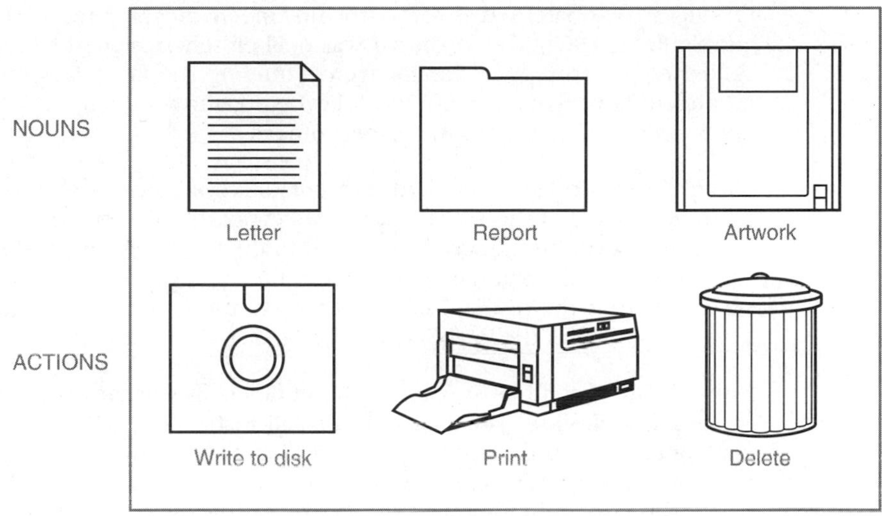

FIGURE 7.8 ICON SCREEN DESIGN

Input Devices

Each of the interface configuration types we have discussed can facilitate a variety of physical input devices for the user. The user can input through the various screen formats by using the keyboard, mouse, light pen, touch-sensitive screen, or voice.

Keyboard. The most common physical input method is the **keyboard**. The user types characters as answers to questions or moves a cursor to a menu and makes a selection by pressing the Enter key. The most commonly used keyboard layout, the

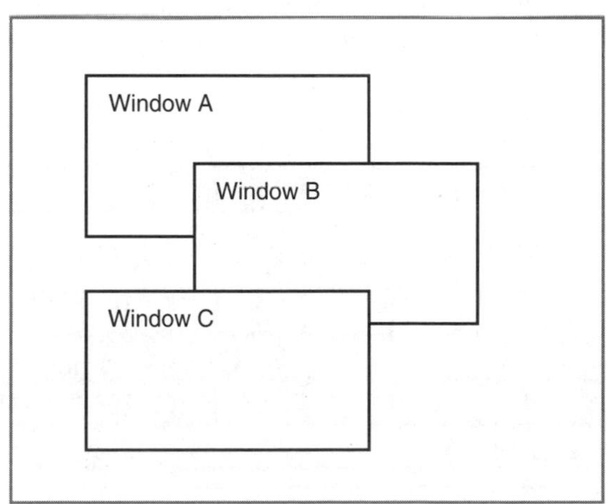

FIGURE 7.9 THE WINDOWING INTERFACE

QWERTY, was patented as part of the first mechanical typewriter by Sholes, Glidden, and Soule in 1864. The keyboard was deliberately designed to reduce the speed of skilled typists to reduce the chance of jamming the keys. Despite its awkward and inefficient key layout, the QWERTY keyboard, shown in Figure 7.10, is still the most commonly used keyboard in business offices.

As an added de facto standard, the cursor keys have been superimposed on the numeric keypad of the IBM PC series of keyboards. The cursor keys enable a user to move the cursor on the screen (in steps) to the right, left, up, or down. The user then can position the cursor over a menu selection, and press the Enter key to transfer the selection from the display screen to the computer program. Thus, through cursor keys and software, the keyboard can be used as a pointing device for menu selection.

Mouse. The second most common input device is the **mouse**, also called a **pointer** or **locating device**. The mouse is a small instrument, about the size of a deck of playing cards, that is attached to the computer by a thin cable. The mouse has one, two, or three buttons for user control. As a user moves the mouse around on a desktop pad or other surface, the cursor on the display moves in a corresponding direction. By using the buttons, the user can convey actions to the computer. For example, by pressing the mouse button while the cursor is on an icon, the user can activate the task indicated by the selected symbol. The mouse is standard equipment with some computer systems, and it is very easy to use. Studies show that the mouse is particularly useful for people who use computer systems over a long, continuous work session. The mouse allows considerable freedom in moving both the hand and the entire body, but it also requires additional desk space.

Touch-Sensitive Screen. Another common input device is the **touch-sensitive screen**. This device is a special screen that allows the user to touch a portion of the screen (icon, menu choice, and so on) and communicate the selection through software to the computer program. These screens are often used in public information systems for the average user in banks, malls, and amusement parks. This type of screen is also common in educational applications. Touch-sensitive screens are best used for applications requiring infrequent use and where coarse pointing is adequate. Their disadvantages include accidental activation and problems caused by dirty screens.

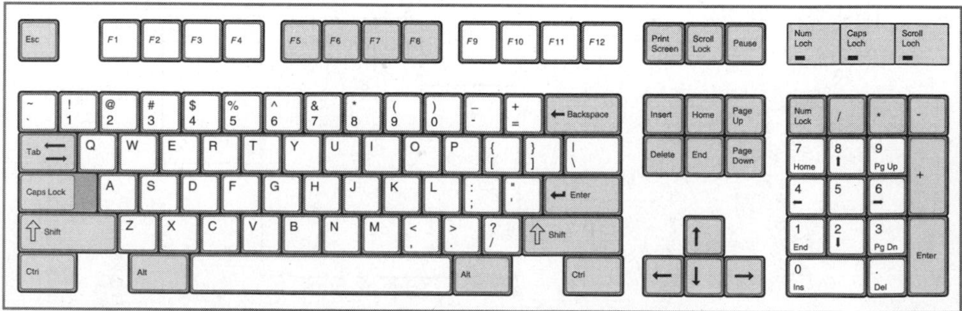

FIGURE 7.10 THE QWERTY KEYBOARD

Light Pen. A fourth input device is the **light pen**, a small wand attached to the computer by a cord. The user positions the pen on a portion of the screen. A small photocell in the pen emits light. This increase in brightness on an area of the screen generates a signal to the computer.

Voice. A new input device, **voice**, exists in varying stages of development. The user speaks and the computer equipment uses the voice signals to control the computer. Even though this is a simple concept, voice-driven systems are not yet used extensively. Voice input is best and most effective in systems in which the user's hands and eyes are not free. Because voice systems require step-by-step confirmation of response, they are not recommended for noisy or stressful environments.

How do we know which input device is the best choice? Each device adds expense, programming overhead, mechanical and/or electrical functionality, and user preferences. The best input device is the one that fits the application, the user, the working environment, and the budget.

Input devices and screen interfaces can be combined. Any type of screen layout could be matched with almost any input device. For example, in today's offices, many computer applications have a WIMP interface: a combination of screen design and input device. A WIMP environment is a user interface composed of windows, icons, menus, and pointer device. This kind of interface is common in the Macintosh and IBM PC microcomputer world under Microsoft Windows. The possible combinations of screen layout and input device(s) are limited only by the designer's imagination, the firm's budget, the user's requirements, and the programmer's development tools (for controlling the video or computer screen).

Integration Interface. A current trend in user interface design for expert systems is the **integration** of the expert system with a spreadsheet or database format. Businesses and management professionals are already acquainted with the tabular format used in spreadsheet and database applications, as shown in Figure 7.11.

Because expert systems contain large amounts of information and facts, connecting existing spreadsheet/database-like formats to the contents of an expert system is a natural link. This is particularly true in the information systems world of both mainframe and microcomputer-based applications. Many users now view the spreadsheet/database format of a table as the front end to an expert system. In some systems, the user can fill in a table and the expert system software will convert the table format to production rules, as illustrated in Figure 7.12.

The expert system software converts the information in each record (row) of spreadsheet/database information to a production rule. For example, record one of ABC blank, is converted to

IF A AND B THEN C.

The first two parts (fields) of each row are interpreted as antecedents (IFs), and the last two fields are considered as actions (THENs). Thus, business expert systems are being constructed on top of the familiar and functional interfaces already common in the business world.

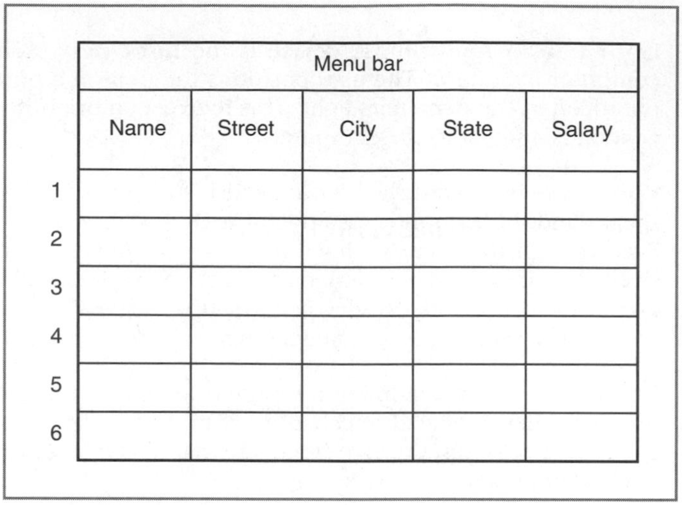

	Menu bar				
	Name	Street	City	State	Salary
1					
2					
3					
4					
5					
6					

FIGURE 7.11 TABLE FORMAT FOR SPREADSHEET/DATABASE INTERFACE

Effective User Interface Guidelines for Expert Systems

This section will summarize some basic guidelines for the development of effective user interfaces. There is no set of magical rules that will guarantee that the user interface will be optimal in design, content, or format. There are, however, basic guidelines that should be considered when designing an expert system. Whether the user is a domain-specific expert (lawyer, accountant, surgeon, or aircraft mechanic), or a more general user (public citizen or consumer), the basic communications between the expert system and the intended user determine the functionality and efficacy of the system. Good screen design and user interaction for a well-designed expert system will enhance the performance of the system.

TABLE FROM THE USER

Record #	Condition	Condition	Action	Action
1	A	B	C	
2	A	D	F	B
3	B		G	H
4	G	D	H	

RULES GENERATED BY THE EXPERT SYSTEM

Rule 1:	Rule 2:	Rule 3:	Rule 4:
IF A	IF A	IF B	IF G
AND B	AND D		AND D
THEN C	THEN F AND B	THEN G AND H	THEN H

FIGURE 7.12 TABLE INTERFACE FOR PRODUCTION-RULE GENERATION IN EXPERT SYSTEMS

We will look at guidelines for the two facets of the user interface: screen display and user interaction via an input device. These two components constitute the user interface. First, let's look at the design and organization of the screen display.

Screen Display Guidelines

Some general guidelines or directions for effective **screen display** in expert systems are described as follows.

1. **Use uncluttered, well-spaced screens.**
   The user screens should be uncluttered with appropriate spacing between fields, lines, and text. Experts advise that only 15 to 20 percent of the screen space should be utilized. Unnecessary items are distracting to the task at hand.
2. **Avoid paragraphs of text on the screen.**
   Paragraphs of text should rarely be used on the screen. Reading from a display terminal is not the same as reading from or browsing through a book, so organize information into vertical lists instead of paragraphs. If paragraphs are necessary, however, display part of the text, then allow the user to press a key to continue with the next part. A mixture of uppercase and lowercase text is more readable than all capitals.
3. **In general, left-justify text and right-justify numbers.**
   Text and number lists should be formatted distinctly. Lists of text should be left-justified; that is, the leftmost letter of each word in the list should be aligned vertically. In contrast, numbers should be right-justified, aligned vertically to the right. When using decimal values with varying decimal places, the list of numbers should be decimal aligned. These examples are shown below:

Left Justify	Right Justify	Decimal Align
Investments	4,300	34.78
Stocks	543	123.6
Bonds	678,413	6.789

   This arrangement of lists is easiest to read.
4. **Limit the number of menu selections to five to seven.**
   In general, a user can retain information from about seven items at a time. Therefore, the number of selections from a single menu or pull-down submenu should be limited to five to seven items. Each one should be numbered (1, 2, 3, 4). Use letters only with a short menu list of three or four selections identified by codes (i.e., D for delete, U for update, and A for add). The items can be listed by time, function, number, importance, or other criteria. A general rule of thumb, however, is to list the menu options by frequency of use, with the most frequently used at the top of the menu screen (breadth), and by the highest level of menu (depth).

5. **Be consistent in wording and location of prompts.**
The wording and location of prompts, fields, and other identifiers should be consistent throughout the screens. Use uniform wording of phrases. For example, do not use *change the record* on one screen and *update the record* on another. With data entry fields, maintain the same formats for name, social security number, date, and so on throughout all screens. The screen formats should match the formats used in printed reports. The prompts to the user should always appear in a corresponding location. If the lower-left-hand corner of the screen is used to display the message on how to move to the next screen, retain that relative positioning in all appropriate screens. Keep the language consistent with the expectations of the user of the system; this means to adhere to the vocabulary of the user. If the user is an accountant, the terms *debit* and *credit* are fine, but they are not appropriate for the general user when referring to money paid in and money paid out.

6. **Use reverse video or color for data entry fields.**
If a user is filling in the various items on a form-like screen, each data entry field should be highlighted. Reverse video (the opposite of the screen background: black letters on a white background or white letters on black background) is one popular method in many business-oriented information systems. Color is another way to draw attention to a particular field. Windowing is still another technique used to highlight fields. The system should not permit the user to alter information or words that are not to be changed on the screen. In other words, the programmer should "protect" these screens when the screen programs are being implemented.

7. **Use color and graphics to add to screen design.**
Color can be a useful asset to screen design if it is available on the video system. It can be used to highlight, to contrast, and to make complex diagrams such as maps more readable. Color on diagrams makes intricate detail more discernible. Care should be taken in selecting the color of the background (blank screen) and the color of the text that appears on the screen. Certain color combinations are hard to read, such as red letters on a blue background, for example. The designer of the system should make sure that if the color screens are displayed on a black and white monitor, the visual effects are still discernible and pleasing to the eye.

8. **Let the user select some characteristics of the screen interface.**
For those applications that users work with extensively, a method of configuring or tailoring the screen environment for the individual should be available. For example, if color is used, the user should be able to select certain colors for his or her working environment. If a beep or sound is used to identify the end of a field, the user should be able to turn the sound off. Different people have different reactions to color, sound, and so on, and these minor characteristics can be very annoying to those who use the system continually. If the users, on the other hand, vary from day to day, the screen characteristics should be determined by the designer of the system. Advising systems used by the general public are a type of expert system that users should not be allowed to modify.

The preceding rules are guidelines for general screen design for expert systems. The more specific and narrow the domain of the expert system, the more these rules may have to be adjusted to fit the functional needs of the specific expert system application.

Guidelines for the Use of Input Devices

In this section, we will look at guidelines for implementing interaction through screen displays. Interaction between the user and the system is sometimes called **dialogue**, similar to a conversation between two individuals. Besides considering rules for implementing effective dialogue, we will explore general rules for using different input devices.

1. **Minimize the number of keystrokes.**
   To perform tasks and move between menus, users should have to make as few keystrokes, or other required actions, as possible. If users are working with codes, the codes should be short. The depth of menus should be taken into consideration at design time.

2. **Use a two-level command or menu selection structure.**
   If the system is driven by commands, or a menu, the instructions should exist at two levels, one for the beginner and one for the more experienced user. A beginning user does not mind selecting "D" for document, "P" for print, and "Go" to start the process of printing. As the user gains experience, this three-step process is abbreviated to entering DPG (in one step). The same principle applies to menu selection when the user is very familiar with the system. A two-level command or menu selection structure accommodates users with different levels of expertise.

3. **Help the user navigate through the system.**
   Effective dialogue and functionality are fostered when the user understands how to progress through the system. This concept is analogous to reading the symbols on a roadmap. To determine which way to go, the symbols (or prompts) must be understood. Similarly, in a computer or expert system, users must recognize whether they are in the mode to look at production rules, change a rule, or observe the execution of a set of rules. They also must understand how to move from one mode to another. Proper symbols and understanding of how to navigate through a system are important for the user.

4. **Provide on-line help.**
   If the user is unsure of backward chaining, he or she should be able to invoke the help system (press the Help key) to find information about backward chaining. If the user does not remember what information is to be entered, an on-line help option could review the material. On-line help supplements the formal training and written documentation for the system. Help systems are considered standard features, not only to expert systems but to any computer information system.

5. **Guarantee quick response time.**
   Response time is the amount of time (in seconds or smaller time units) that the computer system takes to respond to the user input. Users on mainframe or

microcomputer systems are accustomed to a two-second response time. No one likes to wait to get the task completed. Response time is particularly important in multiuser or networked systems. The response time is a function of both the hardware and the software. Response time is important at design time and after the system is implemented because of continual changes in the system and working environment.

6. **Avoid blank screens during any dialogue session.**
   In a dialogue session, the user should see an appropriate response. Many times, searching and sorting techniques take considerable time (or so it seems). Instead of leaving the display screen blank, some message appropriate to the task should appear. For example, if the expert system is firing the rule base, a message such as "Searching the rule base" should appear on the screen. Users are likely to be more patient if they see an explanation for the lapse of time. Hence, for the user, appropriate messages are preferred to blank screens.

7. **Implement an undo feature in the commands.**
   An undo command is one that reverses a command or action that has just been implemented. Not all actions in a system are reversible, though. Users have become accustomed to this helpful system interface, making the implementation of this feature a useful design enhancement.

8. **Simplify error messages.**
   Error messages, warning messages, and other related feedback should be in terms the user understands. They are information for the user, not the programmer. For example, the message, "Check disk drive door to make sure it is closed," is a better user message than "System failure: abort, retry?" Messages like "syntax error" or "Runtime error 256" are not particularly useful to users. In addition, users should be involved in *approving* the messages that are built into the operation of an expert system to ensure that they understand them.

9. **Provide for atypical users.**
   Most computer systems are designed for people who have 20/20 vision, are right-handed, and have full-color vision. Many people, however, have vision problems, are color-blind, or are left-handed. Input devices such as the numeric keypad and the trackball accommodate left-handed users. Color display screens, for color-blind users, should appear readable and discernible on black-and-white monitors. An expert system used by a wide variety of users must take into consideration the atypical user.

10. **Minimize the number of changes.**
    If the keyboard system is combined with another device such as a mouse or light pen, minimize the changes from keyboard to device. When another input device is used with the keyboard, frequent changes between the two styles can be confusing.

These guidelines should be considered during the design stage of the expert system's life cycle. Incorporating user-friendly procedures is easiest and most effective during the design stage. The best judge of a user interface is not a given set of rules, but an evaluation by the user. Attention to the user interface design will reap countless benefits to the acceptability, functionality, and efficacy of the expert system.

Case Study: A User Interface for WPHELP

As more and more people use computers, both in the workplace and at home, the quality of the user interface is becoming increasingly more important for vendors and users. People want computer software that is user friendly. Today's user interfaces have improved, but there still is much to be done. Let's consider what, ideally, would be the user interface for WPHELP.

First, our expert system should have the capability of either keyboard or mouse input. The use of a mouse has become very popular in recent years.

We would like the communication between the user and WPHELP to be as English-like as possible, so we need natural-language understanding capabilities. Voice input/output is a possibility for future versions of WPHELP, but the technology is not well developed now.

Communication with the user should incorporate all four methods of screen design discussed in the chapter: question and answer, menus, icons, and windows.

Question-and-answer screens should be developed in ordinary English. They should allow the user to respond in an open-ended fashion, as though he or she were conversing verbally with a human expert. Voice would be the ideal medium for question and answer, but the keyboard will have to do for some time to come. These screens will allow the user to express his or her word processing needs.

Menu selection and icon screens would be useful for allowing the user to make selections about desired attributes of a potential word processing system. These screens could be easily implemented using a mouse or a keyboard.

The alternate use of different input devices may be bothersome to users who prefer to use only one device. The use of a mouse would not allow the use of the free expression capability of the question-and-answer screens, so we will have to decide whether to use only the keyboard or gamble on combining two devices.

Windows would also be very useful. WPHELP might make use of three windows on the screen simultaneously. One window might show the question-and-answer communication between the user and the expert system (using one-half of the screen).

A second window could be used to show keyword summaries of the user responses. As a user, it is easy to forget what responses have been made; rather than querying WPHELP, the information could be easily retrieved. Perhaps the window size could be one-fourth of the screen.

The third window (also one-fourth of the screen) could be available for help on word processing terminology or on the use of WPHELP. Good help facilities will go a long way toward user friendliness.

Next, WPHELP should have report-generation capabilities for the printer in a format that the user can design. Different users usually have quirks about how reports of findings are to be formatted. Depending on the user's desires, the report could be printed out or stored electronically for later retrieval.

Finally, WPHELP should have a database and a spreadsheet attached to it. The database could store in summary form complete details about the selected word processing package. The spreadsheet could assist the user in analyzing the cost impact of the purchase of the selected word processing package for his or her enterprise.

Key Points

- An expert system is composed of three parts: knowledge base (rules, facts), inference engine (reasoning mechanism), and user interface (communications vehicle between the user and the expert system).
- The user interface is analogous to an atmosphere surrounding a planet. The user interface has physical, perceptive, and psychological components.
- The user interface is important to the overall success of the operation of the expert system.
- Identifying the user's tasks is critical to designing the user interface.
- The user interface is affected by the type of screen design and the method (input device) of implementing interaction (dialogue) between the user and the expert system.
- There are four common types of screen design:
  - Questions and answers
  - Menus
  - Icons
  - Windows
- There are many input devices for effective communication or a dialogue between the user and the expert system. The six most common are:
  - Keyboard-cursor keys
  - Mouse
  - Touch-sensitive screen
  - Light pen
  - Voice
  - Integration interface
- The QWERTY keyboard of 1864 is still the most popular keyboard in offices today.
- A WIMP interface is a term describing a user interface of windows, icons, menus, and pointers (usually a mouse).
- A user interface can be any imaginative combination of screen format and user input device.
- No set of magical rules will ensure an effective user interface, but consideration of good screen design guidelines will enhance the expert system.
- The following are general guidelines for effective screen design:
  - Use uncluttered, well-spaced screens.
  - Avoid paragraphs of text on the screen.
  - In general, left-justify text and right-justify numbers.
  - Limit the number of menu selections to five to seven.
  - Be consistent in wording and location of prompts.
  - Use reverse video or color for data entry fields.

  - Use color and graphics to add to screen design.
  - Let the user select some characteristics of the screen interface.
- Guidelines for effective user communications or dialogue with input devices are as follows:
  - Minimize the number of keystrokes.
  - Use a two-level command or menu selection structure.
  - Help the user navigate through the system.
  - Provide on-line help.
  - Guarantee quick response time.
  - Avoid blank screens during any dialogue session.
  - Implement an undo feature in the commands.
  - Simplify the error messages.
  - Provide for atypical users.
  - Minimize the number of changes.
- An interface for the general user (consumer, citizen, or voter) must be designed differently than the interface for a domain-specific expert (doctor, lawyer, or mechanic).
- A WIMP interface, or touch-sensitive screen, would be the preferable interface for the case study WPHELP because of the diversity of the users of the system.

Exercises

1. Define user interface.
2. What is the most common screen format?
3. What is the most common input device for computer systems today?
4. Given the four pull-down menus in Figure 7.13, identify a heading that would be appropriate to identify each submenu.

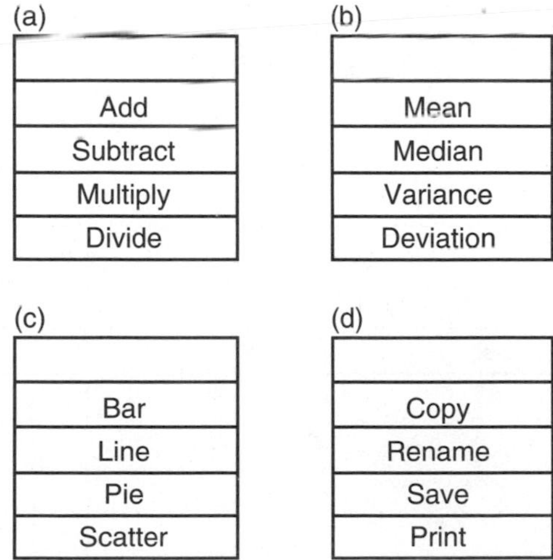

FIGURE 7.13 PULL-DOWN MENUS

5. Place the following topics under the most appropriate pull-down menu in the menu bar in Figure 7.14. Some topics may apply to more than one submenu.

Escrow	Adjustable rates
Cost of checks	Money market
Monthly service charge	Certificates of deposit
Principal	Closing costs
Points	Insurance
Fixed rates	Interest rate

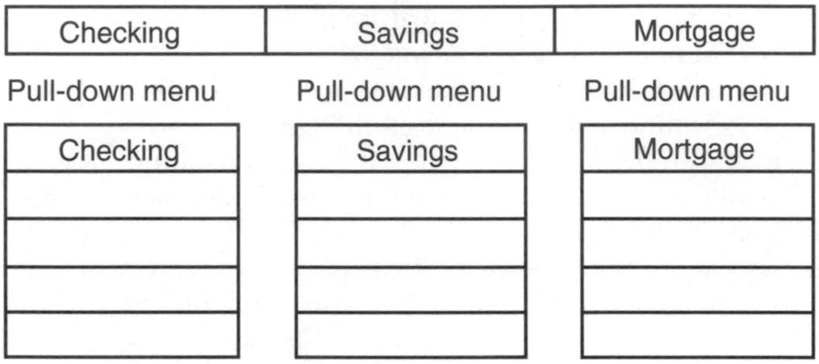

FIGURE 7.14 MENU BAR AND PULL-DOWN MENUS

6. Given the hierarchical menu structure in Figure 7.15, translate the command structure into a tree structure of menu screens.

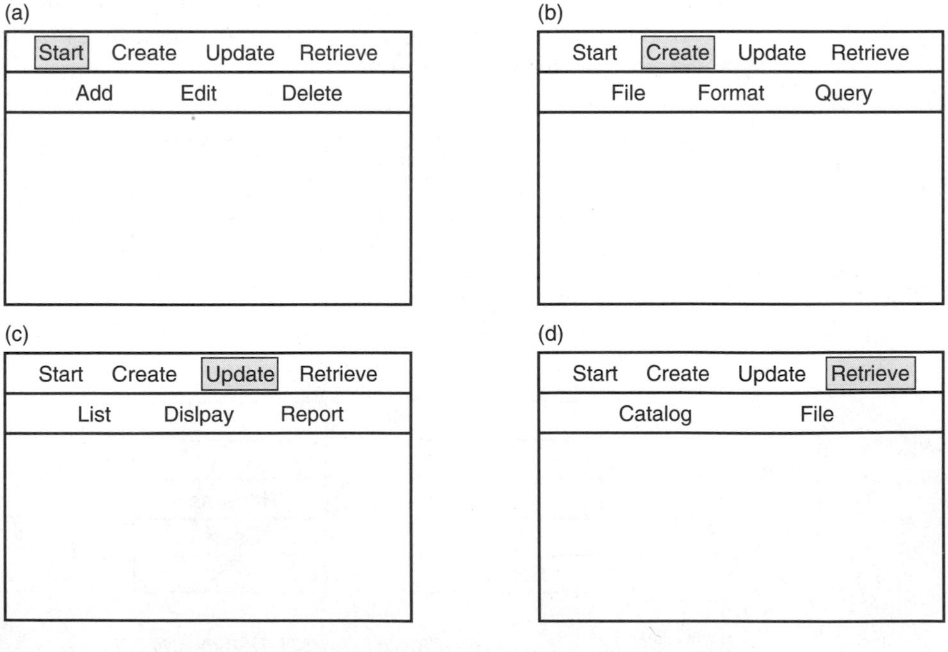

FIGURE 7.15 HIERARCHICAL MENU STRUCTURE

7. Given the tree structures in Figure 7.16, translate the information into a menu bar with pull-down submenus.

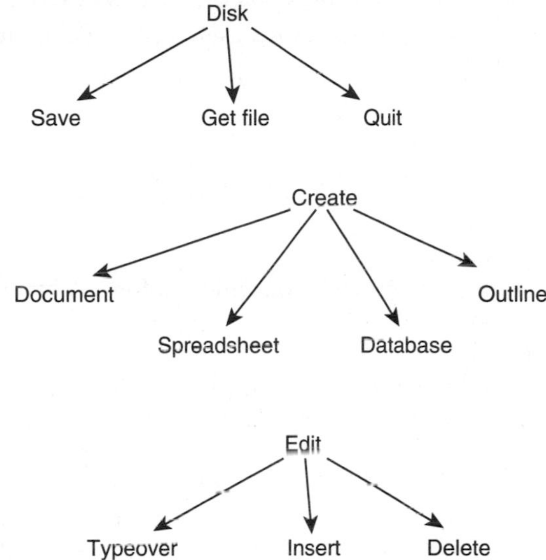

FIGURE 7.16 TREE STRUCTURES FOR MENU INTERFACE

8. Each of the six screens in Figure 7.17 violates at least one user interface guideline. Identify the violations.

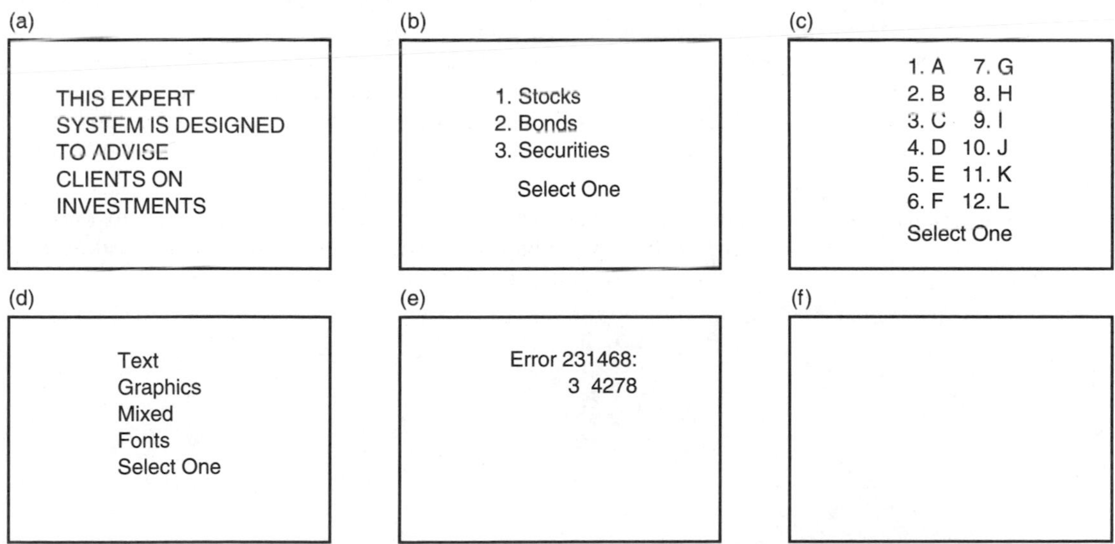

FIGURE 7.17 VIOLATIONS OF USER INTERFACE GUIDELINES

9. What are the three components of the user interface?
10. What two general areas affect the user interface?
11. What are the four most common types of screen design?
12. Give the set guidelines for user/software communication.

Assignments

1. Design icons for a campus information expert system for a nearby community or four-year college. By each icon, write the noun or verb appropriate for the symbol.
2. Expand the list of guidelines for effective interfaces by consulting the book *Human Computer Interface Design Guidelines* by G. Martin Brown Norwood, NJ: Ablex Publishing Company. Chapter 14 is a summary of the guidelines presented and explained throughout the book.
3. Write a report on the desktop interface of the Macintosh microcomputer. A good reference is *Human Interface Guidelines. The Apple Desktop Interface* by Apple Computer, Addison-Wesley Publishing Co., New York.
4. For detailed information about existing interfaces for expert systems, read *Expert Systems: The User Interface* (a collection of works on expert system interfaces) edited by James Hendler (Ablex Publishing Company, 1988).
5. As you probably know, there is no consistency among user interfaces from software package to package. Several vendors have attempted to develop the graphical user interface (GUI) or the common user interface (CUI). Examples are IBM's Common User Access (CUA) and Microsoft Windows/Windows NT. Use your library resources to write a short paper (6–8 pages) describing the GUI or the current state of the art in user interface design. Look especially for materials written by Ben Schneiderman. Other good references are the publications of the Special Interest Group on Computer/Human Interface (CHI) of the Association for Computer Machinery (ACM).
6. Using what you have learned from reading the text and from Assignment 5 above, evaluate the user interface of VP-Expert. What are its strengths and weaknesses? What changes would you make to improve the user interface?
7. Repeat Assignment 6 using the materials gathered in Assignment 5 in Chapter 6.

8 Software Engineering and Expert Systems Development/ Integration

In the last three chapters, we have examined what is involved in creating the three component parts of an expert system. We are now ready to put these pieces together.

In this chapter, we will examine the relationship between software engineering and expert systems, including a discussion of the software life cycle, the expert system life cycle, and appropriate software engineering techniques that should be used in expert system development. We will then discuss the various phases of expert system development. Potential difficulties will be examined. Finally, we will discuss the issues involved in integrating the expert system into the organization.

Software Engineering and Expert Systems

An expert system is a piece of software. In some ways, an expert system is like conventional software, but in other ways, it is quite different. In this section, we will discuss the life cycles for both conventional software and expert systems and how software engineering techniques can be used in the development of expert systems.

The Software Life Cycle

To understand the importance of software engineering, it is helpful to understand the **software life cycle**, which details what happens to a piece of software from the time the idea is born until the time it is discarded. The software life cycle is illustrated in Figure 8.1.

The figure shows that once the software has been developed, it enters a *looping pattern* of use and modification. The software stays in this loop until it is discarded.

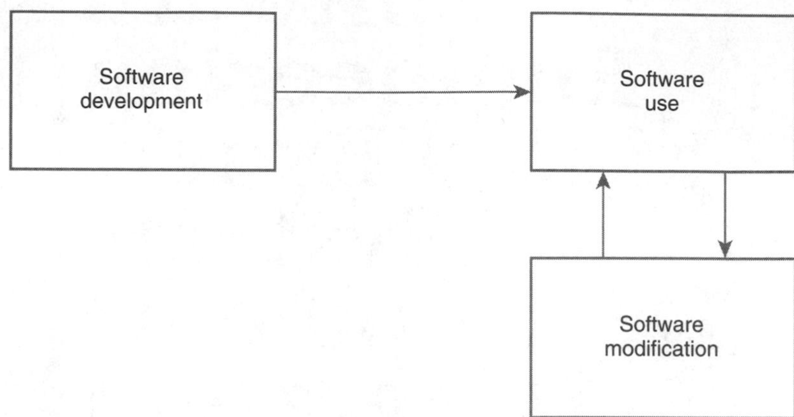

FIGURE 8.1 THE SOFTWARE LIFE CYCLE

Such a looping pattern is common for any product that is manufactured. The product cycles between use and repair/maintenance until it eventually wears out and is replaced.

Unlike other manufactured products, software does not wear out. After its development, it moves into the use phase, where someone may determine that the software contains errors or that changes are required. The software then loops through this pattern because once errors are corrected or changes are made, new problems may arise during the next use phase. Whenever changes are made to software, there is a risk of causing new problems. For example, changes in an accounting procedure will require changes in a program that handles accounts payable or accounts receivable. These changes, in turn, may cause problems that will not be discovered until a later time.

Let's examine the relationship between the development and modification phases of the software life cycle. The amount of time and effort spent in the modification phase is likely to be inversely proportional to the time and effort spent in the development phase. That is, more time and effort spent in the development phase may cause less time and effort to be required in the modification phase. If care is taken early on to ensure proper system design and implementation, the resulting system is more likely to meet most of the user's needs. Also, it is likely that there will be fewer errors to correct in the modification phase.

A great deal of importance must be placed on the development phase of any software project, especially large software systems that involve many people. A methodical approach using software engineering techniques is needed.

In a systematic approach, several steps are required. Our discussion here will be brief; if you want more detail, consult a text such as *Systems Analysis, Design, and Implementation* by John Burch [boyd & fraser publishing company, 1992].

The first step in the development phase is to **formalize** the problem specifications. At this point, the decision is made that software will be appropriate for the given application. The requirements are then isolated and production begins. These requirements are refined and become the specifications of the application. For example, the application might require that data be accessed only by authorized personnel.

The set of specifications thus forms the basis for the software. This process of formalization is analogous to using a blueprint to build a building or CAD/CAM (computer-aided design/computer-aided manufacturing) to manufacture a product.

The formalization process should maximize the understanding of the objective of the application by both the user(s) and the software developer. If the process does not work this way, additional time and effort will be needed later in development to make changes.

Formalization is probably the most difficult part of the development phase. The software developer (who may or may not do the actual programming) must be a skilled communicator who can ensure that requirements are clearly stated. Often, users cannot articulate their application needs. Sometimes a manager will provide limited input to indicate the parameters of the application. Except for a completion deadline, the software developer may get little direction. The software developer is then responsible for most of the formalization and for presenting the specifications to management for review.

The second step involves the **design of the software system structure**. At this point, the application must be divided into a series of tasks or modules. At the same time, techniques for interfacing the modules must be considered to minimize the time spent in future system maintenance. The software developer must ensure that all the specifications have been accounted for in the design of the software system.

Implementation, the third step, involves the actual creation of the software program. As in the design step, implementation is accomplished on a modular basis using the modules defined in the design process. As modules are completed, their interface to other modules in the software system must be tested thoroughly. Testing is tedious and, at times, is considered to be of little importance by system developers. Slighting the testing process in this step, however, usually creates larger and more difficult problems when the entire system is put together; these problems otherwise may have been avoided.

The fourth step involves the **testing** and **debugging** of the entire software system. Do all the modules interact properly with each other? What bugs remain that were not found in the design step? At this time, the overall software system is assessed. Does the software give the user satisfactory performance? Have all original specifications been met? These questions must be resolved now.

When this step of the software system development phase is completed, the system should require minimal time in the modification phase, except for changes requested after the original formalization stage.

The final step in the development phase is the **preparation of documentation** for the software system. This step is very important because poorly documented software will often become unusable. Poorly written documentation greatly complicates the efforts of people assigned to the modification phase; it also makes using the system very difficult. The worst case could result in the software system being seldom or never used. If the latter occurs, a lot of time, effort, and money will have been wasted.

Just like good help, good documentation is hard to find. For a variety of reasons, documentation on most software systems rates mediocre at best. Providing well-written system documentation takes time and effort; documentation is usually the area that falls short of user expectations. Good documentation makes good software better; poor documentation can make good software unusable.

The Expert System Life Cycle

Because an expert system is a software system, its life cycle shares a number of similarities with the general software life cycle. There are, however, some differences that will become clear in our discussion. The **expert system life cycle** is illustrated in Figure 8.2.

In the development phase of the expert system life cycle, the expert system is actually created. This is analogous to the development phase in the software life cycle. The same basic steps are used in the development of the expert system except for some unique differences.

First, a conventional software system is designed to deal with a single application or problem. Its specifications are developed in conjunction with the user, who has a task to be accomplished.

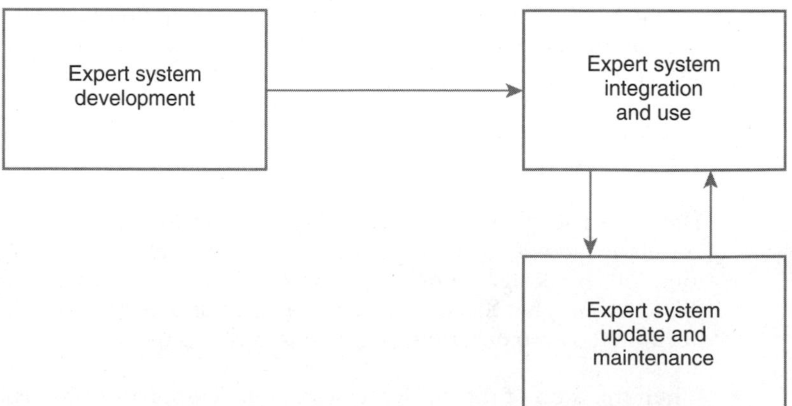

FIGURE 8.2 THE EXPERT SYSTEM LIFE CYCLE

On the other hand, an expert system is intended to solve a range of applications or problems in a narrow domain of knowledge. Knowledge, usually from a human expert, must be captured and put into a form that can be understood by the expert system. While formal specifications are not developed as they are in a conventional software system, the domain of knowledge must be clearly defined. The knowledge engineer who develops the expert system works with the human expert to acquire the knowledge but does not, generally, work with the user. The exception to this occurs when the knowledge engineer must model the user's interactions with the expert system, such as choosing the vocabulary used in system/user interaction.

The use and modification phases of the expert system life cycle are analogous to the corresponding phases in the software life cycle. In the **use phase**, the expert system solves problems in its domain, just as the conventional software system deals with the desired application. In the **modification phase**, the knowledge base is updated or modified and errors are corrected. As with conventional software, the expert system loops between its use and modification phases.

In comparing the differences in the two life cycles, we can best summarize them this way. Conventional software, by its nature, is usually designed to accommodate changing user needs. Modifications are expected from the beginning. Because conventional software development is a very competitive business, there is a great deal of pressure to get the software to market as soon as possible. Often, less-than-adequate testing results in a high rate of errors present in the software as it enters the use phase. Consequently, more time and effort are needed in the modification phase to customize needs and to correct errors. This description of conventional software may be disputed by some software developers, but, sadly, it is reality.

On the other hand, an expert system must be of high quality before it is released for use; it must be considered high-performance software. In general, it is not in a competitive marketing situation so it does not face the pressure to hurry to market.

This discussion can be summarized in Figure 8.3, which compares the time spent during the life of development and modification/maintenance phases for conventional software systems and expert systems. The time spent on the different phases during the life cycle varies dramatically.

The development phase of the expert system life cycle contains several stages, which include:

1. Problem analysis
2. Problem formalization
3. Knowledge acquisition
4. Knowledge representation
5. Prototype development
6. Full system development
7. System evaluation and documentation

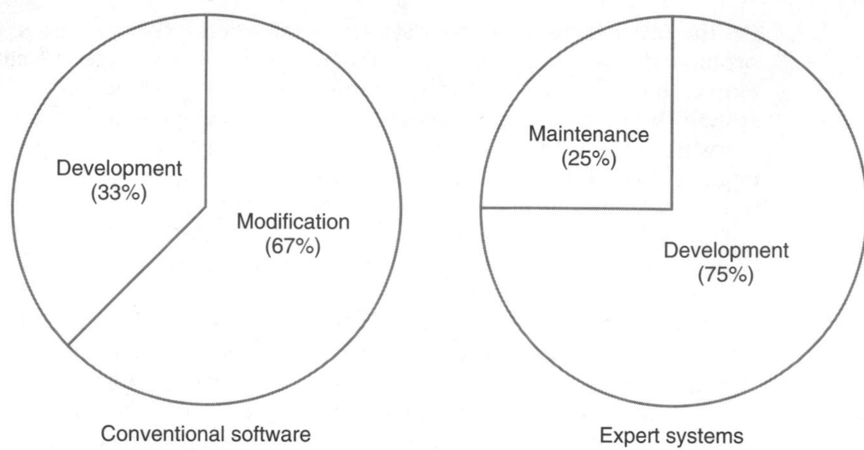

FIGURE 8.3 COMPARISON OF APPROXIMATE TIME SPENT IN THE LIFE CYCLES

We can briefly summarize the objectives of each step as shown in Figure 8.4.

Details of the first four steps were discussed in Chapter 5. We will now discuss details of the last three steps.

STEP	PURPOSE
1. Problem analysis	Determine if the problem is appropriate for expert system development
2. Problem formalization	Design the basic structure of the expert system
3. Knowledge acquisition	Locate an expert and acquire knowledge from the expert
4. Knowledge representation	Determine how the knowledge should be represented and choose a development tool
5. Prototype development	Initially test the system design, knowledge representation, and development tool choices
6. Full system development	Construct the full expert system
7. System evaluation and documentation	Finalize system evaluation and documentation before system goes into use

FIGURE 8.4 STEPS IN EXPERT SYSTEM DEVELOPMENT AND THEIR PURPOSES

The Software Engineering Methodology

The role of software engineering is to provide law and order in the chaotic software development process. This role can be summarized in Figure 8.5.

When a project emerges as a candidate for software development, several conditions are certainly true. First, there is a wide variation of the form of the project, such as its definition and scope. A relatively high cost for the development process is assumed.

The software developer must take this dynamic situation and fit the project into a framework for software development. This is best accomplished using techniques of software engineering.

Software engineering techniques should produce plans for the development process including timetables and structured design for the software. Software engineering will also use the software life cycle as a model for the development of the software.

The result of the application of software engineering techniques is a product that should have certain characteristics. The product should be:

- Fully documented
- Verified, validated, and tested
- Cost effective
- Completed on time
- Easily manipulated and enhanced

These characteristics are certainly reasonable and appropriate. Their attainment must be the goal of every software development project, whether conventional software or an expert system.

Putting the Expert System Together

Prototype Development

The fifth step in the development of the expert system involves the development of a **prototype** of the expected final version of the expert system. The prototype represents a piece of the final system and serves a number of purposes.

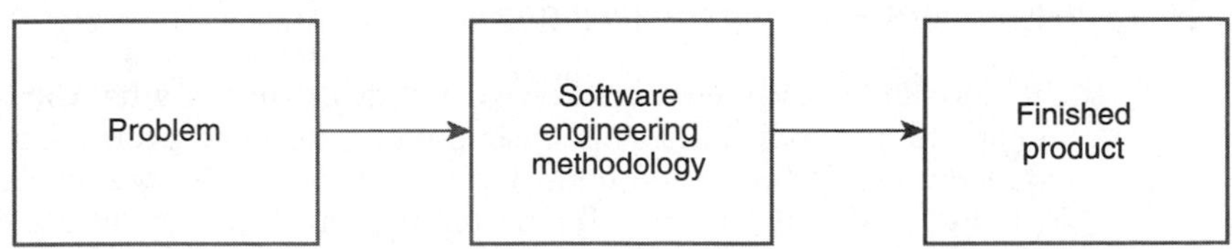

FIGURE 8.5 THE EFFECT OF SOFTWARE ENGINEERING ON SOFTWARE DEVELOPMENT

Before we discuss the purposes of prototyping, we must note that the process of acquiring knowledge should be performed in parallel with the prototyping process. We will not elaborate on the knowledge acquisition process here, as it was discussed in detail in Chapter 5.

Prototyping has several purposes. First, it gives the knowledge engineer an opportunity to *validate* the choice and definition of the problem to be solved. Does the knowledge engineer understand the nature or scope of the problem?

A second purpose of prototyping is to determine the *appropriateness* of the chosen development tool. Of course, not all possible difficulties in the use of the tool would surface at this point; if a gross mistake in the tool choice has been made, it is now likely to be evident. Near the end of the full development process, discovering that the chosen development tool did not, in fact, work as expected would be frustrating. That late discovery would force a change to a new tool.

A third purpose of prototyping is to give the knowledge engineer a chance to *assess* the initial completeness of the knowledge acquisition process. In consultation with the human expert(s), the knowledge engineer assesses the initial results of the creation of the knowledge base. This interaction with the human expert(s) carries a beneficial side effect: active involvement between the human expert(s) and the expert system is likely to sustain their interest in the system. It also fosters communication between the knowledge engineer and the human expert(s).

A fourth purpose of prototyping is to give the knowledge engineer an opportunity to *evaluate* the design of the expert system. If design flaws are encountered, they are more readily corrected at the prototyping stage.

The prototype can also serve as an initial test of the reactions of a group of potential users to its output. Involving potential users in this process can assist in correcting any problems with the user interface. Suppose, for example, that the language used in the user interface is inappropriate for the end user; that problem could be detected and corrected at this time.

A final purpose of prototyping is that it *involves management* in the development process. Viewing initial development results helps keep management informed; because they are part of the development process, managers may be more willing to allocate additional resources should the knowledge engineer need them.

These six purposes obviously indicate a number of measures that will and should occur in the actual prototyping process.

The knowledge engineer should have a clear picture of what the final expert system should do and look like. This clear picture, however, must exist not only in the engineer's mind, but in the minds of all concerned. Tools such as flowcharts of the sequence of events and story boards (scripts) may be very useful; every effort should be made to make the system design available in a visual form, on paper or on a computer, to others (human experts, management, and so on) involved in the development process.

The knowledge engineer also must analyze the knowledge that has been gathered. What is known? What is not known? What relationships among portions of knowledge need to be explored?

Finally, the prototype development should be accomplished as soon as possible to expedite any changes that may need to be made. As the knowledge engineer gains experience in developing expert systems, the prototyping process will likely become easier to implement.

Prototyping is a crucial part of the expert system development process. Time and effort taken at this point will be rewarded as the process moves to completion. We cannot overstate the importance of the prototyping stage.

Full System Development and Evaluation

The sixth step of expert system development involves the **extension** of the prototype into the full expert system. At this stage, initial problems discovered in the prototyping stage should be solved. It is now a matter of putting the components together.

The knowledge engineer continues to expand the knowledge base toward its final form, as indicated by the initial problem identification and expert system specifications. Continuous interaction between the knowledge engineer and the human expert(s) is needed to validate the acquired knowledge. This validation often is performed successfully by presenting new problem situations that are still within the specified knowledge domain to both the expert system and the human expert and comparing their results for consistency.

As new knowledge and rules are added to the knowledge base, their effects must be assessed relative to the inference engine. Does system performance degrade after more knowledge and rules are added? Lower performance levels may be evidenced by, for example, more time needed to solve a given problem, inconsistent or unexpected results, or incorrect results. These situations may be caused by searching techniques used on the knowledge base or conflicting rules. If one or more such problems occur, they should be addressed immediately because they are likely to compound themselves if ignored.

There may also be a problem in the original design of the knowledge base, the choice of the expert system development tool, or the choice of knowledge representation. Even though the prototyping stage should have identified these issues, problems sometimes "fall through the cracks." This situation is more likely to happen with an inexperienced knowledge engineer, but it can occur with even the most experienced knowledge engineer.

The user interface must be continuously evaluated through contact with potential end users. As we mentioned in Chapter 7, the user interface is typically the most neglected component of the expert system, especially if time becomes a major constraint.

Appropriate management personnel should be given periodic progress reports and demonstrations of the expert systems as it is being developed. This progress update should provide information only. The knowledge engineer should not allow management to modify or add to the original specifications.

When the system has been completed, a final testing and evaluation period follows. Testing and evaluation is often done by a person or group of people who did not actively participate in the construction of the expert system because they will give a more realistic evaluation of the expert system. However, potential users of the expert system should also participate in the final testing and evaluation process.

We do not mean to suggest that the knowledge engineer and the human expert should be left out of the final testing and evaluation process. Clearly, they must be included, but a community of people will facilitate the most successful evaluation.

Plans for maintenance must also be formulated and executed during the full system construction process. Documentation on updating the system, handling ongoing maintenance, and using the system, must be provided. Development of this documentation should be a continual process that begins at the start of prototyping and ends with the final phases of construction of the full system.

Difficulties in Expert Systems Development

Companies wishing to apply expert systems technology to solve their problems encounter several types of difficulties: scarce resources, time for development of the expert system, errors in the expert system itself, and its inherent limitations. In this section, we will examine these difficulties and the impact they have on development.

Scarce Resources

Within a company or organization, competition for **resources** is tremendous. There is rarely sufficient time, personnel, or money to accomplish management's goals. The overall objective of every organization is to increase productivity at the same or lower cost.

Freeing resources for expert systems development is difficult for two reasons. First, expert systems, like their parent field AI, represent a new technology. They have not yet established a track record of success and are still relatively unfamiliar to many computer specialists and managers. The decision makers in the organization must be convinced that using expert systems technology will, in fact, increase productivity at a justifiable cost.

The second problem is that a number of companies and organizations are attempting to enter the arena of expert systems development. Because the technology is new, experienced knowledge engineers and other trained personnel are in short supply.

Due to this scarcity, hiring a consultant to do expert systems work may cost more than can be justified. To fill this void of expertise, a number of companies are training in-house personnel. This solution presents additional difficulties during the learning period when employees are not available to perform their assigned job responsibilities. Consequently, the slack must be taken up somewhere else in the organization.

Development Time

Development of expert systems cannot be hurried. Depending on the application, completing the development process may take months or even years. This is especially true when the knowledge engineer is inexperienced.

One of the major obstacles to development is time for the knowledge acquisition process. Executed properly, the process of extracting the knowledge from the human expert to create a complete knowledge base cannot be rushed. Both the knowledge engineer and the human expert must work diligently to complete the knowledge acquisition process successfully.

Many expert system development projects are team efforts, which take extra time simply because of their nature.

From another angle, time constraints can also pose a problem. Often, a time limit is placed on the completion of the expert system development project. When the limitation is enforced, the people involved may believe that increasing the team size will decrease development time. More than one organization has fallen into this trap.

Sources of Errors

In the development process and in the expert system itself, errors can occur in a number of sources, as summarized in Figure 8.6.

The basis of knowledge for the expert system is the human expert's knowledge. If this knowledge is *incorrect* or *incomplete*, the same errors will be present in the knowledge base.

Having another expert or a panel of experts certify the human expert's knowledge may be desirable. This is critical when the expert system is concerned with life-threatening situations, such as medical diagnosis expert systems are.

Semantic errors can occur in the interaction phase between the human expert and the knowledge engineer. The message may become lost in the translation; this is more likely to happen with inexperienced knowledge engineers or engineers who have not become as familiar with the domain area as could be expected. Gaps in the knowledge stream can also cause errors.

Syntax errors in the knowledge base can be caused by using an incorrect form of a rule or a fact. The development tool should flag these errors.

SOURCE	ERROR
Human expert	Incorrect or incomplete knowledge
Knowledge engineer	Semantic errors in communication with the expert, knowledge gaps
Knowledge base	Syntax errors in format, errors caused by incorrect or incomplete knowledge and uncertainty
Inference engine	Bugs in the inference engine and/or development tool, incorrect rule priorities or rule location in the knowledge base
User interface	Incorrect content of communication between the expert system and the end user

FIGURE 8.6 SOURCES OF POSSIBLE ERRORS IN THE EXPERT SYSTEM

Errors may occur in the inference engine due to *bugs* in either the structure of the engine or the development tool. *Rules* may have incorrect priorities or may be inserted incorrectly (in the wrong place) in the knowledge base.

Finally, *content errors* can occur in the user interface. Users may be given incorrect or incomplete instructions, causing a breakdown in the system effectiveness and use.

Most, if not all, of these errors should be corrected during final system evaluation, before the system is released to users. In reality, totally eliminating errors is very difficult and increases in difficulty as system size and complexity grow.

Inherent Limitations

For all the things it does well, expert system technology currently has **inherent limitations.** If these limitations are recognized and accounted for, they are less likely to cause severe problems.

Expert systems now are not capable of acquiring knowledge directly. The acquisition process must employ a knowledge engineer, who then translates the knowledge into a form the expert system can understand. Researchers are working on the development of automated acquisition software, but this software is some years away from commercial use.

Expert systems now are not capable of dealing with knowledge that has a mixed representation. For example, an expert system in one application cannot deal with knowledge that incorporates the forms of rules and of semantic nets. Such mixing capability might be appropriate for applications that merge several representations.

Current expert systems are not capable of refining their own knowledge bases. Refinements might include combining rules with the same conclusion or eliminating redundant or contradictory rules. These actions must be performed externally.

A final limitation is that an expert system cannot learn from experience. This would be a desirable characteristic because learning from experience is an integral part of a human expert's capability. Researchers are avidly pursuing the ability to learn from experience, but expert systems capable of it are still several years away from the commercial world.

Integrating the Expert System into the Organization

The final stage of the expert system development process is its **integration** into the organizational structure. This stage actually puts the final expert system *into production*.

The integration process, however, does not start after the completion of the expert system, as you might infer. It starts with the initial idea to build an expert system. One key to the successful integration of the expert system is the early involvement of management and other personnel whose work will be affected by the presence of the system. As we suggested earlier, their involvement in the planning stages and the effect of their input during the development process are likely to enhance the expert system's acceptance. Those who will work with and around the expert system must recognize its potential to improve productivity and preserve jobs. Successful integration of the expert system depends on how well these two views (increased productivity and job security) are espoused. This task is the job of the project leader (who may or may not be the knowledge engineer).

From the perspective of the organization, the arrival of the expert system cannot be disruptive. The organization must continue to function in an efficient manner with as little interruption as possible.

From the expert system's perspective, the expert system must be seen as helping to achieve the organization's goals. The development team, including the human expert and the knowledge engineer, must feel that the expert system has a well-defined and accepted place in the organization.

Finally, from management's perspective, the expert system must increase productivity and preserve good employee morale.

A Final Remark

The development process for an expert system is complex. The successful completion of an expert system depends on the cooperative efforts of a number of people from beginning to end. Well done, the expert system development process can be a major asset to an organization's future; poorly done, the process can cost the organization a great deal of time, money, and effort.

Case Study: WPHELP

So far, we have attempted to use our sample expert system, WPHELP, to illustrate the theory of expert systems. Some of the illustrations have been actual examples, while others have been hypothetical. We will continue using WPHELP to illustrate our purpose in the modules that follow.

Although WPHELP is a hypothetical system, its development throughout the chapters illustrates the principles of software engineering discussed in this chapter. In its present form, WPHELP is in the prototype stage of development. As more rules are added, it may become a marketable expert system.

Key Points

- An expert system is a piece of software; its development process can use principles of software engineering.
- The software life cycle contains three phases: development, use, and modification.
- In the software life cycle, approximately 67% of the time and effort are spent in the modification phase and approximately 33% in the development phase.
- The steps in the development phase of the software life cycle include:
  - Formalization of the software specifications
  - Design of the software system structure
  - Construction of the software system
  - Testing of the software system
  - Documentation of the software system
- The expert system life cycle includes the following phases: development, use/integration, and updating/maintenance.
- In the expert system life cycle, approximately 75% of the time and effort are spent in the development phase and approximately 25% in the updating/maintenance phase.
- The development phase of the expert system life cycle includes the following steps:
  - Analysis of the problem
  - Formalization of the problem
  - Knowledge acquisition
  - Knowledge representation
  - Prototype development
  - Full system development
  - System evaluation and documentation
- The role of software engineering is to bring law and order to the chaos of software development.
- The software product that results from the software engineering process should be:
  - Fully documented
  - Verified, validated, and tested
  - Cost effective
  - Completed on time
  - Easily manipulated and enhanced

- Knowledge acquisition should be performed in parallel with the prototyping process.
- Prototype development has several purposes:
  - Validate the choice and define the problem to be solved.
  - Determine the appropriateness of the chosen development tool.
  - Assess the initial completeness of the knowledge acquisition process.
  - Evaluate the design of the expert system.
  - Test the reactions of potential users.
  - Allow management to see initial results of the expert system development process.
- Full system development is an extension of the prototype, but new problems may occur during the extension process.
- The final system evaluation must be complete and thorough and include outside evaluations and potential users.
- Difficulties in the development of an expert system include scarce resources, development time, sources of errors, and inherent limitations of expert systems.
- To integrate the expert system successfully into the organizational structure:
  - The integration process must start at the beginning of the expert system development phase.
  - Early involvement of management and other personnel whose work will be affected by the presence of the expert system is important.
  - The arrival of the expert system cannot disrupt the organization.
  - The expert system must be seen by all concerned as making a contribution to the organization.
  - The presence of the expert system must increase productivity.

Exercises

1. What is the relationship between expert systems and software engineering?
2. Describe the software life cycle.
3. Describe the expert system life cycle.
4. Describe the software engineering methodology.
5. Compare and contrast the development phases of the software and expert system life cycles.
6. Name the seven steps in the development of an expert system. What is the purpose of each step?
7. What are the purposes of developing a prototype?
8. Describe the processes of full system development and final system evaluation.
9. Discuss the four major areas of difficulties in the development of expert systems.
10. What is the role of software engineering?
11. What characteristics should the end product of the software engineering process possess?
12. What strategies should be used to integrate the expert system into the organizational structure?

Assignments

1. The final evaluation of an expert system is the last phase of the development. In Chapter 5 (Assignment 5) and Chapter 6 (Assignment 3) you developed a rule-based knowledge base for each of a series of problems. For each knowledge base, outline a plan for final evaluation of the knowledge base. Your plan should include the following:
   a. how the rules will be verified.
   b. how the inference chain(s) will be validated. (Include a complete set of test data.)
2. Read and summarize two articles that deal with the application of software engineering techniques. Discuss how the content of these articles might apply to the development of expert systems. Good sources of articles are *Byte*, *AI Expert*, and *PC AI*; however, there are other sources as well.
3. From your library resources, collect several case studies of actual expert systems. For each expert system described, determine to what extent the expert system life cycle was implemented in the expert system development. Report your findings to your classmates or friends. One example is an article by Snell and Mah, "A Real-Time Expert System for Quality Control," in the October 1992 issue of *IEEE Expert*.

9

The Future of Expert Systems

Expert systems technology is gaining widespread acceptance in the commercial world. Companies are talking about their successes with the technology. Other companies are experimenting with the technology, and new applications continue to surface. The interest has created a demand for new and better expert systems development tools that can be used by less experienced knowledge engineers. Improvements in computer hardware/software technology and vendors' desire to produce quality development tools have led to a greater use of the maturing expert systems technology.

In the future, we will see improvements in several areas of expert systems technology: knowledge representation, knowledge acquisition, development tools, expert system design, and the programming of expert systems. In this chapter, we will briefly examine these areas. We will also discuss the future availability and cost of expert systems and the concept of intelligent systems. Our final case study scenario will outline a future word processor.

Knowledge Representation

As we pointed out in Chapter 3, the choice of a **knowledge representation scheme** is a key to effective expert systems development. Current models of knowledge representation will be improved and extended. For example, the rules model might be extended to include the direct use of *inheritance*, a characteristic of semantic networks that contends that if a node is part of a hierarchy, then all properties of the higher level are also properties of any node connected at a lower level.

An integration of different representation schemes in the same expert system is a likely development. This would allow an expert system to receive and use knowledge from more than one model. Knowledge represented as rules could be combined with knowledge represented as a semantic net or in the form of predicate calculus. Expert systems technology then could be applied to applications in which knowledge is represented more than one way.

Finally, new knowledge representation schemes will be developed that will be easier and more powerful to use; for example, research is being done on a **model-based knowledge representation scheme** that uses a model structure. The knowledge is given to the representation model, and the model converts the knowledge into the scheme. The knowledge is entered through a dialog between the model and the knowledge engineer, thus freeing the engineer from actually converting the knowledge into the representation scheme. Elements of this technique are already being seen in some expert system shells.

Other new knowledge representation schemes will allow the use of a single time dimension. The **script knowledge representation scheme**, currently under development, allows the use of specifications such as done-before and done-after.

Automated Knowledge Acquisition and System Development

The use of **automation** in the future development of expert systems is very feasible and appealing, especially for knowledge acquisition and the development of self-learning systems. In this section, we will examine its use in separate but closely related areas.

As discussed in Chapter 5, the task of knowledge acquisition is a complex one. It requires the full cooperation of both the human expert(s) and the knowledge engineer. The manual methods of knowledge acquisition—the interview and the analysis of verbal conversations between the human expert and the knowledge engineer—can cause a loss of valuable time for both parties. Inaccuracy and knowledge gaps also can occur. Even when the knowledge engineer is skilled and experienced in the knowledge-acquisition process, productive time that the human expert might spend elsewhere is inevitably lost. In addition, the human expert may not be able to detail his or her reasoning in solving the problem explicitly. These problems must be addressed for the knowledge base development process to be successful.

These problems have paved the way for the development of **automated knowledge-acquisition tools** (Parsaye and Chignel 1988). These new tools provide a structure that can be filled through the use of a computer program. Their use diminishes the role of the knowledge engineer as the facilitator of the knowledge-acquisition process.

Another major problem with the manual methods of knowledge acquisition is, as the reader can imagine, the cost involved. Automated knowledge-acquisition tools can decrease costs of acquiring knowledge in several ways. Martin and Oxman (1988) listed some of these ways:

1. Decrease the amount of time required to acquire the knowledge.
2. Decrease the skill level required of the person acquiring the knowledge.
3. Automate some aspects of the knowledge acquisition process.
4. Enable users and human experts to add to the knowledge.

It seems reasonable that these steps would decrease costs. The human expert and a skilled knowledge engineer usually are high-priced personnel. Reducing the human expert's time needed in the knowledge-acquisition areas means that he or she can spend more time doing what he or she does best—being an expert. The knowledge engineer can be freed from some mundane tasks of knowledge acquisition, such as collecting statistical data. Also, the knowledge engineer need not be called back to the site to update or modify the expert system if someone else, such as a user, can modify the expert system's knowledge base.

The cost of knowledge acquisition will decrease if the needed skill level of the knowledge engineer can be decreased. Knowledge engineers can be very expensive. Training a person who, with guidance from the knowledge engineer, can use automated tools for knowledge acquisition or perform lower-level acquisition tasks is one way to reduce costs. This situation parallels the use of paramedics and paralegals in the medical and legal professions. Also, the increasing quality of the knowledge-acquisition capabilities of expert system shells means that a less skilled person perhaps can perform as well as an experienced knowledge engineer. A word of caution is appropriate: investigate the alternatives before implementing this method of decreasing cost.

Automating some aspects of the knowledge-acquisition process can help decrease costs. Automation will also allow other people, both users and human experts, to add to the knowledge base. Some areas of the knowledge-acquisition process, such as the determination and closing of "gaps" in the knowledge base, however, probably cannot be automated. Automated knowledge-acquisition tools are not yet capable of this function.

Research and development work on automated knowledge-acquisition tools is ongoing in the industrial world and at universities. Activities have concentrated on automated tool development, automated consistency-check tools, and self-learning systems (Martin and Oxman 1988).

The development of automated knowledge-acquisition tools allows more people to serve as knowledge engineers at a lower skill level than ordinarily needed. The expanded pool of available people allows more expert systems to be built, making expert systems technology more useful to more applications. Developing cost-effective automated tools to deal with the tasks of knowledge acquisition will eventually be feasible.

Automated consistency-check tools would be used to refine the knowledge base to eliminate duplicate or contradictory rules. For example, if two rules had the same conclusion but different premises, the tool would notify the knowledge engineer so that he or she could make appropriate changes. As another example, two rules might have the same premise but contradictory conclusions. Suppose these two rules were present in the knowledge base:

Rule 1: IF there is a fire,
THEN throw water on it.

Rule 2: IF there is a fire,
THEN don't throw water on it.

The tool would indicate to the knowledge engineer that Rules 1 and 2 were contradictory and the engineer could make needed changes. Automated consistency-check tools also appear to be feasible, but much more work needs to be done.

Most researchers in expert systems believe that **self-learning systems** will be the future model for expert systems. These systems would make use of automated knowledge-acquisition tools in conjunction with automated techniques for refining and updating the knowledge base. These systems would be able to use *inductive techniques* and to learn from experience. Automation would be very much a part of a self-learning system.

The development of self-learning systems would enable expert systems to resemble the activities of a human expert more closely. Most experts increase their expertise through experience. They use inductive reasoning by hypothesizing on a set of examples. Through experience, the set of examples increases or is modified, possibly resulting in modified hypotheses. An inductive, self-learning system would emulate the expert's behavior.

There are a number of applications for *automated induction systems*. Such systems could search large databases and look for possible relationships. For example, marketing strategies for merchandise items such as clothing or children's toys could be formulated with an induction system. Over time, these strategies could be re-examined as new data became available.

It is also hoped that self-learning systems will be able to take knowledge from different forms (representations) and combine it into a single knowledge base. For example, a self-learning system might combine knowledge from a textbook or manual with knowledge from experience to create and update a knowledge base. Diagnostic, control, repair, and educational/training systems would benefit from this induction capability.

System Development Tools

Initially, expert systems were built with high-level languages. While the use of these languages had some merit, they were usually general-purpose languages that did not lend themselves to the inclusion of intelligence as an integral part of the expert system. Informing schemes had to be built from scratch. Even though the high-level languages had limitations, they were used because they were the only tools available.

Expert system shells changed the way that expert systems are developed. Unfortunately, the first shells were, in general, not very powerful because the shell technology was immature. Newer versions of the shells will have much more powerful utilities, such as graphics and natural-language interfaces, that shell users can work with easily. Future shells will appear to have CAD-like capabilities.

Martin and Oxman (1988) note that intelligent design tools will assist the knowledge engineer in a number of ways:

1. Determining problem domain characteristics
2. Ensuring that all important problem attributes are considered
3. Determining which knowledge representation(s) and inferencing mechanism(s) are used
4. Assessing system user capabilities and selecting appropriate user interface design schemes
5. Helping to select appropriate shells for given applications and determining if other software tools are needed

These advances in expert systems tools technologies will result in a higher quality expert system that will be able to deal more effectively with a large number of applications.

User Interfaces

A definite trend in recent years is improving the design and effectiveness of **user interfaces**—both the content of the communication and the physical means of communication between the user and the expert system. The majority of the improvements have come, however, in the content phase.

The term *"user friendly"* is familiar to many users and software developers. Most people who use the term have at least an intuitive, if not formal, understanding of what it means.

The term has had a significant impact on expert systems development. Historically, communication between an expert and a non-expert on the expert's domain of knowledge has not been easy. The expert usually communicates in what might be called the "domains," a dialect not always known by the user. The vocabulary of computers is an excellent example of "domainese."

The expert system must translate the expert's vocabulary for the user. Expert systems are using natural language and help facilities to make using the expert system as easy as possible for the user.

Natural-language interfaces attempt to make the human/machine interaction as natural as possible. A great deal of research is going on in the area of natural-language processing. However, significant breakthroughs have been constrained by the complexity of the problem.

The techniques of **key-word analysis** and **syntactic/semantic analysis** have been the usual one used in natural-language processing work. Because of the nature of the English language, where many words having multiple meanings, key-word analysis has had limited popularity. Syntactic/semantic analysis appears to have a much brighter future.

The physical aspects of the user interface will likely undergo dramatic changes in the future. Traditionally, the physical interface has the user sitting in front of a terminal with a keyboard and the screen as the communication media. Communication takes the form "I speak, you listen, and then you speak, I listen," where "speak" means putting messages on the screen via the keyboard or the computer's character generator and "listen" means seeing a message on the screen or entering a message into the computer's memory. It is a rather inhuman way of communicating.

Expert systems of the future will give "speak" and "listen" new meaning. The user will actually talk to the expert system, and the expert system will talk to the user through *speech recognition and synthesis* (speech production) technologies housed in the computer. And, the communication will be in ordinary English.

Research in speech recognition and synthesis technologies has progressed slowly so far. Success has been limited by the same constraints faced by natural language: memory limitations and computing speed.

In the future, expert systems will have user interfaces that will facilitate communication. The interfaces will be verbal and user friendly, with maximum attention given to the quality of the user interface.

Availability and Cost

The increasing popularity of expert systems and the extended number of possible applications have been evident in recent years. Expert systems technology now is available in some form to almost any business that needs it. The increased number of knowledge engineers and availability of affordable development tools have created a healthy environment for expert systems technology; the availability of people and tools will continue to grow in the future.

We must warn, however, that just because expert system technology is available, it is not the solution for every problem. Any business or individual seeking to implement expert systems technology must perform a careful needs analysis. As we have previously pointed out, expert systems technology is not a panacea.

Currently, the technology has significant inherent limitations. Keeping these limitations in perspective, as well as performing a careful, initial needs analysis, will help a business make a sound decision about whether to implement expert systems technology. A major factor in this decision is the cost to develop the expert system.

The price of computer hardware has decreased dramatically over the years, but the same cannot be said for computer software. Software development requires people, and the cost of people has increased; this trend is likely to continue, even with the advent of automated development tools. Consequently, the cost of developing an expert system is likely to increase. However, the expense of expert system development possibly can be justified. In addition to the initial outlay for development, there is the cost of updating and maintaining the expert system after its development. In each of these processes, people are involved, so costs are not likely to decrease.

The development of automated tools will help bring down the costs of expert systems. The tools will carry a high initial cost which may be offset by a reduction in the number of people needed.

Intelligent Systems

As we consider what is next for expert systems, we will find that they will be integrated with other systems, such as computer or manufacturing systems, to form **intelligent systems.** Up to now, we have considered expert systems in only a standalone mode.

The integration of expert systems with other systems affords the opportunity for more efficient and effective operation of the system. An expert system can be used to monitor or control the system, diagnose problems, or perform a number of other tasks. For example, suppose we have a robot with an expert system embedded in it. If the robot malfunctions, the expert system could diagnose the problem and report its findings to a human, who could then repair the robot. The diagnostic function performed by the expert system could save hours of troubleshooting time for the human. In another situation, the robot may be programmed to perform a particular task. If the environmental conditions in which the task was to be performed changed (such as weather or object misplacement), the expert system could reprogram the robot to meet the changed conditions, thus probably saving time and money.

Expert systems can be very useful when integrated with other systems. In the future, many more applications will make use of expert systems, enabling each application to be performed in an intelligent manner.

Case Study: The Future Word Processor

In our case study for Chapter 9, we would like to consider the future word processor. It will likely be an intelligent word processor with some very useful features. Consider this scenario.

Mary Jones, claims adjuster for the Federated Insurance Company, arrives at her office on Monday morning to face a very heavy day. Severe thunderstorms and a tornado hit the community on Saturday and Sunday, causing a great deal of paperwork to settle the claims. She also has to file a number of reports with the district office.

Mary sits down at her desk with a cup of coffee and switches on her word processor. She can communicate verbally with the word processor, using a header, which is directly linked to the district office so that electronic mail can be used.

Mary begins by filing a report to her boss in the district office recounting her weekend activities dealing with the storm damage. It was a hard weekend for Mary, so she decides to let her word processor write the report. She gives her essential information to the word processor, which then composes the report. After a few seconds of work, the system asks Mary if she would like to review the report to determine if any important details were omitted. The review of the document can be done via audio, visually on the screen, or from hard copy. Mary elects the auditory method; as she listens, she recalls some missing details. At the end of the review, Mary indicates the details to be added. The system constructs appropriate sentences and integrates them into the report. After she reviews the changes, the system sends the report electronically to the district office and files a copy on Mary's hard disk.

Next, Mary must deal with the various claims forms. She verbally enters the necessary information about each claim (name, address, description of loss, and so on). Reports on each claim are then generated using the appropriate template and sent electronically to the district office.

Altogether, it has taken Mary less than an hour to perform a set of tasks that might have taken her and a secretary, without a word processor, well over a day to complete. Since Mary is also an insurance agent, she now has time to do what she enjoys best: leave the office to work with her clients.

The scenario we have just presented is not far from reality. It is easy to see that intelligent systems, such as the word processor in our example, will be more productive in the future.

Key Points

- Expert systems technology is gaining widespread acceptance in the commercial world.
- Expert systems technology will be improved in the areas of knowledge representation, knowledge acquisition, system tools, and user interfaces.
- Expert systems of the future will likely be able to deal with more than one knowledge-representation scheme at the same time.
- New knowledge-representation schemes will be developed to be more powerful and easier to use, and to allow the use of single-time dimension.
- Two problems of manual knowledge-acquisition methods (such as interviews and analysis of verbal conversations) are the loss of valuable time and potential problems with inaccuracy and gaps in the knowledge.
- The problems and uncertainties of time loss and potential knowledge inaccuracies/gaps have paved the way for the development of automated knowledge-acquisition tools.
- Another major problem with the manual methods of knowledge acquisition is the cost.
- There are several ways to decrease the cost of knowledge acquisition:
  - Decrease the amount of time required to acquire the knowledge
  - Decrease the skill level required of the person acquiring the knowledge
  - Automate some aspects of the knowledge-acquisition process
  - Enable users and human experts to add to the knowledge
- There is a trend to sell packaged knowledge for purposes other than use in an expert system.
- Research and development work dealing with automated knowledge-acquisition tools is being done in both the industrial world and universities.
- Automated knowledge-acquisition tools will allow more people to serve as knowledge engineers at a lower skill level; these tools should be cost-effective.
- Automated consistency-check tools would be used to locate duplicate or contradictory rules in the knowledge base.
- Self-learning systems are the hope of the future for expert systems.
- Most experts increase their expertise through experience.
- It is hoped that self-learning systems can take knowledge in different representations and combine it into a single knowledge base.
- Expert systems were initially developed using general-purpose, high-level languages.
- Expert system shells will become more powerful and easier to use.
- Intelligent design tools will aid the knowledge engineer in a number of ways:
  - Determining problem domain characteristics
  - Ensuring that all important problem attributes are considered
  - Determining which knowledge representation(s) and inferencing mechanism(s) are used
  - Assessing system user capabilities and selecting appropriate user interface design schemes
  - Helping select appropriate shells for given applications
- A definite trend in recent years has been to improve the design and effectiveness of user interfaces in expert systems.

- The user interface is concerned with the content of the communications and the physical aspects of the communication.
- The term *user friendly* has meant many things to many people in the past; it will take on new meaning in the future for expert systems.
- There has always been a problem in the communication between an expert and a nonexpert in the expert's domain of knowledge.
- The use of natural-language interfaces is an attempt to make the human/machine interaction as human-like as possible.
- Key-word analysis and syntactic/semantic analysis are the techniques used in natural-language processing.
- Natural-language processing is constrained by computer memory limitations and computing speed.
- Physical communication between the user and the expert system has traditionally been through the keyboard and the terminal screen.
- Speech recognition and synthesis technologies are good possibilities for the physical interface, but they suffer from the same constraints as natural-language processing.
- User interfaces in expert systems of the future will be verbal and user friendly, as expert systems developers pay more attention to the quality of the interface.
- Expert systems technology is now available, in some form, to almost any business that needs it.
- A careful needs analysis must be performed by any business or individual seeking to implement expert systems technology.
- Cost is a major factor in the decision to implement expert systems technology.
- While computer hardware costs have decreased, software development costs have tended to rise.
- Automated tools, although costly initially, will bring the cost of expert systems down.
- Expert systems can be integrated with other systems, such as computer or manufacturing systems. The result of the integration of an expert system with another system has come to be called an intelligent system.
- Intelligent systems, such as the intelligent word processor, will boost productivity in many future applications.

Exercises

1. What improvements will be made in knowledge-representation schemes?
2. What are the problems of manual methods of knowledge acquisition?
3. How can the cost of knowledge acquisition be decreased?
4. What are the characteristics of automated knowledge-acquisition tools?
5. What is the purpose of automated consistency-check tools?
6. What are the characteristics of self-learning systems?
7. What improvements will be made in system development tools?
8. Discuss the two major aspects of the user interface.

9. What is the effect of natural-language processing on the user interface? What is the current status of this technology?
10. What is the effect of speech recognition and synthesis on the user interface? What is the current status of these technologies?
11. Describe the user interface of the future expert systems.
12. What factors will affect the availability of expert systems in the future?
13. How will cost affect the future of expert systems?
14. Describe the concept of an intelligent system.
15. What will be the effect of automated tools on the knowledge engineer? Why is this important?
16. Initially, what tools were used in expert systems development?

Assignments

1. Locate recent articles (published within two years) that deal with each of the major sections (excluding the Case Study) of the chapter. Read and summarize one article for each section. *The Readers' Guide to Periodical Literature* in the library might be a good place to start.
2. You are familiar with the concepts of database and spreadsheet. Design an intelligent system that could be integrated with each. What characteristics would your designed systems have? How would productivity be increased?
3. Design your expert system of the future. What characteristics would it have?
4. The Japanese have undertaken the development of a model for the future of AI and expert systems, called the Fifth-Generation Project. The project has a ten-year duration; it began in 1984 and is a cooperative effort between Japan's public and private sectors. The project is described in *The Fifth Generation* by E. Feigenbaum and P. McCormick (Addison Wesley, 1983). With the help of your reference librarian, locate a copy of this book. What are the implications for business, assuming a successful completion of the project?
5. If you have been compiling a list of vendors of expert system development tools, take your list and contact each vendor. Ask the following questions:
   a. What do you see as the future of expert systems?
   b. What development tools will be developed?
   c. How will your product line reflect your responses to the first two questions?
6. Fault-tolerant systems have had a significant impact on manufacturing and government systems. What are they and how do they relate to expert systems? Write a short paper (8–10 pages) summarizing your research.

The Applications

Module 1

Expert Systems and VP-Expert

In today's business and scientific worlds, interest in and emphasis on expert systems are growing. Expert systems, computer programs that mimic the knowledge of a human expert, are already being used efficiently and profitably in business, manufacturing, training, government, and other areas. In this module we will explore expert systems in general, provide overview examples, explain the purpose of using VP-Expert (a tool to develop expert systems), and guide the reader through the preliminary tasks of making backup disks, formatting disks, and making data disks.

Introduction to Expert Systems

Expert systems are part of a technology called artificial intelligence. Artificial intelligence (AI) seeks to create intelligent systems of hardware and software that perform certain tasks or solve problems. Other fields of artificial intelligence are robotics, machine vision, natural-language processing, and general problem-solving. We will concentrate on understanding expert systems for the business world.

Expert systems are designed to behave like an expert in a particular field or domain of knowledge. Like a human expert, an expert system draws information from a user and then, by using its own "internal experience and resources," gives advice, makes suggestions, delivers an answer, predicts an outcome, or exhibits similar behavior. The purpose of having an expert system is to manage and run a business (or other activity) more economically and efficiently. Expert systems are also useful because they can be more permanent than a human expert. Human experts change companies, go on vacation, retire, and so on. With proper maintenance of hardware and software, expert systems can remain on the job permanently.

Creating expert systems involves several steps. First, a problem domain must be identified, such as diagnosing a medical disease, repairing computer networks, writing mortgage loans, or determining credit approvals. Once the problem area is determined, a team of knowledge engineers, domain experts, and users can work to create an expert system that will help solve the problem. The expert provides the knowledge about the problem. The knowledge engineer identifies the needed knowledge and translates it into structures that computers use; the user of the proposed system gives advice and suggestions about the usability of the system in the normal work environment. With effective teamwork, the chance of producing a usable expert system increases.

The expert system is actually a software program that runs on a computer (supercomputer, mainframe, minicomputer, or microcomputer). It may be written in a language (Pascal, C, BASIC, and so on) or with a special tool (shell). To use a shell, you do not have to be a programmer. We will confine our attention to producing expert systems with a tool (shell) called VP-Expert.

Expert system software has three conceptual parts: the knowledge base, the inference engine, and the user interface. The knowledge base is often a set of rules (or other representation) and facts about the problem area. The inference engine is the mechanism that reasons or infers conclusions after examining the knowledge base. The user interface is the communication between the user and the actual software. A typical interface is the way that users communicate through a video display terminal. These three logical components are pictured in Figure Ml.1.

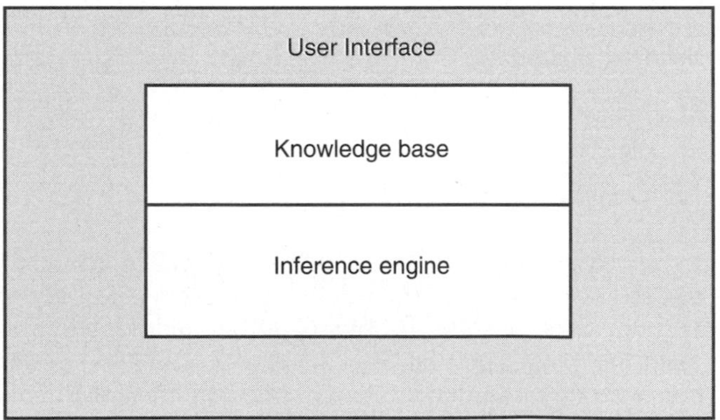

FIGURE M1.1 THE THREE LOGICAL PARTS OF AN EXPERT SYSTEM

Examples of Expert Systems

Expert systems can help in various applications. They can assist in the functions of controlling, debugging, designing, diagnosing, instructing, simulating, interpreting, planning, predicting, and repairing. An expert system can give advice on investments, determine the acceptability of a credit application, predict back orders, control inventory, design investment portfolios, or give advice on the purchase of a car. Managers could use an expert system to simulate personnel situations. A traditional computer program can determine a mortgage balance, but it cannot give advice to a client on whether to buy or what mortgage plan to follow. A traditional computer system may ticket a passenger for a vacation, but the system cannot advise the client on the best vacation for the least money. Expert systems can help in these areas.

Expert systems are also used in other application areas. An expert system may advise a mechanic on repair procedures. Car-trouble analysis, microcomputer troubleshooting, or airplane maintenance can be improved with the use of expert systems. In the field of medicine, special expert systems can diagnose ailments and suggest possible treatments. In the manufacturing field, an expert system can control a production process. In the field of education, expert systems can be used as effective tutors. Clearly, we have a wide range of potential expert systems across many areas of application.

Emphasis of the Modules: VP-Expert

In this wide range of applications, you will learn how to create an expert system. The purpose of the modules of this book is to give you step-by-step instructions in creating an expert system with the tool or shell, VP-Expert. This particular tool combines economy, ease of use, and the ability to use spreadsheet and database files already known to business. The step-by-step procedures to be used are detailed in the modules. The theory and underlying principles of expert system technology are given in Chapters 1 through 9. The reader may choose to do the hands-on activities (Modules 1 through 8) first or to read the conceptual information (Chapters 1 through 9) first or combine theory with practice (read Chapter 1, then work through Module 1, read Chapter 2, then work through Module 2, and so on). The purpose of the modules is to give you a simple, sequential "recipe" for using and creating expert systems with VP-Expert. The modules, exercises, and assignments are designed to give you hands-on experience with VP-Expert. The exercises are designed as a guided review of material in the modules, whereas the assignments are *extended* hands-on exercises from the module. The expert systems used in the applications section are chosen for educational use rather than performance. Small expert systems are used at first so you can get a clear overview of an entire expert system.

VP-Expert is a software tool or shell from Wordtech Systems, Inc. of Concord, California. The software available for use with this book is an educational version, 3.1. The major differences between the educational version and the regular version are capability and cost. The educational version has three limitations:

1. The size of each knowledge base is limited to 16K (16,384 bytes).
2. Only three separate knowledge bases may be chained together.
3. The maximum number of database records that can be used to create an expert system is 150.

Given that the objective is to learn how to create expert systems with VP-Expert, these limitations are not serious.

Getting Started with VP-Expert

Before you can use VP-Expert, you must perform some preliminary but important tasks. The VP-Expert shell must be backed up, the Sample Files Disk must be backed up, and the system requirements must be checked. In particular, you will need two 3.5-inch disks to do the following:

1. Make a backup copy of VP-Expert and label it as a backup.
2. Make a copy of the Sample Files Disk and label it as a backup.

To run VP-Expert, you will need the following:

1. IBM-compatible microcomputer.
2. 512K or more RAM (640K total needed to use all of VP-Expert's features).
3. A hard disk and one 3 1/2-inch disk drive.
4. DOS version 3.1 or higher.

(If you do not have a hard drive, see the special instructions that came with your software.) To use the graphics features, you will need a graphics card such as CGA, EGA, or VGA, Hercules monochrome, or other compatible video card. A mouse is also recommended for graphical applications.

Start by first creating the backups and directories (explained in the next section). Upon successful completion of these tasks, you will be able to use VP-Expert in Module 2.

Creating Backups and Installing VP-Expert

We need to provide for backup and recovery before we use the new tools. In any computer application, we must protect ourselves against the accidental loss of data and damage to programs. Thus, the first thing to do is to back up the software that comes with this book. The conceptual steps to do this are as follows:

1. Produce new labels for the backups to be made.
2. Give the computer the DISKCOPY and COPY commands from DOS (operating system) to copy all files from the originals to backup disk 5.
3. Place the new labels on the new disks (backups).
4. Store the original disks in a safe, dry place.

The first procedure in diskette backup is to create the labels for the disks. Using a pen or felt-tip pen, fill out two labels with the information shown in Figure M1.2. Careful attention to this simple task may save you time and effort in the long run.

VP-Expert Educational Version 3.1 Backup Your Name Date	Sample Files Disk Backup VP-Expert Educational Version 3.1 Your Name Date

FIGURE M1.2 LABELS FOR BACKUP DISKS

The version number is necessary to avoid problems caused by the use of different software versions. Your name is needed in case you lose the disk, and the other information is needed because the number of disks grows exponentially; you will soon forget which disk contains what files. Have the labels ready before you make the backups.

To create the backups of VP-Expert and the Sample Files Disk, you may use the DISKCOPY command. From the C drive prompt (c:\ ), enter the following DISKCOPY command:

```
DISKCOPY A: A: / V
```

After creating your backup copy, place the appropriate label on the new disk. Repeat the same procedure for making a backup of the Sample Files Disk. Store the original disks in a safe, dry place and use the backups to install the VP-Expert work environment.

To create this environment, we will place all VP-Expert files in a subdirectory called VPX3ed and sample (or program) files in a subdirectory called VPX_samp (see Figure M1.3). We will execute an Install program to create the VPX3ed environment, but the

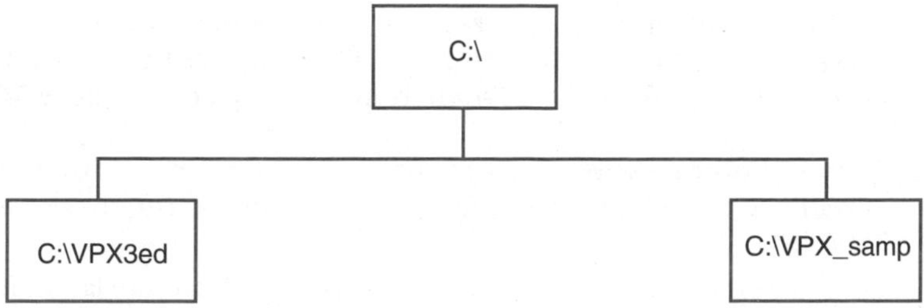

FIGURE M1.3 CONCEPTUAL VIEW OF SUBDIRECTORIES

VPX_samp subdirectory will be created with DOS commands. The next paragraph will step you through the Install procedure. If you do not use the standard subdirectories, VPX3ed and VPX_samp, parts of VP-Expert will not work. See assignment 11 at the end of the module to handle this situation. Place the (backup) disk labeled VP-Expert Educational in drive A. At the C prompt, type

A:install

This will execute install.exe in drive A and bring up an opening screen, as shown in Figure M1.4.

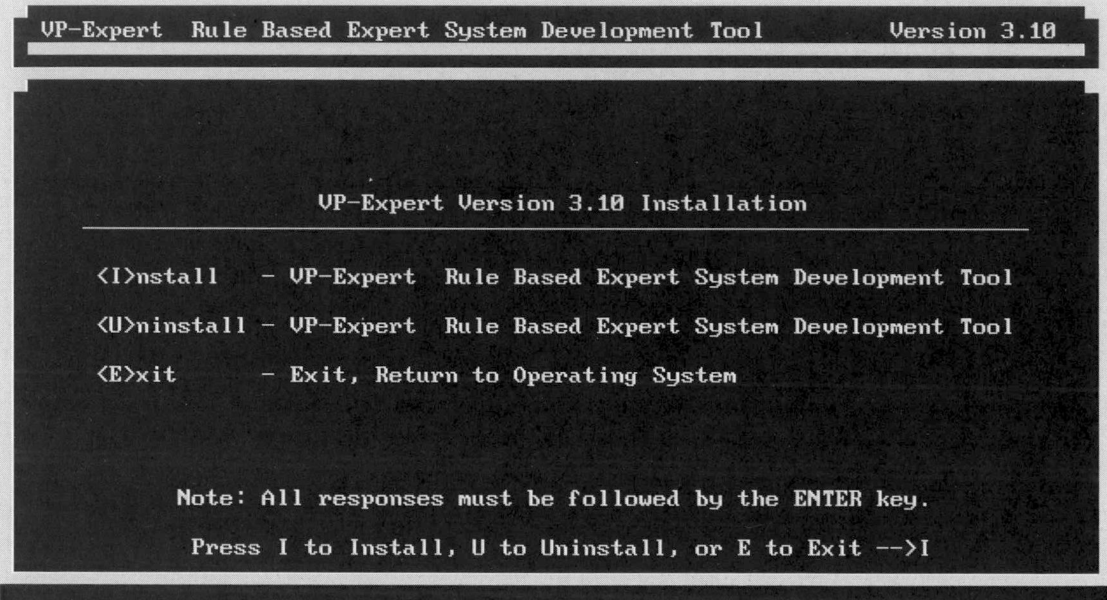

FIGURE M1.4 VP-EXPERT VERSION 3.10 INSTALLATION

The default option is I for Install. Press Enter to execute this option; otherwise type U for Uninstall or E to Exit. The next screen will ask for the drive on which you wish to place the VP-Expert system files (see Figure M1.5).

Press Enter to select the default drive, the C drive. The next screen that appears will prompt you for the subdirectory name on the drive just selected (see Figure M1.6).

The default selection is VPX. To use VPX_samp, move the cursor to the space after x and type 3ed. The name VPX3ED will appear on your screen. Press Enter

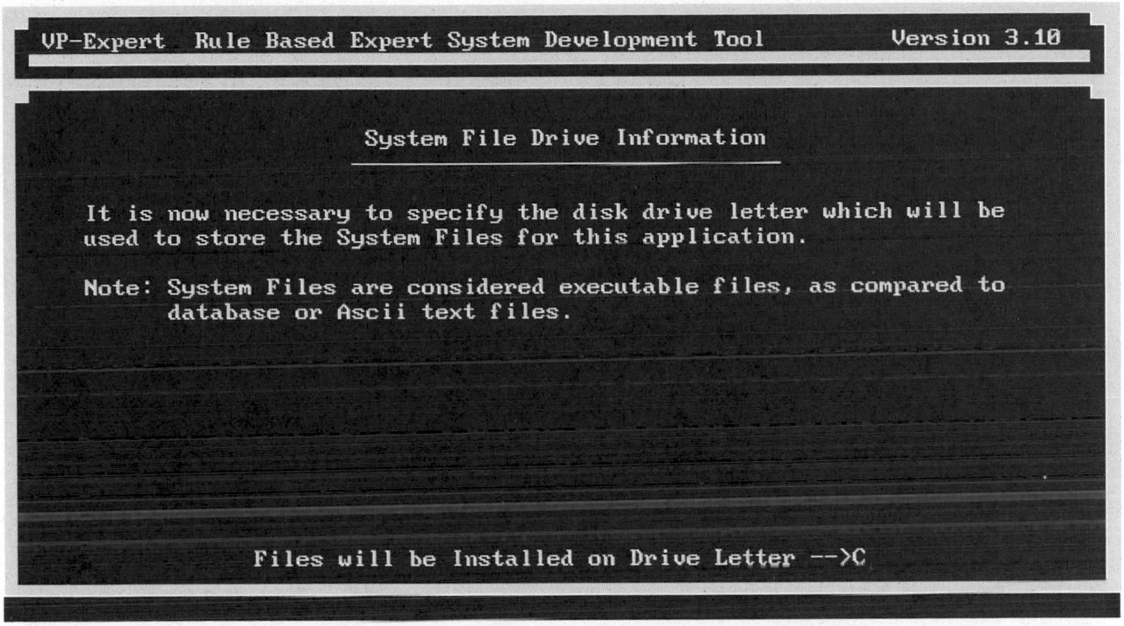

```
VP-Expert  Rule Based Expert System Development Tool          Version 3.10

                      System File Drive Information

   It is now necessary to specify the disk drive letter which will be
   used to store the System Files for this application.

   Note: System Files are considered executable files, as compared to
         database or Ascii text files.

               Files will be Installed on Drive Letter -->C
```

FIGURE M1.5 SYSTEM FILE DRIVER SPECIFICATION

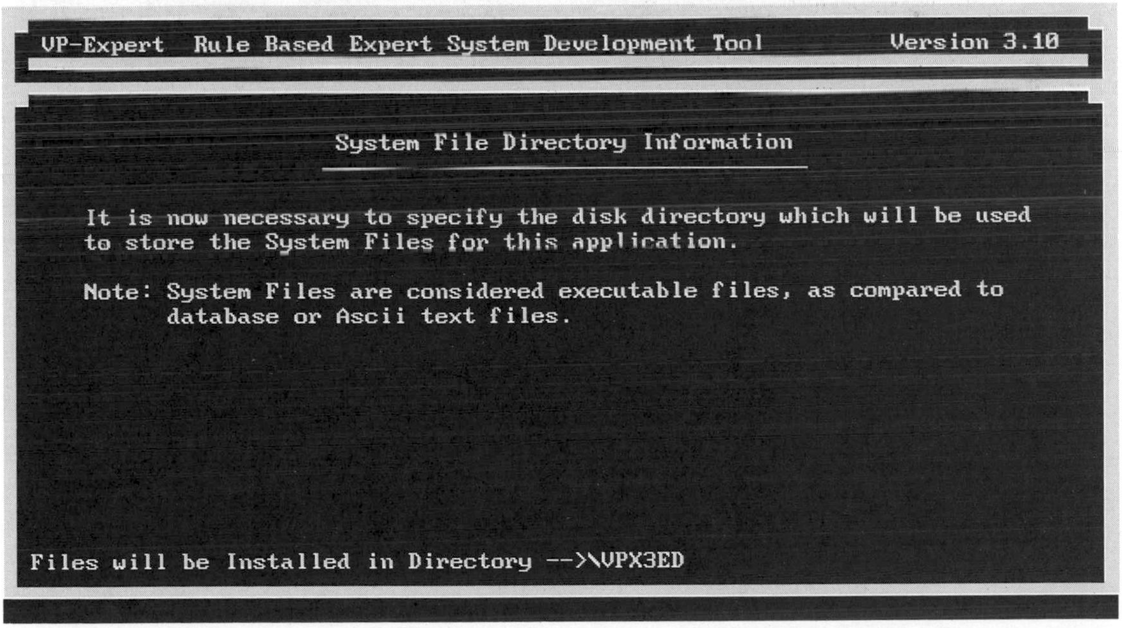

```
VP-Expert  Rule Based Expert System Development Tool          Version 3.10

                    System File Directory Information

   It is now necessary to specify the disk directory which will be used
   to store the System Files for this application.

   Note: System Files are considered executable files, as compared to
         database or Ascii text files.

Files will be Installed in Directory -->\VPX3ED
```

FIGURE M1.6 SYSTEM FILE DIRECTORY INFORMATION

to designate this subdirectory name. The next screen from the Install program asks if you are ready to proceed (see Figure M1.7). Select Y for yes. The program will now prompt you to wait as it creates the subdirectory c:\ vpx3ed. Then it will display the names of VP-Expert system files as they are copied from drive A to c:\ vpx3ed.

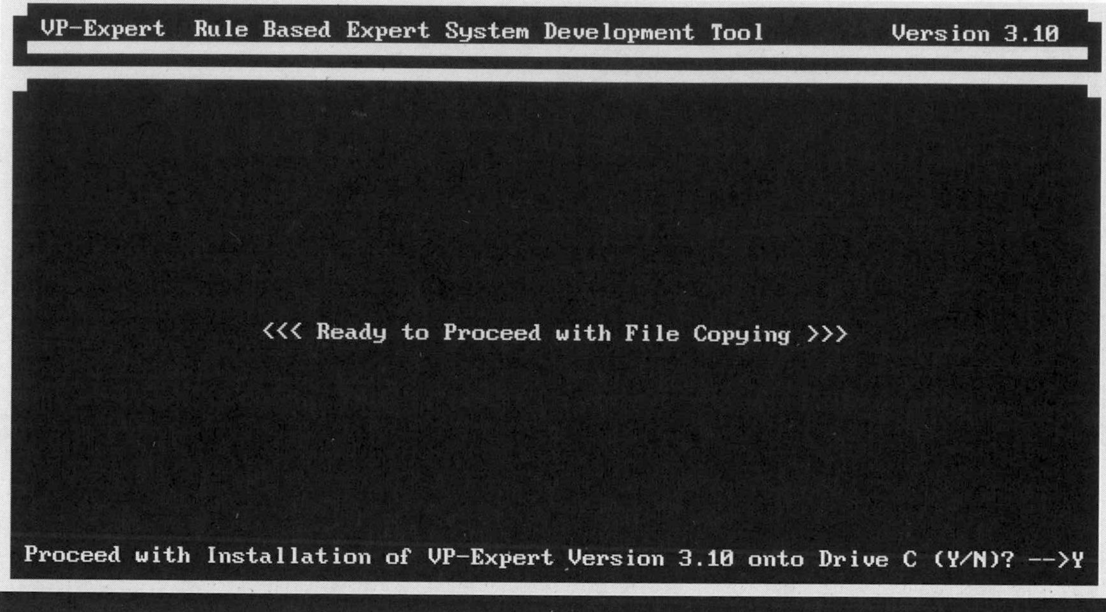

FIGURE M1.7 PROCEEDING WITH INSTALLATION

The next screen from the Install program will check your existing DOS config.sys file to make sure that the operating system can manage enough open files and buffers for VP-Expert to run efficiently (see Figure M1.8).

The VP-Expert environment requires 32 files and 10 buffers. If your config.sys file meets this requirement, select the option to proceed. Otherwise select the option to automatically update your config.sys file. The next screen (Figure M1.9) will show you the status of the Install program.

To make sure that all worked properly, move to the new subdirectory and check the contents by typing the following at the C prompt:

CD VPX3ed

Then, at the c:\vpx3ed> prompt, type

DIR

```
VP-Expert  Rule Based Expert System Development Tool          Version 3.10

                       CONFIG.SYS Contents Report

Parameter          Required  Found
_______________    ________  _____
FILES                 32       40
BUFFERS               10       40


       The CONFIG.SYS file checks out okay, no changes are required...



                       Press any key to continue...
```

FIGURE M1.8 CONFIG.SYS CONTENTS REPORT

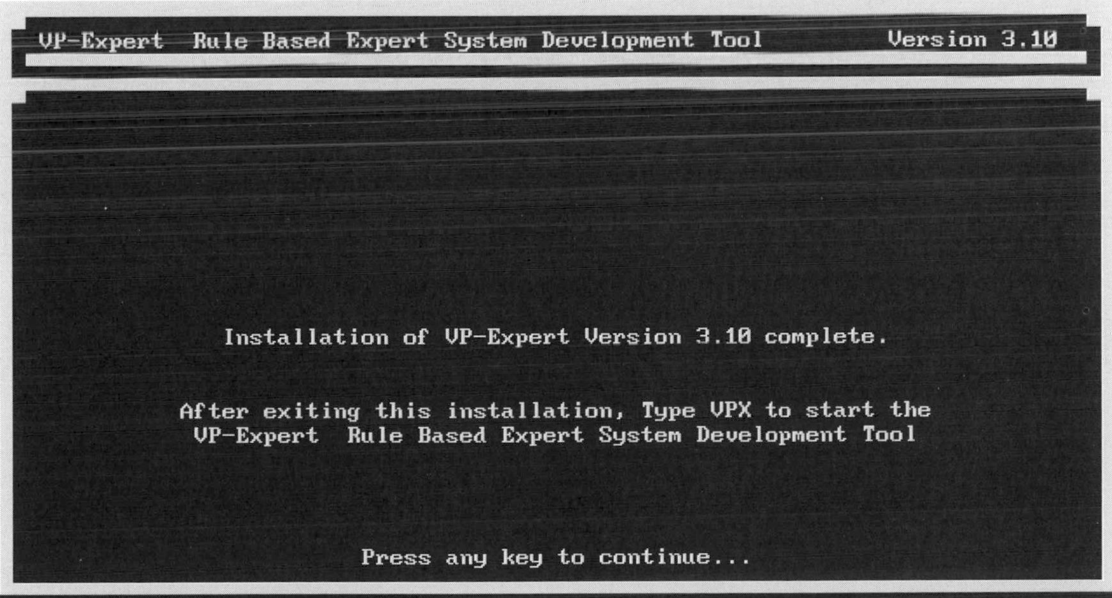

FIGURE M1.9 INSTALLATION STATUS FOR VP-EXPERT

The results should show that there are 17 VP-Expert system files occupying approximately 578,097 bytes (see Figure M1.10).

```
 Directory of C:\VPX3ED

.            <DIR>      09-11-94   1:57p
..           <DIR>      09-11-94   1:57p
MAKEHELP EXE     16416 01-31-94   3:00a
EZI      HST       231 09-11-94   1:57p
VPX      EXE    203582 01-31-94   3:10a
VPXE     EXE     63643 09-01-91   3:00a
VPXED    EXE       678 09-01-91   3:00a
VPXH     EXE     20480 09-01-91   3:00a
VPXHELP  EXE     19120 09-01-91   3:00a
VPXI     EXE     34440 09-01-91   3:00a
VPXPCX   EXE     16459 09-01-91   3:00a
VPXTE    EXE     57020 09-01-91   3:00a
VPXHELP  CHP    122389 01-31-94   3:10a
INDEXER  COM       978 09-01-91   3:00a
VPXHERC  COM      6749 09-01-91   3:00a
README   VPX      7234 01-31-94   3:10a
VPX      CFG        99 02-09-94  11:01a
VPX      IMG       985 01-31-94   3:10a
VPX      MSG      7175 01-31-94   3:10a
       19 file(s)      577678 bytes
                      5939200 bytes free

[MICRO1] C:\VPX3ED>
```

FIGURE M1.10 SUBDIRECTORY CONTENTS OF VPX3ED

Files that end in .EXE are executable files. The file readme.vpx is an ASCII file that contains advanced and recent information on VP-Expert. To display this file to the screen, at the c:\vpx3ed prompt, enter

TYPE readme.vpx ¦ more

To obtain a printout, type the following:

COPY readme.vpx prn

You ideally have stored the VP-Expert system files in the subdirectory c:\VPX3ed.

The next step is to create your working directory called VPX_samp to hold your sample files. Return to the root or top directory by typing CD.. and create the new subdirectory by typing the DOS MAKE DIRECTORY command:

MD VPX_samp

Move to this new subdirectory by typing the DOS change directory command:

CD VPX_samp

Place the Sample Files Disk in drive A and copy the files to this new subdirectory by typing the following at the c:\VPX_samp prompt:

COPY A:*.* /V

The destination defaults to c:\VPX_samp. Use the CD and DIR commands to verify that your sample files have been copied. VP-Expert can now be found in the VPX3ed subdirectory; the default setting or path to the work files is VPX_samp. This will be our standard environment for working with VP-Expert in the rest of the modules.

Starting VP-Expert

Let's start VP-Expert to make sure that everything is ready in order to use VP-Expert in subsequent modules. Get to the subdirectory by typing the DOS CHANGE DIRECTORY command, CD \VPX3ed. At C:\VPX3ed, start VP-Expert by instructing DOS to execute vpx.exe. Type vpx to get to the Opening screen. Press any key to get to the Main menu like the one shown in Figure M1.11.

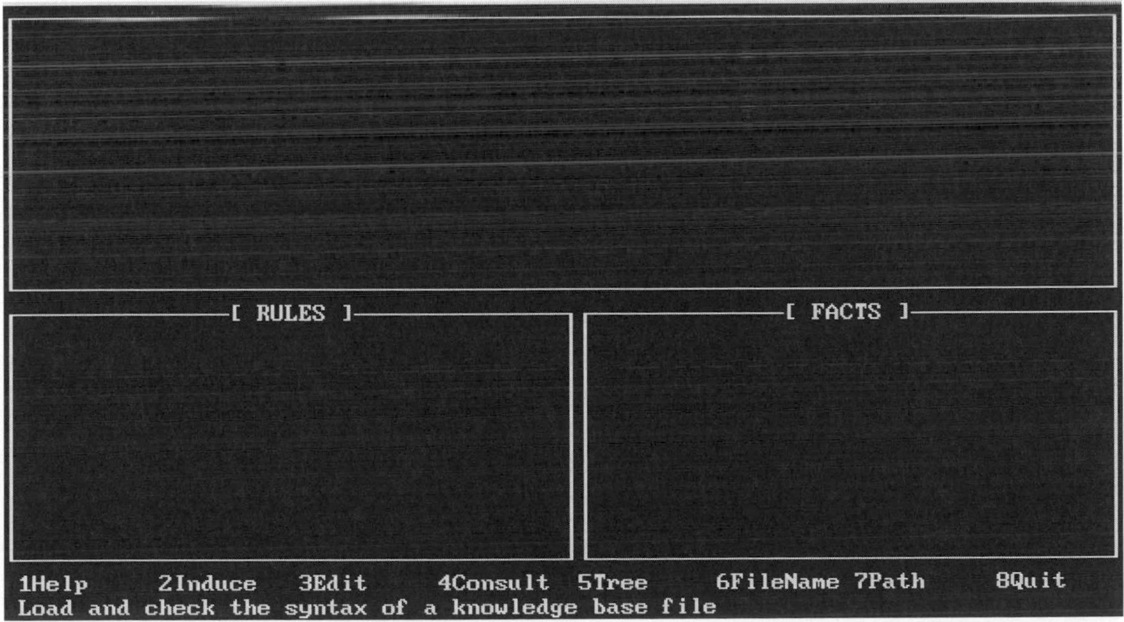

FIGURE M1.11 STARTING VP-EXPERT

Now move the light bar from 4Consult to 8Quit by pressing the Right Arrow key (found on the far right of your keyboard) and then pressing the Enter key. You will be returned to the C:\VPX3ed prompt.

If all went well, you accessed VP-Expert and exited the software. This process is represented schematically in Figure M1.12, which tells us that to enter VP-Expert we use vpx and to exit, we use 8Quit.

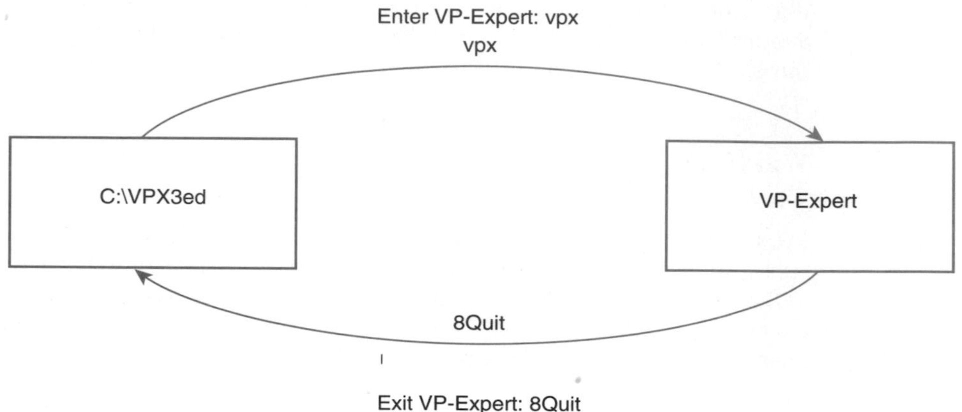

FIGURE M1.12 USING VP-EXPERT

Getting "in, out, and around software" (navigating) is important for productive sessions with the computer. If you have any problems, check to make sure that VP-Expert is in the correct subdirectory. Other problems may be associated with the system requirements; a user has no way of looking at a computer and knowing its memory size. To check the version number of DOS or the operating system, type VER at the C prompt. In any computer environment, it is useful to have the system hardware specifications (memory size, graphics board, and so on) listed on an index card attached to the front or side of the computer. This list will save endless hours of frustration in trying to run new microcomputer software. It might be a good idea to ask the person in charge of your computer environment to provide this information.

Now, with the original software put aside, our backups made, and a short test of VP-Expert executed, we are ready to begin work with VP-Expert.

Key Points

- Expert systems are part of a technology called artificial intelligence, or AI.
- Expert systems are software programs that emulate the intelligence and knowledge of a human expert.

- An expert system can be used in business, manufacturing, government, and education to control, debug, design, diagnose, instruct, simulate, interpret, plan, predict, or repair.
- Expert systems in business can perform specific tasks such as controlling inventory, managing an investment portfolio, advising on mortgage loans, planning a new building, predicting interest rates, and so on.
- Programmers can implement expert systems with a programming language. Non-programmers can produce expert systems by using a tool called an expert system shell.
- The emphasis of this book's modules is to present easy, step-by-step instructions on how to create and use expert systems with VP-Expert.
- You may read all the modules first (applications), all the chapters first (theory), or combine them in pairs as you go.
- To use VP-Expert (or any other microcomputer software), you should create backups of each original disk, create your working environment (subdirectories for hard disk systems), label the new backups properly, and store the originals in a safe place.
- The DISKCOPY command of DOS is used to create the backups. The syntax to copy all files on the source drive to a destination drive is: DISKCOPY source drive: destination drive. If the destination drive is left out, the destination defaults to the current drive. The /V option verifies that all files are copied correctly.
- The system requirements for running VP-Expert are an IBM-compatible microcomputer, DOS 3.1 or higher, 512K of memory (640K to run all the features of VP-Expert), and (a hard disk system) with one 3 1/2-inch drive.
- The recommended working environment for using VP-Expert is to place the system files in a subdirectory called VPX3ed and the sample files in VPX_samp.
- To create a subdirectory, use the DOS command MD subdirectory-name. The MD command stands for make directory.
- To switch to a subdirectory (beneath where you are), use the change directory DOS command or CD subdirectory-name.
- Use the DOS command CD.. to move up from a subdirectory.
- To start VP-Expert, type vpx and press Enter. To exit VP-Expert, select 8Quit and press Enter.

Exercises

1. a. Name the three parts of an expert system.
   b. Name five business applications using expert systems.
2. Name the expert system shell or program to be used with this book.
3. What is the purpose of a backup disk?
4. Identify the syntax or "spelling" error(s) in each of the following DISKCOPY commands:
   a. DISKCOPY A : B
   b. DISKCOPY A:B:
   c. DISKCOPY A B
   d. DISKCOPY A A
   e. DISKCOPY A;B

5. Match the numbered command to the lettered task.

1. DISKCOPY	a. Formats and copies
2. VER	b. Makes a duplicate
3. FORMAT	c. Makes a disk usable
4. COPY	d. Reveals the version number of the operating system

6. Mark the following statements T (true) or F (false). VP-Expert:
   a. is a programming language like COBOL or Pascal.
   b. is a product of Wordtech Systems of Concord, California.
   c. requires 200K of memory.
   d. runs under any DOS.
   e. is a microcomputer software development tool.
7. To create a subdirectory, enter
   a. CS
   b. CD
   c. MS
   d. MD
   e. SD
8. To run VP-Expert, enter
   a. go
   b. vp
   c. vpe
   d. vpx
   e. vpxt

Assignments

1. From your computer system, fill in the information below.

   Type of computer ______________________________

   Version of DOS ______________________________

   Memory size ______________________________

   Graphics adapter ______________________________

   Mouse ______________________________
2. a. Check the contents of your VP-Expert backup by placing the disk in drive A, typing DIR, and pressing Enter. The names of your files should appear on the screen.
   b. To obtain a printout of what appears on the screen, press the Shift key and the Print Screen key at the same time. Use Print Screen to obtain a list (directory) of the files on your backup disks.
   c. How many files are on the disk?
3. The Samples Files subdirectory vpx_samp has so many files that the screens scroll faster than you can read. To remedy this, enter at the C:\vpx_samp prompt:

   DIR/W

Columns of file names appear, but dates and file sizes are missing. To send the complete directory to the printer and screen, enter at the c:\VPX_samp prompt:

CTRL PRTSC

DIR

CTRL PRTSC

The CTRL PRTSC command is a toggle to send screen output to the printer. The command is given again to turn off the printing.

a. Obtain a printout of the files on the sample files subdirectory.
b. How many files are on the disk?
c. How many free bytes are left on the disk?

4. When you enter vpx to start VP-Expert, you are executing a file called vpx.exe. Using the printout in Assignment 2 or the command

   DIR vpx.exe

   find out how large the file is.

5. If a file has an extension of .kbs, it is an expert system file created by VP-Expert. In the sample files subdirectory, enter at the C:\VPX_samp prompt:

   DIR *.kbs

   to display all files that end with .kbs.
   a. How many .kbs files are on the Sample Files Disk?
   b. Obtain a printout by using Control-Print Screen before and after the DIR *.kbs command.

6. Most disks have a text file that gives information about the disk.
   a. Use the DIR command to see if a Readme or similar file is on the Sample Files subdirectory. Try DIR R*.*.
   b. If there is, display the contents of the file by typing the command

      TYPE file.ext ¦ MORE

   c. To obtain a printout of the file, enter:

      COPY file.ext>PRN

      OR

      COPY file.ext LPT1

   You must type the exact name of the file. When you have typed in the instruction, press the Enter key.

7. Repeat Assignment 6 for a file called vpx.img in the VPX3ed subdirectory. What are the contents?

8. DOS has two types of commands: internal and external. DISKCOPY is an external command, whereas VER and COPY are internal commands.
   a. Look up the difference between internal and external commands in the DOS manual.

b. How does this explain the problem and frustration that users have with the error message "file not found"?

9. Obtain a printout of the opening screen of VP-Expert by following these steps:
   a. Get into VP-Expert.
   b. Use Print Screen (see Assignment 2).
   c. Exit VP-Expert.
10. Get into the Sample File directory, VPX_samp:
   a. Type DIR ¦ SORT and explain what happens.
   b. Type DIR ¦ SORT ¦ MORE and explain the results.
   c. Type DIR ¦ SORT > PRN and explain what happens.
11. In order to use subdirectory names that are different from the ones used in Module 1, you must edit a VP-Expert system file called vpx.cfg. Insert your VP-Expert Educational Version 3.1 disk in drive A and at the C:\prompt type

    TYPE A: VPX.CFG

    This is a system file that VP-Expert reads. To access .kbs files from a directory different from VPX_samp, edit the .kbs statement from this file with the subdirectory name of your choice. To use a different system directory rather than VPX3ed, change the subdirectory name in the EDITOR statement in this same file. Use an ASCII editor such as DOS EDIT or DOS EDLIN to achieve these changes.

Module 2

Examples of Expert Systems

We want to run our first expert system, WPHELP, to select a word processor. To do this, we must learn to use the Main Menu to set the path, select the file, and then run the expert system with the Consult Menu. We will then run another version of the same expert system that will show us the parts of the system (knowledge base, inference engine, and user interface) and examine the software that produced the interactive expert system. After using the system, we'll explore the Help facilities within the VP-Expert system environment.

Using VP-Expert

We will use VP-Expert to run our first expert system or consultation. The expert system will advise us on selecting a word processor for IBM-compatible microcomputers used in business. Our first example will be small and somewhat limited in its advice, but a small system rather than a large one is easier to conceptualize at first.

The Main Menu

To get started, call up the directory containing the VP-Expert system files (cd VPX3ed), type vpx, and press Enter. Your screen should look like Figure M2.1.

Another name for this screen is the Opening screen. The information on the screen identifies the software (VP-Expert), the version (Educational 3.1), the programmer (Brian Sawyer), and the publisher (Wordtech Systems, Inc.). Press any key to leave the

FIGURE M2.1 VP-EXPERT OPENING SCREEN

opening screen. The next screen is the Main Menu. The bottom of the screen shows the Main Menu choices. The menu or menu bar is arranged horizontally and shows eight choices:

1Help 2Induce 3Edit 4Consult 5Tree 6FileName 7Path 8Quit

The words underneath the menu bar identify the options. For example, the opening screen in Figure M2.2 identifies the menu choice 4Consult. The message underneath 4Consult explains that this option will "load and check the syntax of a knowledge base file." By pressing the Right or Left Arrow key, you can move the **light bar** (lighted box) right or left to change the message line.

Our purpose is to run a sample expert system. VP-Expert calls this a **consultation**, or **consult** for short. To run a consultation, we must first do two things: (1) tell the computer where to find the sample file, and (2) tell the computer the name of the sample file. Hence, we will use the following three Main Menu choices in this order:

Step 1. Identify where the sample files are by using 7Path.

Step 2. Select the sample file name by using 6FileName.

Step 3. Run the chosen expert system file by using 4Consult.

It is a good idea always to start with Step 1. Tell the computer where to find the files (path) whenever using VP-Expert, regardless of the task.

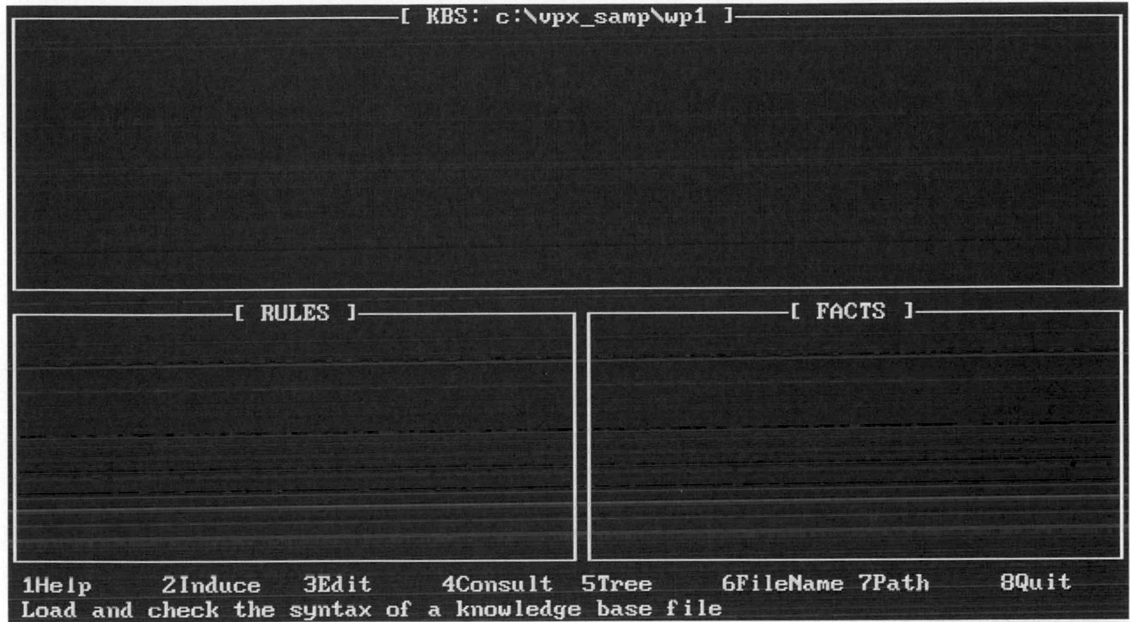

FIGURE M2.2 VP-EXPERT MAIN MENU

Let's select c:\VPX_samp as our path and wp1.kbs as our expert system, and run or consult with this file name. Follow these steps:

Step 1. Press the Right Arrow key until the light bar is on 7Path of the Main Menu. Press the Enter key. The phrase "What is the new pathname?" will appear in a box on the screen. Type C:\VPX_samp and press the Enter key. Use the Backspace key to correct any typing errors before pressing Enter.

Step 2. Using the Right Arrow key, move the light bar to 6FileName and press Enter. The message "Choose a file:" will appear on the screen, followed by a list of file names. This list may extend beyond what appears on the screen. Using the arrow keys, move the light bar to wp1.kbs. When the light is on the bottom row, press the Down Arrow key to scroll through the file names. When the light bar is on wp1.kbs, press the Enter key. If you know the file name, you may instead type the file name, wp1.kbs, then press Enter. (If no file names appear or if you can't find the right file, you did not set the path correctly.)

Step 3. With the path identified as drive C:\VPX_samp, the file as wp1.kbs, and the light bar on 4Consult, press the Enter key to enter the Consult Menu.

The opening screen for the Consult Menu is shown in Figure M2.3.

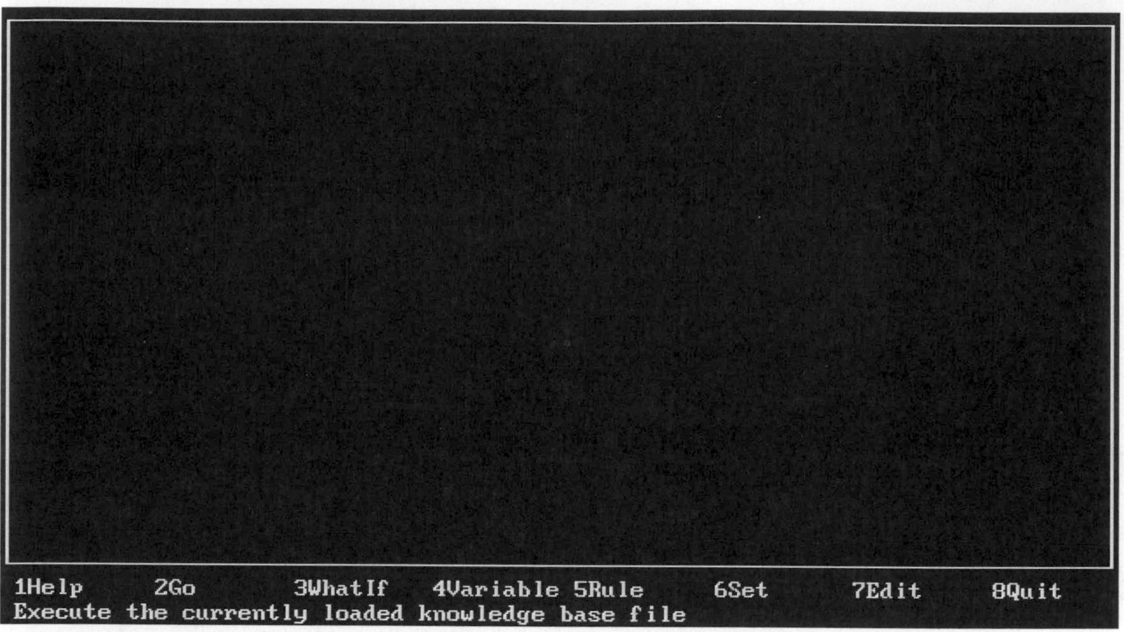

FIGURE M2.3 THE OPENING SCREEN FOR THE CONSULT MENU

If you do not see this screen, retry the preceding steps. If the message "File not found" appears, you may not have specified the proper path. If the message "Not enough memory" appears, VP-Expert does not have enough internal memory (RAM or random access memory) to load the file. Correct this deficiency by trying a higher version of DOS or using a machine with 640K of RAM.

Consultation Sessions

The only identification for our Consult Menu of Figure M2.3 is a blank screen with a menu at the bottom, with 2Go highlighted. The menu bar at the bottom of the screen tells us that we have eight options for consulting with the (sample) expert system file, now loaded in memory. Notice that the menu bar is on option 2Go. When the 2Go option is selected, control will pass to the loaded expert system or .kbs file. In other words, VP-Expert will "execute the currently loaded knowledge base file." You will then follow the instructions from the expert system software until the end of the consultation. For example, press the Enter key when the light bar is on 2Go. You should see Figure M2.4, which shows the start of the sample expert system wp1.kbs.

Follow the instructions on the screen and press the Spacebar to start the advising example. The screen will be cleared, and a question will appear on the display. You are asked to identify the type of documents, simple or complex, that you plan to produce.

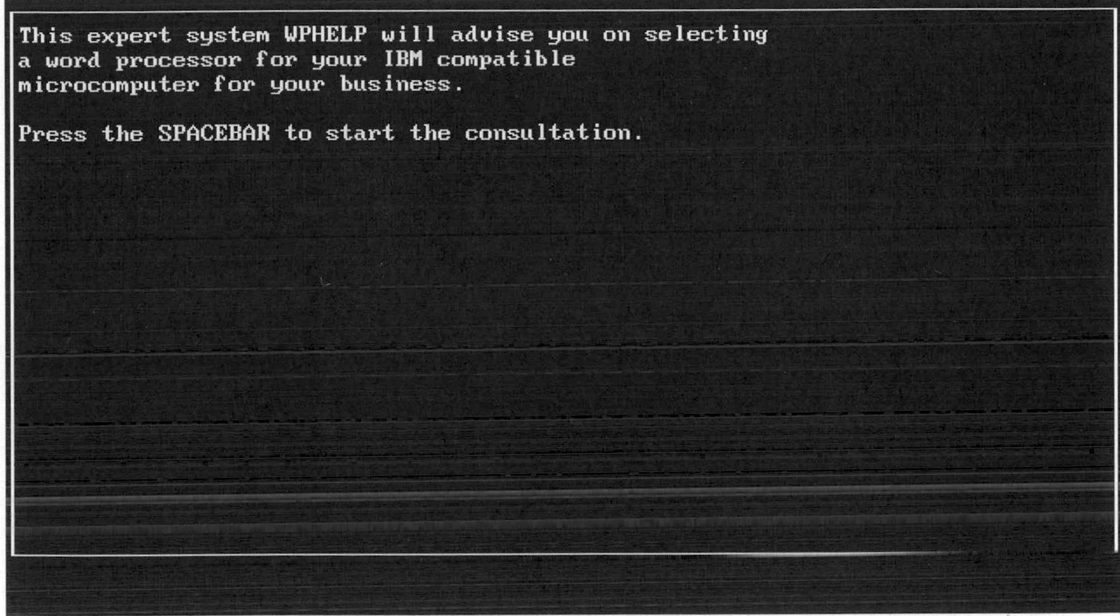

FIGURE M2.4 OPENING SCREEN FOR RUNNING A CONSULTATION WITH wp1.kbs

You may answer a question by following these steps:

Step 1. If neccessary, move the light bar to your selection (Simple).

Step 2. Press the Enter key. A triangle will appear on the screen to mark your choice.

Step 3. Press the End key (on the right of your keyboard). This completes the selection for a given question.

Note that these instructions (and additional ones) are listed on the bottom of the screen, as shown in Figure M2.5.

Select Simple by pressing Enter and then the End key. Your display should look like Figure M2.6, which now asks a second question concerning available money.

Select Low as the answer to the second question by pressing Enter and then the End key. Your screen should now look like Figure M2.7, which shows that the expert system replies "WPHELP advises Product A."

In other words, the expert system advises that Product A is a word processor for simple documents at a low cost. The message "Press the Spacebar to exit WPHELP" tells you how to exit the consultation and return to the Consult Menu. What would the WPHELP system advise if you had given different choices? Rerun the expert system (by selecting 2Go), choose Complex and Low, and verify that the expert system gives the advice shown in Figure M2.8. To start the system, select 2Go from

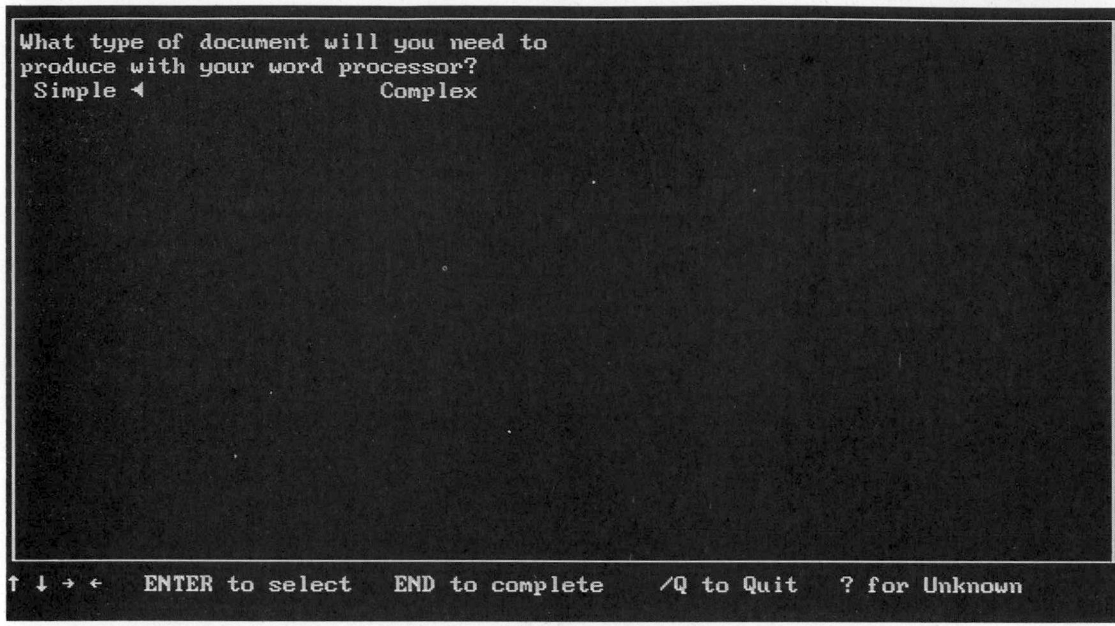

FIGURE M2.5 SELECTING THE TYPE OF DOCUMENT IN wp1.kbs

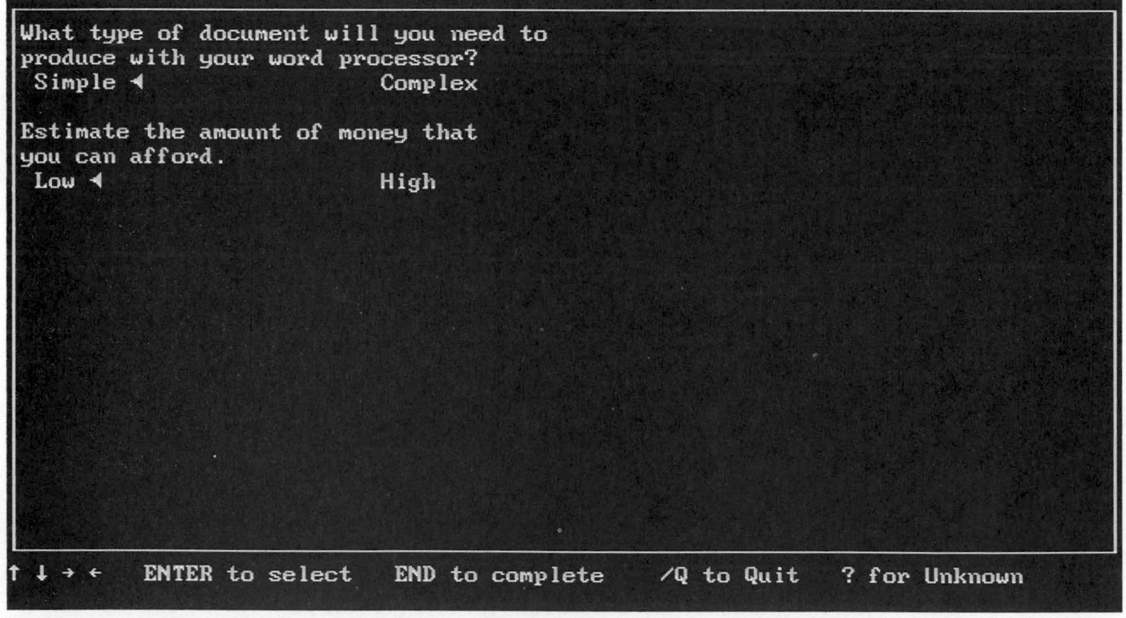

FIGURE M2.6 SELECTING THE COST IN wp1.kbs

```
What type of document will you need to
produce with your word processor?
 Simple ◄                    Complex

Estimate the amount of money that
you can afford.
 Low ◄                       High

WPHELP advises Product A.

Press the SPACEBAR to exit WPHELP.
```

FIGURE M2.7 THE ADVICE FROM WPHELP IN FILE wp1.kbs

```
What type of document will you need to
produce with your word processor?
 Simple                      Complex ◄

Estimate the amount of money that
you can afford.
 Low ◄                       High

WPHELP advises none.

Press the SPACEBAR to exit WPHELP.
```

FIGURE M2.8 DIFFERENT ADVICE FROM WPHELP

the menu bar and make selections by using the arrow keys, Enter, and End. Figure 2.9 summarizes possible answers to both questions and WPHELP's advice.

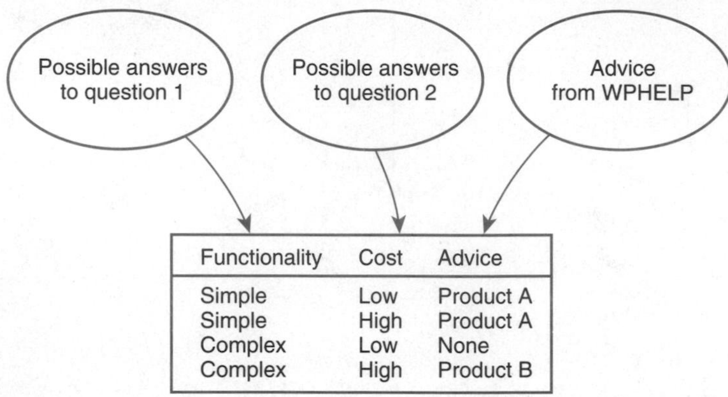

Functionality	Cost	Advice
Simple	Low	Product A
Simple	High	Product A
Complex	Low	None
Complex	High	Product B

FIGURE M2.9 SELECTIONS AND ADVICE FROM WPHELP (wp1.kbs)

Note that the expert system is called WPHELP, but the sample expert system file name is wp1 (wp1.kbs to DOS). You now have completed your first consultation session with a small expert system. When you have finished, select the 8Quit option and return to the Main Menu.

Alternate Methods for Menu Selection

VP-Expert allows you to enter commands in other ways. There are four ways to make menu choices in VP-Expert. One way is to move the light bar with the arrow keys and press the Enter key. Three alternate ways are as follows:

1. Type the number corresponding to the command.
2. Press the function key corresponding to the number by the desired option.
3. Type the first letter of the command.

Thus, in the Main Menu, to select the consult option (4Consult) we can type a 4 to get to the Consult Menu, or press the F4 function key to the left or top of the keyboard, or type an uppercase or lowercase c. All these methods will achieve the same objective: to select the 4Consult option from the Main Menu. You are free to use any method you like, but we will use the light bar method and names to make the words meaningful as you navigate through VP-Expert.

Identifying the Three Parts of an Expert System

We stated that an expert system is composed of three logical parts: the knowledge base (facts and rules), the inference engine (reasoning mechanism), and the user interface (communications). The components are shown in Figure M2.10.

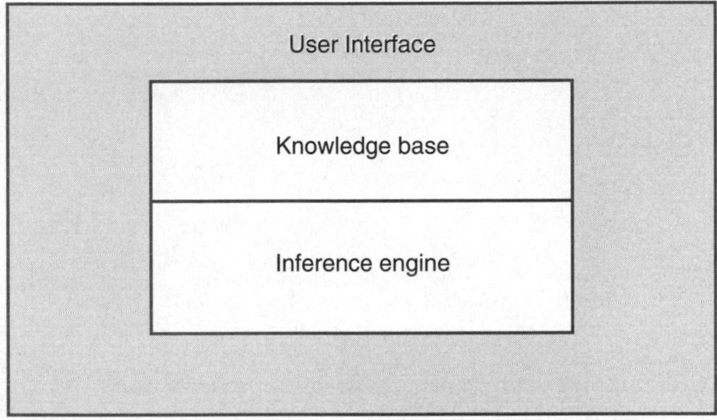

FIGURE M2.10 LOGICAL COMPONENTS OF AN EXPERT SYSTEM

Where are these three parts in the WPHELP system? We will run a slightly different version of our example to show the three components.

Viewing the Component Parts

To expose the three components of an expert system, load and execute the file named wp2.kbs. Assuming that the path is still b:\VPX_samp, follow these steps at the Main Menu:

Step 1. Move the light bar to 6FileName and press Enter. Select wp2 by typing wp2 or use the arrow keys to move the light bar to wp2. Press Enter.

Step 2. With the light bar on 4Consult, press Enter.

Your display should now look like Figure M2.11.

Your screen should now be divided into three parts or **windows**. When we start the consultation, we will see the purpose of the windows as we run the knowledge base or expert system file.

Select the option 2Go to start the consultation with wp2.kbs. Follow the message on the screen and press the Spacebar to start the session with WPHELP. When the first question appears on the screen, your display should look like Figure M2.12, with messages within the lower-left window, called the **Rules Window**. This window tells us what the inference engine is doing.

The comment "Finding Value for Word_processor" tells us that the inference engine is looking for the value of an unknown variable named Word_processor. The inference engine goes to the first rule containing that variable. In this case, it is Rule 1. The comment, "Finding value for Functionality," means that the inference engine wants to

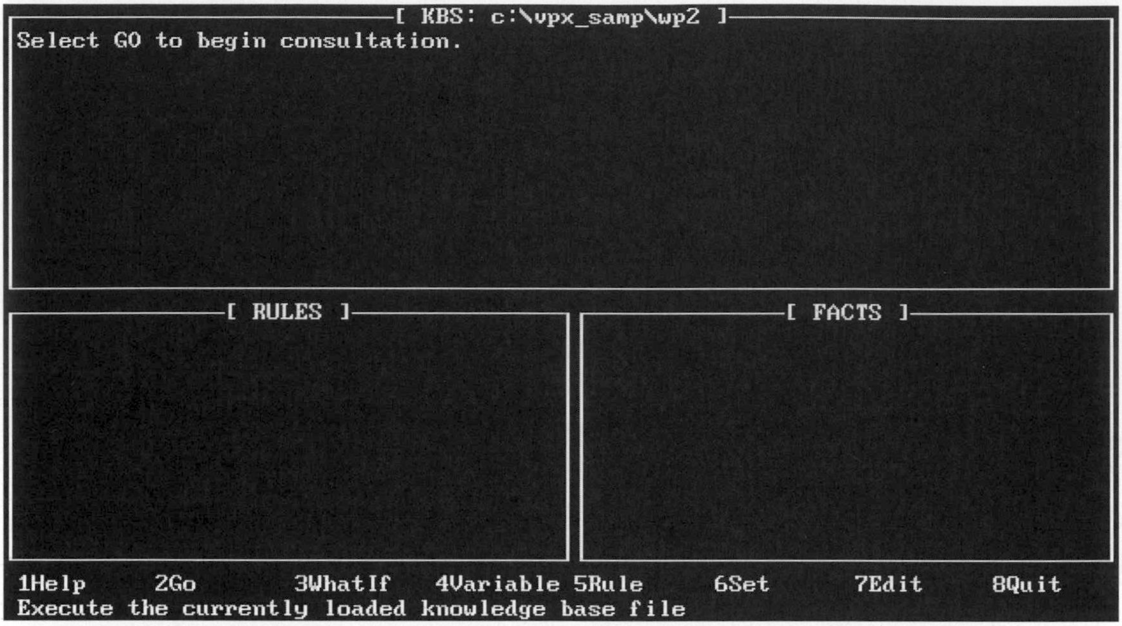

```
[ KBS: c:\vpx_samp\wp2 ]
Select GO to begin consultation.

[ RULES ]                                [ FACTS ]

1Help   2Go   3WhatIf   4Variable 5Rule   6Set   7Edit   8Quit
Execute the currently loaded knowledge base file
```

FIGURE M2.11 OPENING SCREEN FOR LOADING wp2.kbs

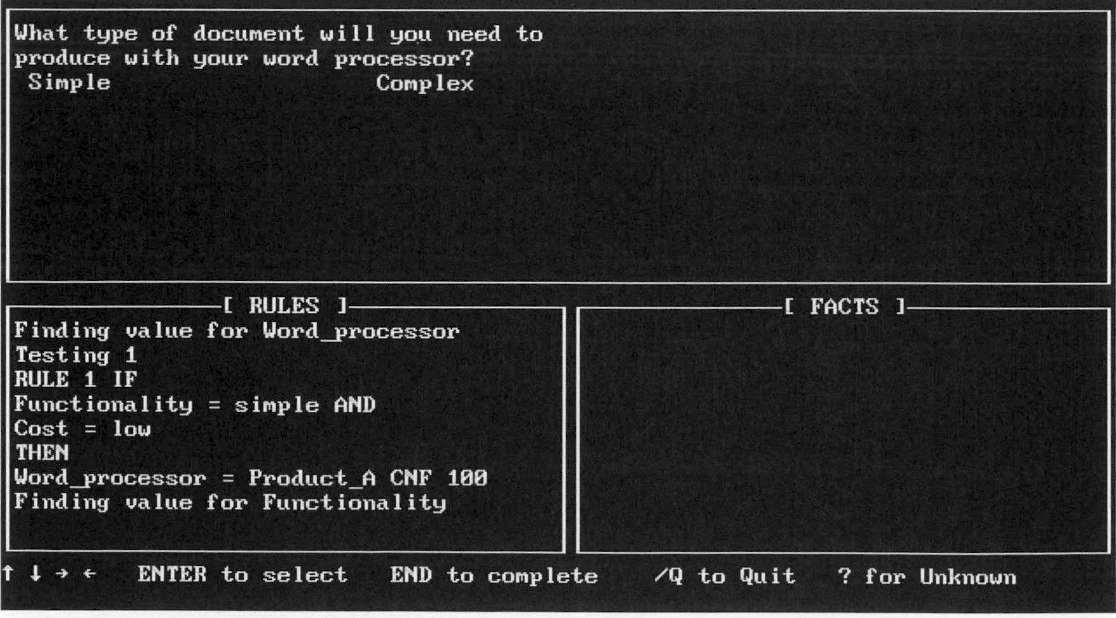

FIGURE M2.12 FIRST QUESTION FROM wp2.kbs BEFORE SELECTION

know the functionality or type of document to be produced. In the Rules Window, the inference engine shows what rules it is testing and what facts it is trying to find from the user.

Now select Simple by pressing Enter and End; your screen now should contain the information shown in Figure M2.13.

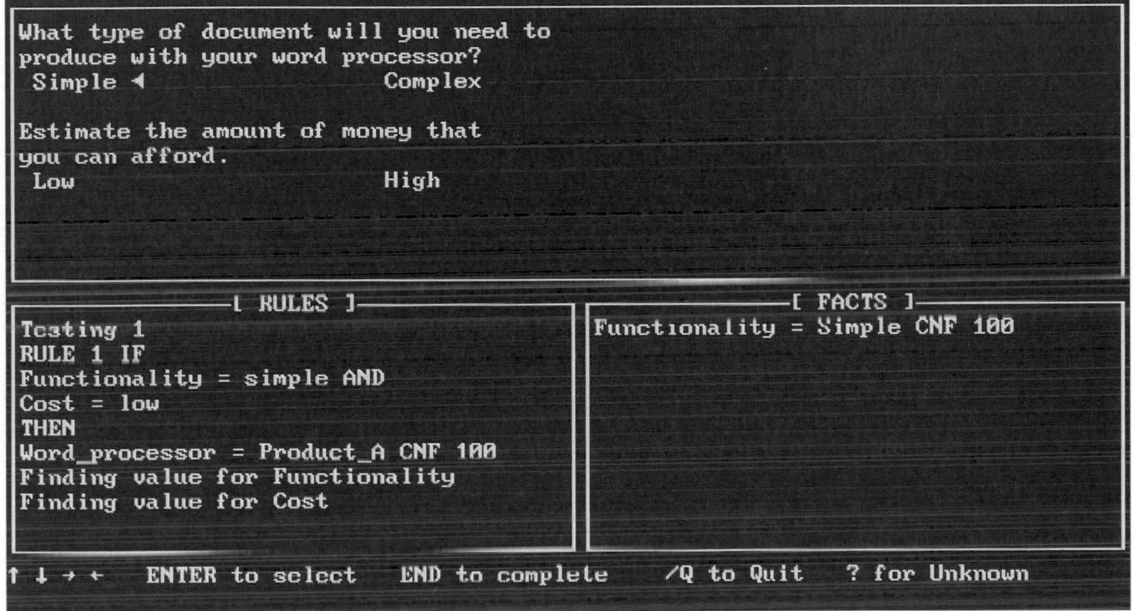

FIGURE M2.13 THE CONSULTATION SCREEN AFTER ANSWERING THE FIRST QUESTION IN wp2.kbs

The window in the lower-right corner shows that the value of Functionality (variable) is Simple. The Rules Window on the left now adds the information that the inference engine is now trying to find the value for Cost. Notice that Cost and Functionality are in the IF part of Rule 1, displayed in the left window. If Functionality had a value of simple and Cost were low, then the Word_processor (variable) would be assigned a value of Product A. In other words, Rule 1 from the knowledge base would be "fired" or considered to be true. Make this happen by selecting Low with the Enter and End keys. Now the system identifies Product A as the selection in Figure M2.14.

The message Word_processor = Product_A CNF 100 is in the lower-right window, the **Facts Window**. The top window is the **Consultation Window,** where the user interacts or communicates with the expert system.

Another name for the Consultation Window is the **user interface.** The lower-left window is the **Rules Window,** where the inference engine displays the name of the

```
produce with your word processor?
 Simple ◄                    Complex

Estimate the amount of money that
you can afford.
 Low ◄                       High

WPHELP advises Product A.

Press the SPACEBAR to exit WPHELP.

[ RULES ]
Testing 1
RULE 1 IF
Functionality = simple AND
Cost = low
THEN
Word_processor = Product_A CNF 100
Finding value for Functionality
Finding value for Cost

[ FACTS ]
Functionality = Simple CNF 100
Cost = Low CNF 100
Word_processor = Product_A CNF 100
```

FIGURE M2.14 THE CONSULTATION SCREEN AFTER ANSWERING THE SECOND QUESTION IN wp2.kbs

variable it is trying to find a value for and the rule from the knowledge base it is testing. The lower-right window, the Facts Window, is where the inference engine displays the values of the variables that it knows. We are close to viewing the three parts of an expert system: the user interface, the knowledge base, and the inference engine. (The windows appear or disappear depending on the absence or presence of a special command, RUNTIME, in the software. RUNTIME causes the Rules and Facts Windows to disappear (close). Otherwise, all three Windows appear. In general, a user doesn't want the Rules and Facts Windows to appear, but while learning about expert systems, we do want them to appear.)

Press the Spacebar to end the consultation and return to the 2Go option of the Consult Menu.

Interaction with a Consultation

When you are viewing and using an expert system, VP-Expert allows you to control the session. The 2Go option has a submenu of six options available during the consultation. These submenu options are not apparent in Figure M2.15. To use these submenu options during a consultation, press the Slash key (/).

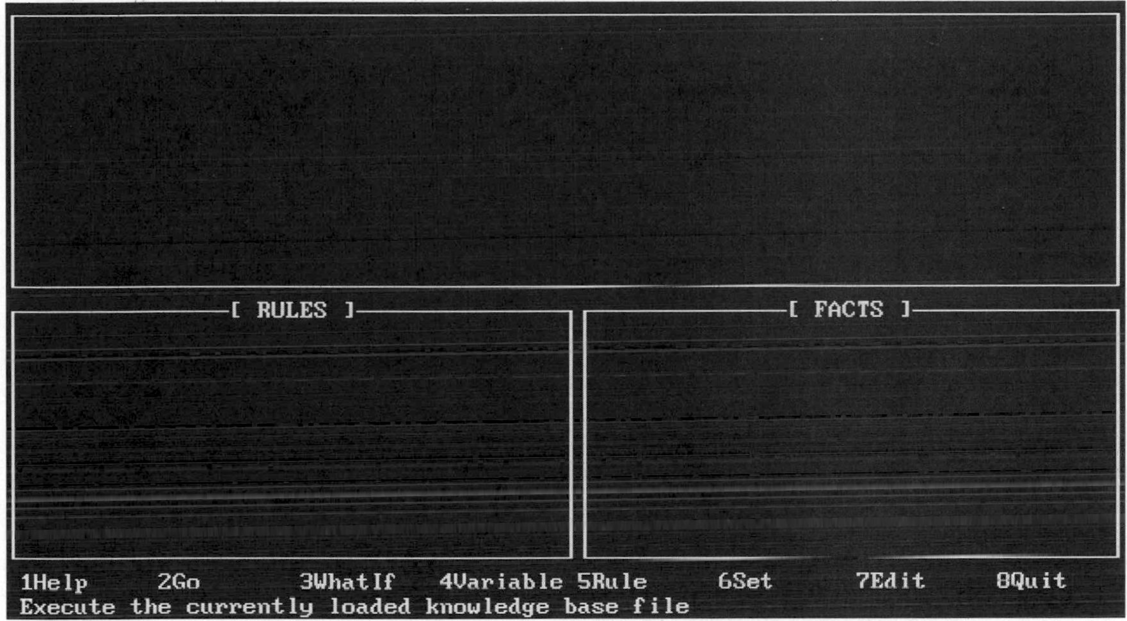

FIGURE M2.15 THE CONSULT MENU WITH THE 2Go MENU

Pressing the Slash key temporarily suspends the consultation so that you can move the light bar with the arrow keys to the option you want and press Enter. The options are summarized in Figure M2.16. The first option is 1Help; the second choice, 2How?, will display the reason for the assignment of a value to a variable. The 3Why? option shows why a question is being asked. The 4Slow option will slow down the scrolling of the rules and variables in the windows so that you can easily read the text. The more often the 4Slow option is selected, the slower the consultation operates. The 5Fast option, the reverse of 4Slow, increases the speed of the consultation. The 6Quit from VP-Expert option returns you to the Consult Menu; it allows you to exit a consultation at any time. Remember that you can use these options during a consultation by pressing the Slash key and making a choice with the light bar and the Enter key.

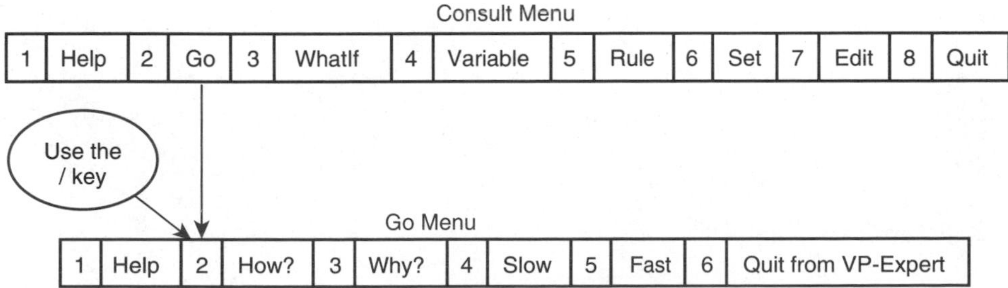

FIGURE M2.16 THE CONSULT MENU WITH THE GO MENU

Rerun the file wp2.kbs using the same selections, complex and low, identified in Figure M2.8. Interrupt the session with the Slash key and experiment with the options. Verify also that what WPHELP advises, Product A, is the same as given in wp1.kbs, as shown in Figure M2.9.

After the consultation is executed, there are additional ways to find out about the workings of the sample expert system file. You may use the 3WhatIf, 4Variable, or 5Rule options available after 4Consult is selected. After you have run a consultation, you may ask what the advice would be if one variable response changes. Instead of rerunning the system, move the light bar to the 3Whatif option and press Enter. A window will appear and ask you what variable you wish to change (see Figure M2.17). You may enter the variable name cost, and VP-Expert will return you to that question in the consultation. To find the value of a particular variable, select the 4Variable option on the Consult Menu.

As shown in Figure M2.18, the 4Variable option will ask you to select the variable you want to see. Select Word_processor and press Enter. The value of the variable Word_processor from the consultation will be displayed in the Consultation Window.

To look at a given rule, select the 5Rule option. Again, a window of information will appear, as in Figure M2.19. Use the arrow keys to select the number of the rule you want displayed, then press Enter.

The difference between the "Slash menu options" and the three options we have just discussed is that the Slash option works *during* a consultation and the three new options (3Whatif, 4Variable, 5Rule) are used *after* a consultation is run. These commands help you to find specific information about a rule or variable in the expert system.

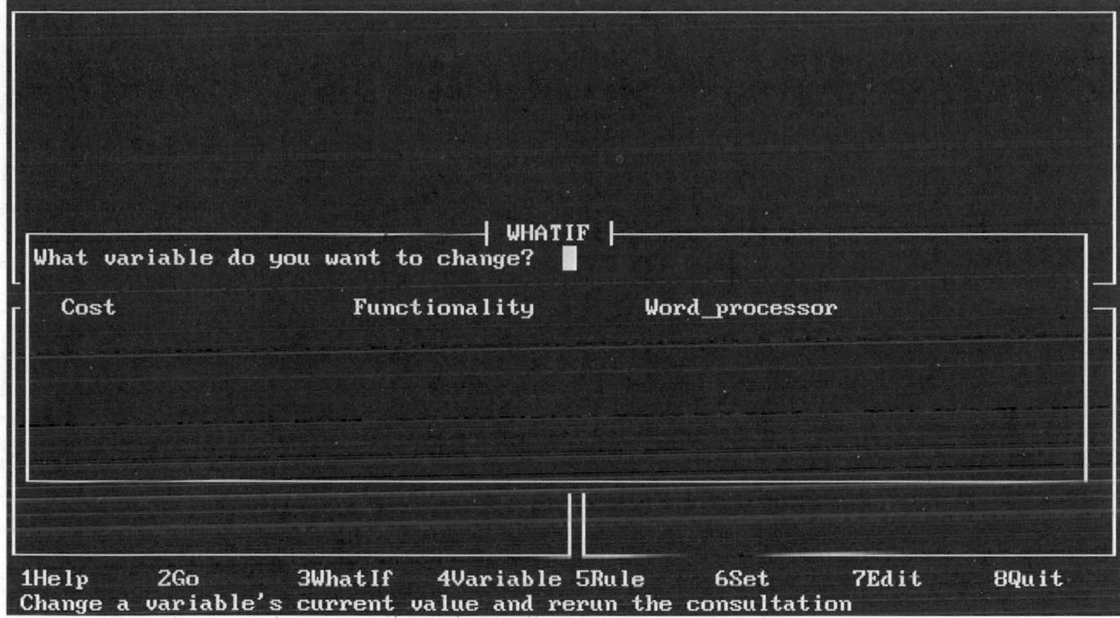

FIGURE M2.17 WHATIF OPTION OF THE CONSULT MENU

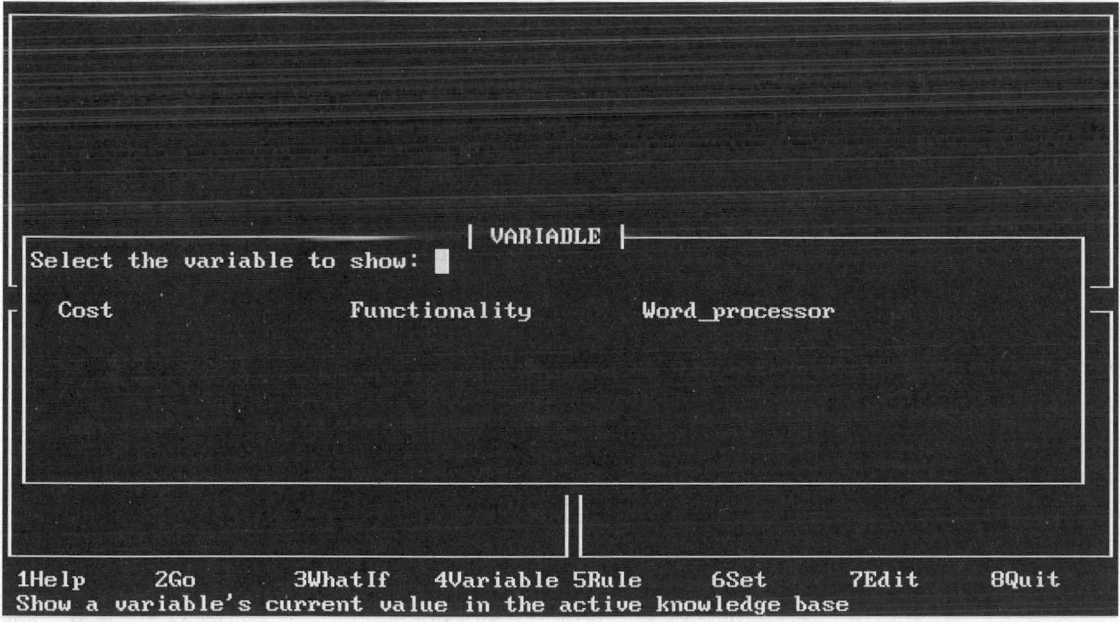

FIGURE M2.18 VARIABLE OPTION OF THE CONSULT MENU

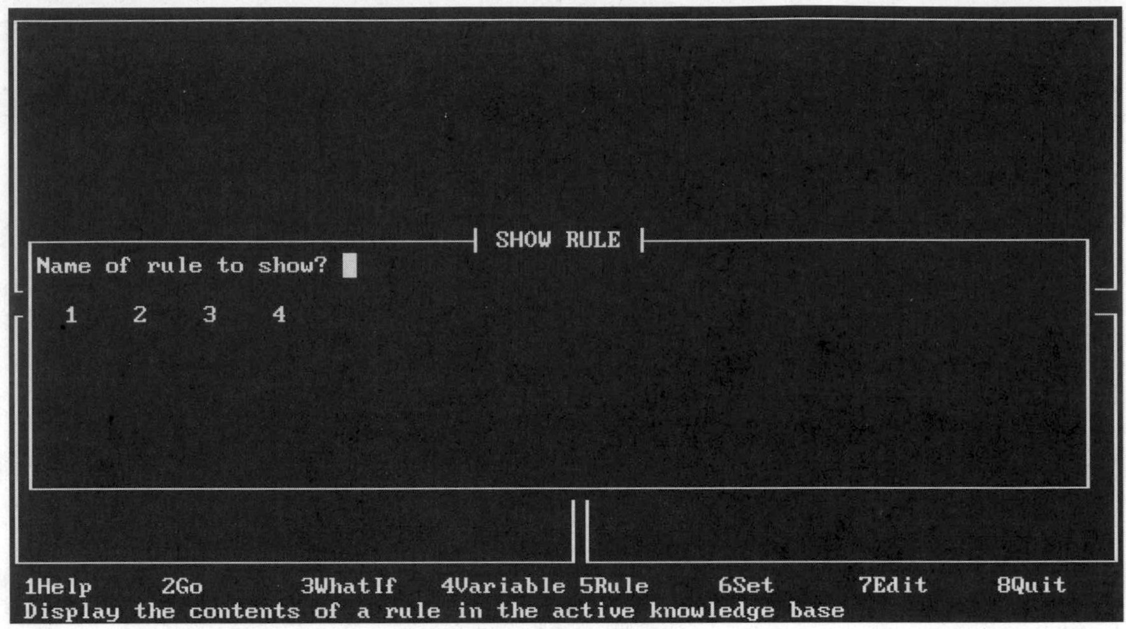

FIGURE M2.19 RULE OPTION OF THE CONSULT MENU

The Software Behind the Consultation

By running WPHELP, in wp2.kbs we obtained three windows that helped us to see the user interface, the knowledge base, and the activity of the inference engine (reasoning mechanism). Now let's take a look at the software behind the expert system again to identify the three parts of an expert system. We want to examine the actual **software** at the screen and obtain a hard-copy printout of the (sample) expert system program.

Obtaining a Printout. We assume that you are in the Consult Menu after running wp2.kbs. We will use the editor's screen facility and printer facility to obtain a printout. Select the 7Edit option from the Consult Menu. While holding down the Shift key, you should see the screen shown in Figure M2.20.

All the text on the screen is the software program for wp2.kbs. The name of the file displayed is shown in the upper-right corner. Note also that the path name is included with the file; the complete name is C:\VPX_samp\wp2.kbs. The C:\VPX_samp tells us that the file is stored in the VPX_samp directory on drive C, that wp2 is the first part of the file name, and that .kbs is the second part of the file name (extension).

Certain editor conditions such as insert on, document (mode) off, boldface off, and underline off are indicated at the bottom of Figure M2.20. The next line is another menu bar of function keys. The 10Print option means to use the F10 or function key

```
                                               Editing: Old File c:\vpx_samp\wp

!The file wp2.kbs shows the three parts◄
!          of an expert system through windows.◄
◄
ACTIONS◄
       DISPLAY "This expert system WPHELP will advise you on selecting ◄
a word processor for your IBM compatible micro-◄
computer in your business.◄
◄
Press the SPACEBAR to start the consultation.~"◄
        CLS◄
        FIND Word_processor◄
        DISPLAY "WPHELP advises {word_processor}.◄
"◄
        DISPLAY "Press the SPACEBAR to exit WPHELP.~◄
"◄
        CLS; ◄
◄
RULE 1 ◄

+    ▲    ▲    ▲    ▲    ▲    ▲    ▲    ▲    ▲    ▲    ▲    ▲    ▲    ▲    ▲
Insert On     Document Off                             Boldface Off Underline Off
 1       2         3Srch↓  4          5Rplc↓  6        7GbRpc↓ 8         9Rept↓ 10
```

FIGURE M2.20 SCREEN DISPLAY WHEN THE EDITOR LOADS wp2.kbs

10 to print the entire document (even though you may only see part of the document on the screen). This function is available only after you release the Shift key. You can see the rest of the document on the screen by using the Page Down (PgDn) and Page Up (PgUp) keys on the right side of the keyboard. To obtain a printout, follow these steps:

Step 1. Make sure your printer is on and has paper.

Step 2. Press the F10 key.

Step 3. When a printer options screen comes up, leave the settings alone; simply press the End key.

The computer screen will display a message that it is printing and then return you to the editor mode. Because all we wanted to do is print a file, we can exit the editor now.

There are many ways to exit the editor, depending on what changes we want to make to the file. Holding down the ALT key displays different options with the function keys in the menu bar at the bottom. We will use ALT-F8 to abandon the editor without making changes to the program. Therefore, to exit the editor, hold down the ALT key and press the F8 key. The editor will ask you (at the bottom of the screen) if you save changes. Type y and VP-Expert should take you back to the Consult Menu. Because we want to discuss the printout, move the light bar to the 8Quit command and press Enter to exit the Consult Menu; press 8Quit again to exit the Main Menu. You should be back at the C:\VPX3ed> prompt. (Users who do not have a hard disk will see a different prompt.)

Examining the Printout. Figure M2.21 provides a copy of the file wp2.kbs. We have divided our VP-Expert program into three parts called Actions, Rules, and Statements.

The ACTIONS block contains the strategic reasoning mechanism of the inference engine; it tells the engine what to look for. From Comment 1 or C1 in Figure M2.21, we observe that a DISPLAY command sends an opening message to the user. Whatever variable is listed after the FIND clause (C2 in Figure M2.21) is what the inference engine is trying to find or conclude. In our case, the inference engine is looking for the value of the variable Word_processor.

The Rules section contains the knowledge base or expertise of the system. In our case, we have four rules, expressed in the IF-THEN format. The THEN parts are determined by the values of the variables, Functionality and Cost, in the IF parts. The STATEMENTS block at the end of the program helps control the user interface so that the user can select values for the type of document (Functionality) and values for Cost.

The ASK statement (C3 and C4) is followed by a variable and then words in quotes, which appear on the display screen. The CHOICES statement after the ASK statement sends the possible values for a specific variable to the screen. After the inference engine has completed its search, it displays the value of the Word_processor and an exit message for you. See C5 in Figure M2.21.

What the user sees on the screen comes from the three DISPLAY commands in the ACTIONS block (C1 and C5) and from the ASK statements (C3 and C4) in the STATEMENTS block. This is shown as two partial screens in Figure M2.22. The comment numbers from the code of Figure M2.21 match the resulting comments on the screen display of Figure M2.22.

Note that the first statements on the screen come from the first DISPLAY statement in the ACTIONS block (C1 from Figure M2.21). The next two questions with choices (C3 and C4) come from the Statements block, but the last sentences on the screen come from the last two DISPLAY statements in the ACTIONS block (C5). The command DISPLAY "WPHELP advises {word_processor}" is used to tell you the value of the word processor after the inference engine has examined the rules and obtained facts. We will discuss the syntax (spelling and special rules) that relate to programs in later modules.

The only difference between wp1.kbs and wp2.kbs is that the RUNTIME instruction in the ACTIONS block of wp1.kbs closes the Facts and Rules windows on the user's display when the consultation is running.

Obtaining Help

Besides learning how to run a consultation and how to obtain a printout of the software, you should know how to obtain any help that VP-Expert might provide. To do this, perform the following steps:

Step 1. At the C:\VPX3ed> prompt, type vpx and press Enter.

Step 2. Press any key to dismiss the Opening screen and the Main Menu will appear. Move the light bar to 1Help and press Enter.

```
!The file wp2.kbs shows the three parts
!             of an expert system through windows.
```

Actions

```
ACTIONS
          DISPLAY "This expert system WPHELP will advise
you on selecting a word processor for your IBM compatible
microcomputer in your business.


Press the SPACEBAR to start the consultation.~"
          CLS
          FIND Word_processor
          DISPLAY "WPHELP advises {word_processor}.
"
          DISPLAY "Press the SPACEBAR to exit WPHELP.~
"
          CLS;
```


Rules

```
RULE 1
IF        Functionality = simple AND
          Cost = low
THEN      Word_processor = Product_A;

RULE 2
IF        Functionality = simple AND
          Cost = high
THEN      Word_processor = Product_A;

RULE 3
IF        Functionality = complex AND
          Cost = low
THEN      Word_processor = none;

RULE 4
IF        Functionality = complex AND
          Cost = high
THEN      Word_processor = Product_B;
```

Statements

```
ASK Functionality: "What type of document will you need
to produce with your word processor?";

CHOICES Functionality: Simple, Complex;

ASK Cost: "Estimate the amount of money that
you can afford.";

CHOICES Cost: Low, High;
```


FIGURE M2.21 THREE PARTS OF A VP-EXPERT PROGRAM

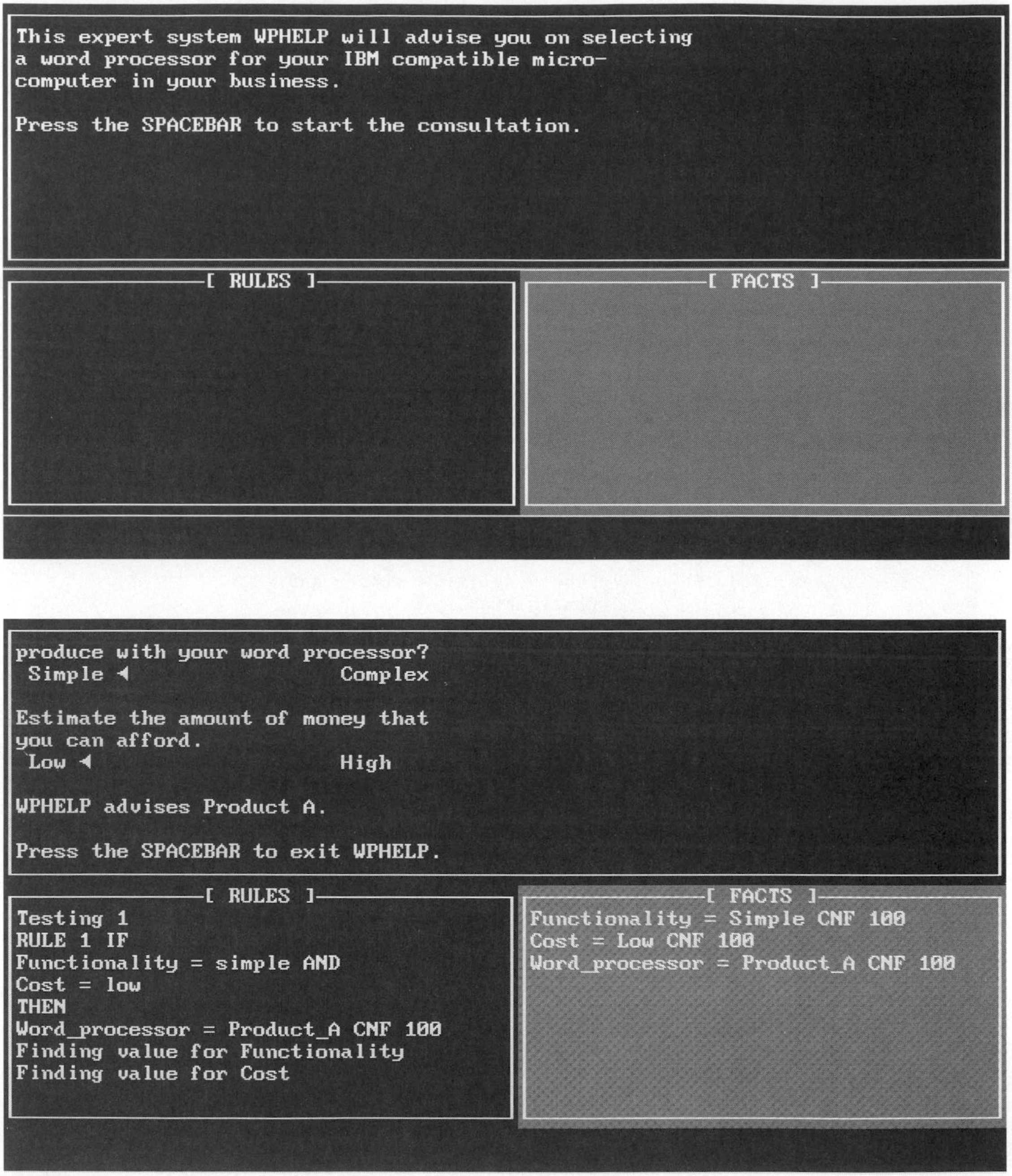

```
This expert system WPHELP will advise you on selecting
a word processor for your IBM compatible micro-
computer in your business.

Press the SPACEBAR to start the consultation.

[ RULES ]                                [ FACTS ]
```

```
produce with your word processor?
 Simple ◄                   Complex

Estimate the amount of money that
you can afford.
 Low ◄                      High

WPHELP advises Product A.

Press the SPACEBAR to exit WPHELP.

[ RULES ]
Testing 1
RULE 1 IF
Functionality = simple AND
Cost = low
THEN
Word_processor = Product_A CNF 100
Finding value for Functionality
Finding value for Cost

[ FACTS ]
Functionality = Simple CNF 100
Cost = Low CNF 100
Word_processor = Product_A CNF 100
```

FIGURE M2.22 USER INTERFACE SCREENS FOR CONSULTATION WITH WPHELP (FROM wp2.kbs)

The Main Menu Help screen should look like Figure M2.23. This screen, just one of many information screens in the Help system, is a summary of the options of the Main Menu. Notice that we have used all of the options listed except Induce, Edit, and Tree. The Edit option shown in the Main Menu is the same as 7Edit from the Consult Menu.

```
                              MAIN MENU
 Related: INDUCE MENU  CONSULT MENU  TREE MENU

   Induce    Creates a working rule base from information organized as a table
             contained in text, database, or worksheet files.

   Edit      Invokes the Editor using the current knowledge base file, or
             will prompt you for a file name to use.

   Consult   Prepares to run the knowledge base file most recently loaded, or
             will prompt you for a file name to use. If no syntax errors are
             detected, VP-Expert displays the Consult Menu.

   Tree      Displays a menu of options for viewing the results of a previous
             consultation inference logic trace.

   FileName  Displays a list of KBS files available in the current directory

   Path      Displays the current drive letter and subdirectory name. Allows
             you to specify a different path if desired.

   Quit      Ends VP-Expert, saves your configuration and returns you to the
↑/↓-Scroll  PgUp/PgDn-Page  Home-Top  End-Bottom  Esc-Topics  P-Print  F1-Keys
```

FIGURE M2.23 THE MAIN MENU HELP SCREEN

Additional available facilities are found on the instruction bar at the bottom of the menu. Press the Escape key (ESC), and your screen should be identical to Figure M2.24. A list of alphabetical topics appears on the left; instructions appear on the right. The instructions tell you to press a single letter key and the arrow keys to locate your desired topic.

Let's find the Help screens for the Consult Menu and get a printout. Follow these steps:

Step 1. Type lowercase c.

Step 2. Using the Down Arrow key, move the light bar until it is on CONSULT MENU.
Press the Enter key.

Step 3. Press P to obtain a printout.

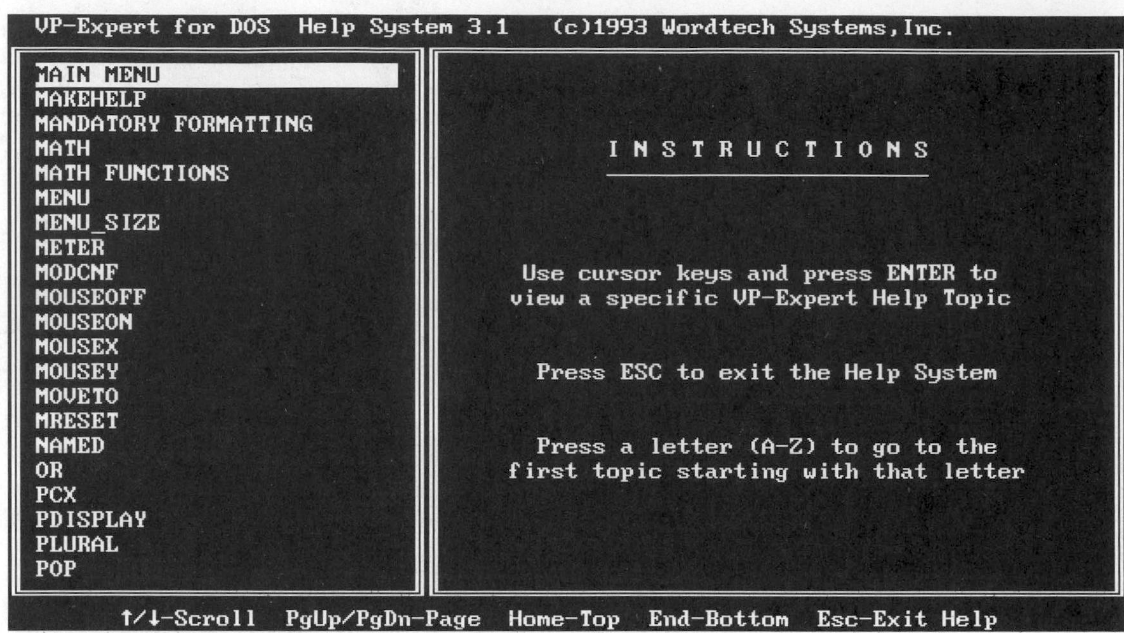

FIGURE M2.24 THE OPENING SCREEN FOR THE VP-EXPERT HELP SYSTEM

The VP-Expert Help facility has a reference feature. Look at the long strip at the top of the screen in Figure M2.25. Topics related to the Consult Menu are listed across the top. You can use the arrow keys to find information on a related menu. (The reference list may scroll off the screen.) A summary of the keystrokes used to engage these options is displayed by pressing the F1 key. Press the Enter key while the light bar is on SET MENU, and the Set Menu Help screen will appear. Obtain a copy by typing P for print. Press ESC to return to the alphabetical topics list. Press ESC again to exit the Help system and return to the Main Menu. The VP-Expert Help facility can be activated before, during, or after a consultation by selecting the 1Help option from any menu. The Help facility contains information on menus, statements, clauses, and other topics, in alphabetical order.

Our work with VP-Expert for this module is now complete. Choose 8Quit and press Enter to return to the DOS system prompt.

We have concluded our use of the microcomputer for Module 2; but let's discuss the printout of the Consult Menu Help screen shown in Figure M2.25. We have used all of these options except the Set option. The Go menu starts a consultation, and Whatif, Variable, and Rule options help find information about the system after a consultation. What the Help screen does not show us, however, is that the Go Menu can be accessed with the Slash key. These Go Menu options could be activated while a consultation is running. At the top of Figure M2.25 we see that related topics are Go Menu, Set Menu, and Main Menu.

```
                              CONSULT MENU
Related:

  Help            Calls the VP-Expert Help System.

  Go              Starts a consultation. If no file is currently loaded,
                  VP-Expert displays a list of files in the current
                  directory

  WhatIf          Allows you to see how the outcome of the consultation would
                  be affected if you assigned a different value to one of the
                  consultation's variables.

  Variable        Shows you the values assigned to one or all variables in
                  the most recent consultation.

  Rule            Displays one or all rules in the current knowledge base.

  Set             Displays an additional menu of options, including the Trace
                  Slow, and Fast commands.

  Quit            Returns to the VP-Expert Main Menu.

↑/↓-Scroll  PgUp/PgDn-Page  Home-Top  End-Bottom  Esc-Topics  P-Print  F1-Keys
```

FIGURE M2.25 CONSULT MENU HELP SCREEN

Note that you will find Edit as an option on both the Main Menu and the Consult Menu (Figures M2.23 and M2.25), even though this option is not included in the Help screen. You can get to the editor mode from either menu. Often, using the Go and Edit features from the same Consult Menu is more convenient. If you experience any unusual "out of memory" errors when using the 7Edit option from the Consult Menu, use the 3Edit option from the Main Menu.

In summary, we have conducted consultations with a simple expert system, identified the three parts of an expert system, obtained a printout of the expert system program (wp2.kbs), discussed the program, and examined the Help facilities in VP-Expert. In the next module we will look at the knowledge base (production rules, or simply rules) in more detail.

Key Points

- VP-Expert is accessed by typing vpx at the DOS system prompt.
- The opening screen or Main Menu has eight options.
- Menu options may be selected in any of four ways:
  - Move the light bar to your choice and press Enter.
  - Enter the number by the menu choice.
  - Press the function key with the same number.
  - Type the first letter of the menu choice.

- The extension name for VP-Expert files is .kbs (for knowledge base).
- 4Consult on the Main Menu is the choice for running an expert system or knowledge base file (.kbs). Then select 2Go from the Consult Menu.
- Before running an expert system, use the 7Path option to indicate where the file is located (C:\VPX_samp) and the 6FileName option to select the file.
- When running an expert system (consultation), make choices by pressing the Enter key followed by the End key.
- The file wp1.kbs is the program for advice on selecting a word processor; the file wp2.kbs, the same expert system, allows the user to view the Rules Window and the Facts Window as the expert system runs.
- When using wp2.kbs, you can "see" the three parts of an expert system:
  - The user interface is at the top.
  - The inference engine communicates its actions, rules, and values of variables via the two bottom windows, the Rules and Facts Windows.
- When running an expert system, you may press the Slash key to bring up the Go Menu of 1Help, 2How?, 3Why?, 4Slow, 5Fast, and 6Quit from VP-Expert. These options will help you understand how the inference engine is working during a consultation.
- After running an expert system, you can select the options, 3WhatIf?, 4Variable, and 5Rule from the Consult Menu to find information about the variables and the rules.
- To obtain a printout of an expert system program (.kbs), or file, select C:\VPX_samp on 7Path, the desired file in 6FileName, and then 3Edit from the Main Menu. Once inside the editor, press the F10 key, followed by the End key. Exit the editor by pressing the ALT and F8 keys simultaneously. (ALT-F6 will save the file and exit at the same time.) You may also use the 7Edit option from the Consult Menu.
- A VP-Expert program is composed of three parts:
  - Actions (strategy or plan of the inference engine)
  - Rules (IF-THEN rules)
  - Statements (commands to obtain choices or information from the user)
- The FIND command in the ACTIONS block of a program identifies the work to be done by the inference engine. What the inference engine is trying to find or conclude is identified by the variable after the FIND command.
- The text within quotes in a DISPLAY command will be directed to the screen as information for the user.
- To obtain help from VP-Expert, select the 1Help option from the Main Menu or other menus. Follow the instructions on the screen to look up a term for more information.

Exercises

1. Which of the following options are on the Main Menu?
   Exit Quit Go Stop Edit Directory Path Help
2. Name the command from the Main Menu to:
   a. get out of the menu.
   b. select a knowledge base program.
   c. tell the computer where files are located.
   d. actually run a program.

3. Which of the following options are on the Consult Menu?
   Run Go Variable Rule Edit Print Stop Help
4. What are the four different ways to select 4Consult from the Main Menu?
5. Select the two successive commands needed to run an expert system:
   a. Consult, Run
   b. Consult, Go
   c. Run, Consult
   d. Go, Consult
6. What are the other three sets of commands that you could use to run an expert system? Assume you start at the Main Menu.
7. Which of the following options are part of the Go Menu?
   Help Go How? Why? Slow Fast Print Quit from VP-Expert
8. Name the options from the Consult Menu to:
   a. find a rule in the knowledge base
   b. see what advice the system gives if one variable is changed
   c. run a knowledge-based program
   d. display the value of a variable
9. Which of the following options from the Main Menu do you use to print a copy of a .kbs file?
   a. 1Help
   b. 2Induce
   c. 3Edit
   d. 4Consult
   e. none of these
10. Name at least one way to access the Edit option.

Assignments

1. Run wp2.kbs again. During the consultation, use the Slash key to select the 3Why? option. Obtain a copy of the answer that appears on the screen by pressing the Shift and Print Screen keys at the same time.
2. Using wp1.kbs or wp2.kbs, do the following:
   a. Identify the variable names.
   b. Tell how many rules are in the knowledge base.
   c. Identify Rule 3.
3. The only difference in the programs wp1.kbs and wp2.kbs is that wp1.kbs has the RUNTIME command and wp2.kbs does not. Verify this by obtaining a printout of wp2.kbs. Remember to:
   a. Set the 7Path to c:\VPX3ed (for hard disk users).
   b. Select 6FileName (wp2.kbs).
   c. Select 3Edit (to use the F10 print option).
   d. Draw a box around each of the ACTIONS, RULES, and STATEMENT blocks on the printout. Label each block.

4. Using the 1Help option of the Main Menu, obtain a printout of the information about these topics.
   a. Edit commands
   b. Cursor command
   c. Delete commands
   d. Function keys
   e. Go Menu
   f. Runtime
   g. Shift-Function keys
   h. ALT-Function keys
5. Using the Taxguide.kbs file from your sample file directory, run the Taxguide expert system.
   a. What is the purpose of the Taxguide system?
   b. How many rules are in the Taxguide system?
   c. Obtain a printout of Taxguide.kbs.
   d. On the printout, draw a box around each of the Actions, Rules, and Statement blocks. Label each.
6. Use VP-Expert Editor to display files. Set the 7Path to C:\VPX_samp and the 6FileName to wp2. Now load wp2.kbs with 3Edit. Within the editor, you can display files on a specified drive. Follow these steps:
   1. Press ALT-F7.
   2. In reply to the message, "Which files do you want to list?", type *.kbs.
   3. Press Enter.
   4. Press Shift and Print Screen simultaneously to obtain a printout. (F10 and End do not work here.)
   5. Press ESC.
   6. Press ALT-F8 to exit without saving changes. Answer Y to the message.
7. There is an alternate way to display a list of files from DOS. At the c:\VPX3ed> prompt, type

   DIR *.kbs

   to display all the knowledge base files. Press Shift and Print Screen simultaneously to obtain a printout.
8. As we use the sample files, the modules will instruct us to create certain files. To make sure that the sample files subdirectory has enough room, find out how much space in bytes is free on the disk. Follow these steps:
   1. Enter the DOS command CD VPX3_samp.
   2. Type dir/w or dir.

   After the file names are listed, DOS will tell you the number of files and the number of bytes free. Record this information.
9. The file vpx.exe in C:\VPX3ed executes when you type VPX. Find out how large the file is by typing DIR VPX.EXE at the system prompt.
10. To obtain a listing of VP-Expert system files, get to the C:\VPX3ed > prompt.

a. To obtain a listing at the screen, type

DIR v*.*

b. To redirect the output to a printer, type

DIR v*.* > prn

11. To obtain information about your working environment, get to the C:/> prompt.
    a. To see the directory structure (tree), type

    TREE C:

    b. To see the tree structure and files, type

    TREE C: /f

    c. To redirect the information (from 11b) to a printer, type

    TREE C: /f. prn

12. Let's explore the VP-Expert Help facility more. You may enter this Help facility from within VP-Expert or by typing VPXHELP at the c:\VPX3ed> prompt. Get to the Help facility and obtain a printout (for study purposes) of each of the following topics:
    a. DISPLAY
    b. FIND
    c. ASK
    d. CHOICES
    e. IF
    f. ACTIONS
    g. STATEMENTS
    h. CLAUSES
    i. RULE
13. DOS can be used to display the contents of certain files. In our VP-Expert environment, files with a .kbs extension can be displayed. (Filenames containing an .exe extension can't be displayed on the screen.) At the C:\vpx_samp prompt, try these DOS commands for displaying text.
    a. To display file contents to the screen, type

    TYPE wp2.kbs

    b. To redirect the screen output to a printer, type

    TYPE wp2.kbs > PRN

    c. To display a long file, one screen at a time, type

    TYPE taxguide.kbs | MORE

Module 3

The Knowledge Base: Production Rules in VP-Expert

The knowledge base is a key part of an expert system. We will concentrate on understanding how VP-Expert structures its knowledge base. We will take a detailed look at production rules (IF-THEN), variables in production rules, optional parts of rules called ELSE and BECAUSE, and rules with compound parts (AND, OR). We will discuss the concept of a PLURAL variable. Using the INDUCE command, we will use a table file to generate a "bare bones" knowledge base without programming. We will conclude with ways to modify our induced knowledge base program to improve the user interface.

Forming Rules

VP-Expert can give us a format to write rules, but it cannot guarantee that the rule represents knowledge. Three terms—data, information, and knowledge—are sometimes confused. *Data* is simple facts or measurements. *Information* is data organized so that it is useful to a user. *Knowledge* is the refinement of information so that productive and efficient problem-solving can occur. Assuming that you have the knowledge, VP-Expert has a certain format in which it must be placed.

Use of IF-THEN Rules

VP-Expert uses the most common model, production rules, to represent or structure knowledge. VP-Expert does not use semantic nets, frames, or the predicate calculus models. *Production rules* are IF-THEN sentences: the IF part is the condition (premise) and the THEN part is the conclusion. For example, in the rule

If the document is complex, then use Product B.

the condition (premise) is "If the document is complex," and the conclusion is "use Product B." The condition and conclusion are both assertions; each may be a true assertion or a false assertion. When using IF-THEN rules, we can say that the conclusion is true only when the condition is true. If the premise is false, then we can't conclude the truth value of the conclusion. We sometimes say that a rule fires; by this we mean that we know that the condition is true, so we can conclude that the conclusion is true (the rule fires). If a condition is false, then the rule does not fire.

Optional Parts of IF-THEN Rules: ELSE and BECAUSE

VP-Expert allows two optional parts to a rule: ELSE and BECAUSE. The format of the rule is as follows:

RULE label

IF condition

THEN conclusion 1

ELSE conclusion 2

BECAUSE " "

A rule in VP-Expert will always have a rule label (name or identifier), condition, and a conclusion 1. The ELSE conclusion and BECAUSE may or may not be included. An IF-THEN rule may or may not fire; in an IF-THEN-ELSE form, however, the rule always fires. If the condition is true, then conclusion 1 is true. If the condition is false, then conclusion 2 is true. Both conclusions (separated by an ELSE) cannot fire in the same rule at the same time.

The BECAUSE option only adds information and has nothing to do with firing the rule. The BECAUSE option is followed by a sentence or words enclosed in quotes. When the user selects the 2How? or 3Why? options during a consultation, the words following a BECAUSE are displayed. It is up to the system's designer to make sure the BECAUSE statement is clear and readable. For example,

IF the document is complex

THEN select Product B

ELSE select Product A

BECAUSE "Product B will process more complicated documents than Product A."

If the condition (document is complex) is true, then conclusion 1 (select Product B) is true. If the condition is false, then conclusion 2 (select Product A) is true. In either case, the rule is fired because of the ELSE option. If the rule did not have the ELSE option, it would fire only if the condition were true. Moreover, the text "Product B will process more complex documents than Product A" would appear if the user selected the 2How? or 3Why? options of the Go Menu. If a rule does not have a BECAUSE part, then the rule itself is displayed for the 2How? or 3Why? options. The BECAUSE option adds to the understanding of the rule but does not affect the firing of the rule.

Variable and Relational Operations for Rules

Any condition and conclusion must be in a particular format for VP-Expert to understand. The sentences are usually composed of relationships between variables and constants (or other variables). For VP-Expert to understand this last IF-THEN rule, we would have to reword it in terms of variables and constants, and the relationships between them. In the rewording, we have:

> RULE 7
>
> IF Functionality = complex
>
> THEN Word_processor = Product_B
>
> ELSE Word_processor = Product_A
>
> BECAUSE "Product B can process more complicated documents than Product A.";

RULE, IF, THEN, ELSE, and BECAUSE are called *key words* or *reserved words*; the 7 is called a rule name or label. The variables are Functionality and Word_processor. The values or constants are complex, Product_B, Product_A, and the text in quotes after BECAUSE. The first equal sign is a relational operator: it indicates the relation, the same as, in the rule. Figure M3.1 shows the parts of an IF-THEN-ELSE-BECAUSE rule. VP-Expert has its own rules for forming or spelling variable names.

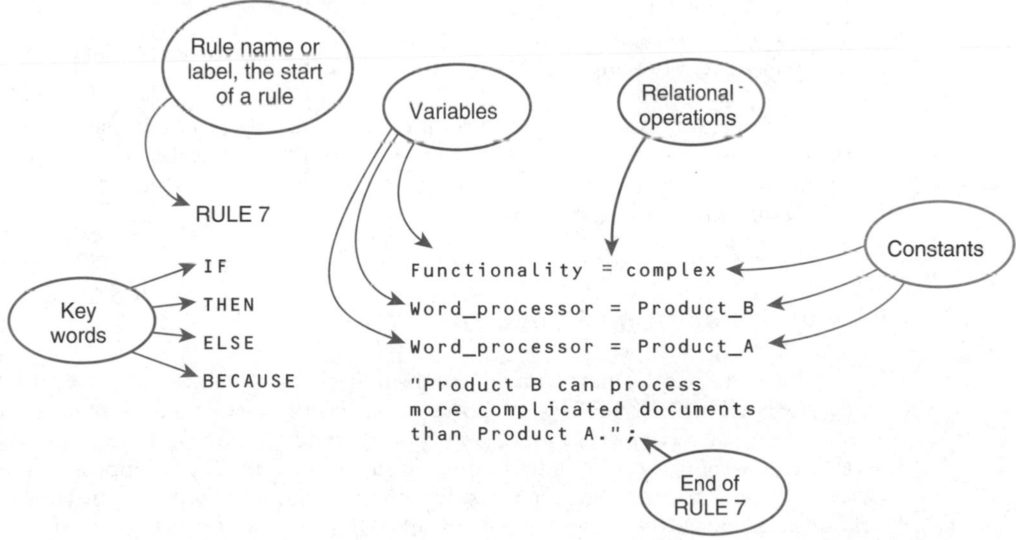

FIGURE M3.1 PARTS OF AN IF-THEN-ELSE-BECAUSE RULE

Other relational operators are allowed in the premise, but only the equal sign, used as an assignment, is allowed in the conclusion. The start of a rule is marked by the word RULE and the end of the rule by a semicolon(;).

Let's discuss the variables used in the condition and conclusion of our rule. A VP-Expert variable can have up to 40 characters, which can include letters of the alphabet (uppercase, lowercase, or mixed), numbers, and the special characters _ $ % ^ ¦. Spaces are not allowed, so the underscore symbol is used to separate words. This is why the variable, **give advice**, is illegal in VP-Expert; instead, we use **give_advice** for readability. Key words or reserved words in VP-Expert cannot be used as variable names. Moreover, each rule must start with the key word RULE, then a space, and then a label name. Label names and variable names follow the same requirements. We will use capital letters for key words and certain commands, but only as a matter of choice. We can use several comparisons between variables and constants in conditions. They are:

= equals

< less than

< = less than or equal to

> greater than

> = greater than or equal to

<> not equal to

If we want to compare the value of one variable to the value of another variable, we must put the second variable in parentheses. For example, to compare Balance and Savings, we could write

IF Balance = (Savings), THEN Advice = save_more;

VP-Expert requires the variable on the right of the equal sign to be in parentheses. It directs the system to find the variable Savings and compare the value of Savings to the value of Balance. Each of our rules in the knowledge base must adhere to all of these rules of spelling and formatting (syntax).

Rules with Compound Conditions (AND, OR)

The *condition* of a rule may contain compound parts that may be joined by two connectives (AND, OR). VP-Expert allows up to 20 different conditions to be connected. When you have two or more parts connected by an AND or OR, you need to find the truth value of the entire compound condition. The compound part is considered true for an AND statement only when all the parts are true; however, the truth value of an OR statement is false only when all parts are false. Figure M3.2 summarizes compound AND and OR statements.

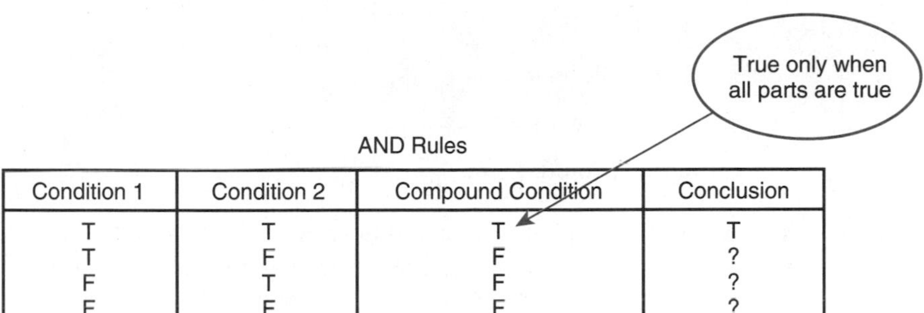

AND Rules

Condition 1	Condition 2	Compound Condition	Conclusion
T	T	T	T
T	F	F	?
F	T	F	?
F	F	F	?

IF condition 1 and condition 2
THEN conclusion

OR Rules

Condition 1	Condition 2	Compound Condition	Conclusion
T	T	T	T
T	F	T	T
F	T	T	T
F	F	F	?

IF condition 1 or condition 2
THEN conclusion

False only when
all parts are false

FIGURE M3.2 COMPOUND CONDITIONS IN RULES

The inference engine fires a rule only when it knows the truth value of the combined condition to be true. If there is a compound condition (AND, OR), the inference engine has to ascertain the value of each part of the compound condition to determine if the rule fires. Note that if the compound condition is false, we can't determine the truth value of the conclusion. The rule does not fire, and the truth value of the conclusion is unknown.

Does VP-Expert allow compound conclusions, as it does compound conditions? VP-Expert does *not* allow the OR connective between conclusions but allows implied AND conclusions. For example, the following rule is invalid:

RULE 4

IF condition 1

THEN conclusion 2

OR conclusion 3;

VP-Expert does not allow OR between conclusions. The following rule, however, is valid:

RULE 4

IF condition 4

THEN conclusion 5

conclusion 6;

The AND connective is not stated between conclusion 5 and conclusion 6, but it is implied. If condition 4 is true, then both conclusion 5 and conclusion 6 will be true. Implied compound conclusions can also be found after an ELSE.

Let's run an example of each compound rule.

The file wp3.kbs contains a small expert system of one compound AND rule. The rule is:

RULE 1	
IF Afford = Yes	(condition 1)
AND Help = Yes	(condition 2)
AND Time = Yes	(condition 3)
AND Need = Yes	(condition 4)
THEN Advice = Proceed	(conclusion 5)
ELSE Advice = Wait;	(conclusion 6)

This rule has four conditions (Afford = Yes, Help = Yes, Time = Yes, and Need = Yes). Conclusion 5 is Advice = Proceed and conclusion 6 is Advice = Wait. If all four conditions are true, then the rule fires and Advice = Proceed.

To verify this, open VP-Expert in your working environment. At the Main Menu, follow these steps after typing vpx:

Step 1. Set the 7Path to c:\vpx_samp.

Step 2. Set the 6FileName to wp3.

Step 3. Select the 4Consult option.

Step 4. Select the 2Go option.

Try running wp3.kbs with the values in rows 1, 4, 10, and 16 from the following table to verify some of the 16 possible results. You probably noticed something interesting when you tried to verify row 4. The inference engine's search ended (you were asked no more questions) when the first No was entered. As soon as one condition in an AND rule is false, VP-Expert ceases to do anything more with the rule.

	CONDITION 1	*CONDITION 2*	*CONDITION 3*	*CONDITION 4*	*CONCLUSION*
Run	Afford	Help	Time	Need	Advice
1	Yes	Yes	Yes	Yes	Proceed
2	Yes	Yes	Yes	No	Wait
3	Yes	Yes	No	Yes	Wait
4	Yes	Yes	No	No	Wait
5	Yes	No	Yes	Yes	Wait
6	Yes	No	Yes	No	Wait
7	Yes	No	No	Yes	Wait
8	Yes	No	No	No	Wait
9	No	Yes	Yes	Yes	Wait
10	No	Yes	Yes	No	Wait
11	No	Yes	No	Yes	Wait
12	No	Yes	No	No	Wait
13	No	No	Yes	Yes	Wait
14	No	No	Yes	No	Wait
15	No	No	No	Yes	Wait
16	No	No	No	No	Wait

Now let's try an OR rule with an ELSE option. If all parts of a compound OR condition are false, then the condition (after the ELSE) is fired. If one or more conditions in the IF part are true, the rule will fire and the conclusion before the ELSE will be true. The file wp4.kbs is a small system advising you if you need or don't need a word processor. The single compound OR-IF-THEN-ELSE rule in wp4.kbs is:

```
IF Letter = Yes                  (condition 1)
      OR Resume = Yes            (condition 2)
      OR Notice = Yes            (condition 3)
      OR Paper = Yes             (condition 4)
THEN Advice = need               (conclusion 5)
ELSE Advice = do_not_need;       (conclusion 6)
```

When one or more of the conditions is true, conclusion 5 is true. When all four conditions are false, the ELSE part (conclusion 6) is fired. The possible results are as shown on the following page.

Note again that there are 16 possible selections for the four variables. To access wp4.kbs, follow these steps:

Step 1. Exit the Consult Menu with 8Quit.

Step 2. Select 6FileName and enter wp4.

Step 3. Select 4Consult.

Step 4. Select 2Go.

	CONDITION 1	*CONDITION 2*	*CONDITION 3*	*CONDITION 4*	*CONCLUSION*
Run	Letter	Resume	Notice	Paper	Advice
1	Yes	Yes	Yes	Yes	need
2	Yes	Yes	Yes	No	need
3	Yes	Yes	No	Yes	need
4	Yes	Yes	No	No	need
5	Yes	No	Yes	Yes	need
6	Yes	No	Yes	No	need
7	Yes	No	No	Yes	need
8	Yes	No	No	No	need
9	No	Yes	Yes	Yes	need
10	No	Yes	Yes	No	need
11	No	Yes	No	Yes	need
12	No	Yes	No	No	need
13	No	No	Yes	Yes	need
14	No	No	Yes	No	need
15	No	No	No	Yes	need
16	No	No	No	No	do_not_need

Verify the results (advice) for runs 1, 5, 15, and 16 from the above table.

What happens, moreover, if an AND and an OR appear in the same rule? This rule

RULE 1

IF condition 1

AND condition 2

OR condition 3

THEN conclusion;

contains both AND and OR between three conditions. Which connective is evaluated first, AND or OR? Figure M3.3 shows a possible problem.

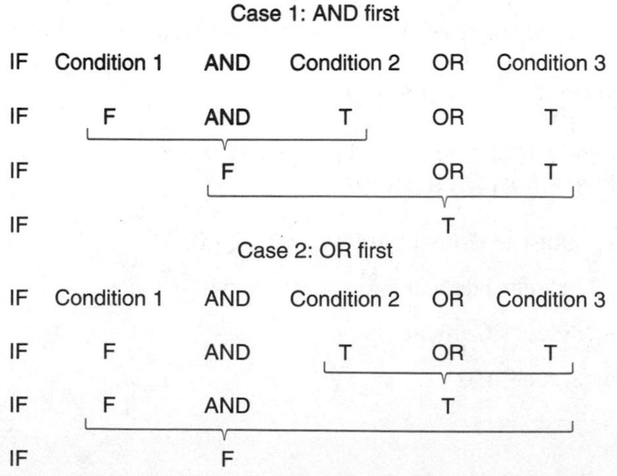

Case 1: AND first

```
IF   Condition 1   AND   Condition 2   OR   Condition 3
IF       F         AND       T         OR       T
IF                  F                  OR       T
IF                                      T
```

Case 2: OR first

```
IF   Condition 1   AND   Condition 2   OR   Condition 3
IF       F         AND       T         OR       T
IF       F         AND                  T
IF                  F
```

FIGURE M3.3 PROBLEM IN THE ORDER OF EVALUATING COMPOUND CONDITIONS

Let's assume that condition 1 is false, condition 2 is true, and condition 3 is true. If we evaluate AND first, followed by OR (Case 1), the final result is a true (compound) condition. If we evaluate OR before AND, as in Case 2, the final result is a false (compound) condition.

To fix this possible problem, VP-Expert assumes Case 2: that is, it evaluates OR first even though it doesn't appear first. VP-Expert does not allow you to change the order by inserting parentheses. To view a variation of WPHELP using compound conditions and various relational operators, run wprelate in Assignment 12 at the end of the module. (Parentheses are allowed only for numerical evaluations, not logical evaluations.)

Mathematical Calculations in Rules

VP-Expert has built-in mathematical operators and functions for use in expert systems programs. The four basic arithmetic operators are implemented with the following symbols:

Operation	*Symbol*
Addition	+
Subtraction	−
Multiplication	*
Division	/

Special mathematical and trigonometric functions are also implemented. In the list below, X stands for a numeric value or a variable with a known numerical value. The functions are as follows:

Mathematical Functions.

@ABS(X)	Returns the absolute value of X.
@EXP(X)	Returns the value of the exponential function.
@LOG(X)	Returns the natural logarithm of X.
@SQRT(X)	Returns the square root of X.

Trigonometric Functions.

@COS(X)	Returns the cosine of angle X.
@SIN(X)	Returns the sine of angle X.
@TAN(X)	Returns the tangent of angle X.
@ACOS(X)	Returns the arccosine of angle X.
@ASIN(X)	Returns the arcsine of angle X.
@ATAN(X)	Returns the arctangent of angle X.

For trigonometric functions, the X represents radian values rather than degree values.

There is a difference in the syntax of arithmetic operators and mathematical functions. The basic arithmetic operators simply use the symbol between two numeric values or variables. The mathematical and trigonometric functions must start with the @ symbol and involve only one number (parameter). All numeric calculations, however, must be contained in parentheses. This signals to VP-Expert to suspend the work of the inference engine and perform the calculation. Moreover, when the calculation is assigned to a variable, the calculation must be to the right of the equal sign, and the variable must be to the left of the equal sign. In other words, variable = (6*price) is a legal statement, but (6*price) = variable is not a legal statement. To make the calculation, however, we must first know the value of price.

Calculations can be performed in both the condition and conclusion parts of rules, but in both cases, the calculations must be on the right side of any **equal** or **assignment** statement. We will explain the concept of an assignment statement by analyzing the following three statements:

Statement 1 IF A = 6

Statement 2 IF A = (B-6)

Statement 3 IF A = 3

THEN B = (A*4)

In Statement 1, the value of the variable A is compared to six to determine if the relationship is true or false. Six is not being assigned to A. Moreover, no math calculation is performed. In the second statement, the subtraction of six from the value of B is compared to A to determine if the two numbers are equal. In Statement 3, the IF part is still a comparison, but the THEN part involves an assignment. The value of A will be multiplied by 4, and the result will be assigned to the variable B. In other words, the equals in the IF part is a comparison, but the equal sign in the THEN part is an assignment. Here is an example in which the same symbol, =, has two different meanings in our rule construction. We can summarize the use of mathematical calculations by the following points:

1. Place the calculations inside parentheses.
2. Special functions must start with @.
3. The value of variables must be known before calculating.
4. Assignment statements must have calculations only on the right side of the equal sign.

We will find many uses for arithmetic operators and math functions, such as computing ratios, percents, averages, and standard deviations. For additional examples of mathematical commands, run wpmath in Assignment 13 at the end of this module.

Variables in VP-Expert

Let's discuss variables in VP-Expert before continuing with production rules. VP-Expert can handle variables that contain text (characters) or numbers. Assignment of values of variables can be achieved by any of the following:

Direct assignment (=) in the ACTIONS block

Direct assignment in the conclusion of a rule

Input from the user (ASK, CHOICES)

As discussed in the previous section, mathematical calculations are assigned by enclosing the calculation in a set of parentheses and using the equal sign. Special math functions must be preceded by the @ symbol. This leads to a discussion of assignment or addressing schemes in VP-Expert. In addition to direct assignment of values there are two additional methods, indirect addressing and double indirect addressing techniques.

Indirect addressing assigns the value of one variable to another variable. The technique is represented below as

```
variable_1 = value
variable_2 = (variable_1)
```

This method of indirectly assigning a value to variable_2 via variable_1 is called indirect addressing. In the following ACTIONS block, the variable best_product is assigned the value of the variable product:

```
ACTIONS
FIND product
best_product = (product)
     :
     :
```

In the following rule, the value difference is assigned to the variable total:

```
RULE 45
IF          difference > 0
THEN total = difference;
```

Indirect addressing with variables can also be used in rule conditions. For example:

```
RULE 12
IF          difference > (last_month)
THEN strategy = successful;
```

Indirect addressing can also be used with the GET clause with database access, which will be discussed in Module 4.

Another method of assigning values to a variable is called **double indirect addressing**. Consider the following:

PET = Cat

CAT = Sport

ANIMAL = (@PET)

VP-Expert would assign the variable ANIMAL with the value Sport. To explain this, we will consider how variables are assigned in memory. The first expression, PET = Cat, creates a place or "mailbox" in memory called PET and stores the value Cat in it. Likewise, the second expression, CAT = Sport, creates another place or mailbox in memory at location CAT with value Sport. The third expression, ANIMAL = (@PET), creates a place in memory called ANIMAL. To assign ANIMAL a value, VP-Expert first looks at the variable location called PET. Because the location called PET contains the value Cat, the location called CAT is examined next. The value contained in the variable CAT is Sport, so Sport is assigned as the value of the variable ANIMAL. This is an example of double indirect addressing. Note that when CAT is on the right side of the equal sign it is a value, but when it is on the left side of the equal sign it is a variable name (location). The parentheses and the @ symbol achieved this double indirect addressing.

As we move into more features of VP-Expert, we will encounter PLURAL and dimensioned variables. The dimensional variable will be discussed in Module 4, but we will now turn our attention to a special type of VP-Expert variable, the PLURAL variable.

Plural Variables with Rules

VP-Expert allows a special type of variable with rules, called PLURAL, which means that the variable may take on more than one value during the consultation. When the inference engine encounters a PLURAL variable as part of a condition or conclusion, it will check for more than one value of that variable. If a variable is not known as PLURAL, then the inference engine will stop its work after finding one value for that variable. One practical reason for using a PLURAL variable is that the user can make more than one selection from a question. For example, in wp4.kbs, we had to ask four questions to get the information we needed. If we change the program, create a variable called Activity, and declare it PLURAL, then we need only to ask one question that can have more than one answer. This change makes the consultation shorter for the user. At this point, exit wp4.kbs by using 8Quit from the Consult Menu. The next file we will use is wp5.kbs, so let's load that file from the Main Menu using 6FileName.

The file wp5.kbs is a version of wp4.kbs that has a new rule with a PLURAL variable called Activity. To view this file, select Edit. A copy of wp5.kbs is shown in Figure M3.4.

The OR rule now contains one variable called Activity as part of the four conditions. The ASK statement has been changed to direct the user to make more than one

```
!The file wp5.kbs uses a PLURAL variable,
!         Activity, in a compound OR IF-THEN-ELSE-BECAUSE.
!
!ACTION BLOCK

ACTIONS
              Display "WPHELP will advise you on whether you need a
word processor or not.

Press the Spacebar to start the consultation.~"
          CLS
          FIND advice
          DISPLAY "You {advice} a word processor at this time.

          Press the Spacebar to exit WPHELP.~"
          CLS;

!RULES BLOCK

RULE 1
IF        Activity = Letters
          OR Activity = Resumes
          OR Activity = Notices
          OR Activity = Papers
THEN      Advice = need
ELSE      Advice = do_not_need
BECAUSE   "If you generate frequent letters, resumes, notices, or papers,
          then you need a word processor.";

!STATEMENTS BLOCK

ASK Activity: "Which of the following documents do you need
to write for your business? Select your answers by pressing
the Enter key. After you have made all your selections, press
the End key. If you don't need to write any of these documents,
select NONE_OF_THESE (with Enter, End).";

CHOICES Activity: Letters, Resumes, Notices, Papers, None_of_These;

PLURAL: Activity;
```


FIGURE M3.4 A PROGRAM WITH A PLURAL VARIABLE (wp5.kbs)

selection if it applies. The instructions for accomplishing this are included in the message on the screen. The CHOICES command lists all the possible documents, along with NONE_OF_THESE. At the very end of the program there is a command PLURAL: Activity. This tells us that the one variable Activity can have more than one value. VP-Expert will allow up to 10 PLURAL variables in a program.

When you run wp5.kbs, note how you select more than one answer to a question. Move the light bar to your choice and press Enter. A mark will appear by your selection. Move the light bar to another choice and press Enter; another mark will appear. When you have made all your selections for the question, signal VP-Expert that you have finished by pressing the End key.

Run a consultation with wp5.kbs by selecting 4Consult and 2Go. Try the 5Rule option after the consultation is over to see the knowledge base. Now, rerun wp4.kbs to compare it to wp5.kbs.

Comparing wp4.kbs and wp5.kbs is interesting. They both give the same advice on whether you need a word processor. The knowledge base in each consists of one rule, but the number of variables in the rule differs. The second program, wp5.kbs, is probably easier to use because there is just one question, with multiple sections, to answer. This is the use of the PLURAL variable. As you have seen from the examples, the same knowledge base can be expressed in different ways, depending on the construction of the IF-THEN rules and the dialogue needed with the user. Now return to the Main Menu with 8Quit.

So far we have discussed the format of IF-THEN rules, relational operators, compound rules, mathematical calculations, and the PLURAL variable. All of these have been parts of pre-existing programs. Now we will explore a way to create a knowledge base from a simple table, without writing the program.

Creating an Expert System from a Table

There is another way to create or induce not only a knowledge base, but an expert system, from a table. Even though an expert system can be created from a database (table), or a spreadsheet (table), we will now concentrate on creating an expert system from a table of text.

Logic Behind Table Files

VP-Expert uses a format to convert the rows and columns of a table to rules. Consider the table in Figure M3.5, which shows the logical placement of condition headings, conclusion headings, and values of the headings.

	Heading A	Heading B	Heading C
Row 1	Value A-1	Value B-1	Value C-1
Row 2	Value A-2	Value B-2	Value C-2
Row 3	Value A-3	Value B-3	Value C-3
•	•	•	•
•	•	•	•
•	•	•	•

RULE 2

IF Heading A = Value A-2
AND Heading B = Value B-2
THEN Heading C = Value C-2

FIGURE M3.5 LOGICAL TRANSLATION FROM ROWS TO RULES

The table is divided into rows and columns. Each row of the table (except the headings) will become a rule in the knowledge base. The contents of the first two columns will become the conditions of the IF part, and the last column will become the conclusion of the rule. For example, row 2 gives the rule "IF Heading A = Value A-2 AND Heading B = Value B-2, THEN Heading C = Value C-2."

Let's take our word processing example. Consider the table file shown in Figure M3.6.

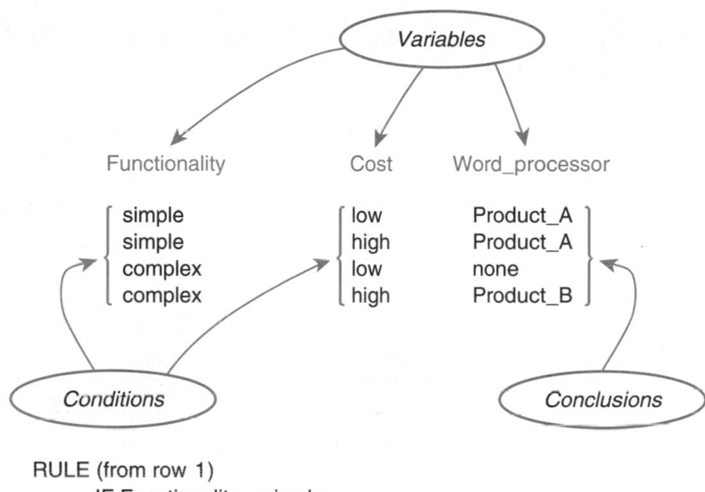

FIGURE M3.6 TABLE FILE (wp1.tbl)

We will assume that the first row identifies the variables: Functionality, Cost, and Word_processor. Because the table has four rows under these headings, there will be four rules in the expert system created from this table. From the first row underneath the headings of Functionality, Cost, and Word_processor, the following rule will be created:

RULE 1

IF Functionality = simple

AND Cost = low

THEN Word_processor = Product_A;

The data in the fourth row under the headings will give us:

RULE 4

IF Functionality = complex

AND Cost = high

THEN Word_processor = Product_B;

Moreover, note that an AND is inserted between the first two column values and that the third column value is the conclusion. The nice part of this method is that the resulting expert system will run in a fixed format without any programming. Let's take a table and **induce** or create an expert system from it.

Translation of a Table File to a Knowledge Base File

Select the 2Induce option from the Main Menu. You will see the new options as shown in Figure M3.7.

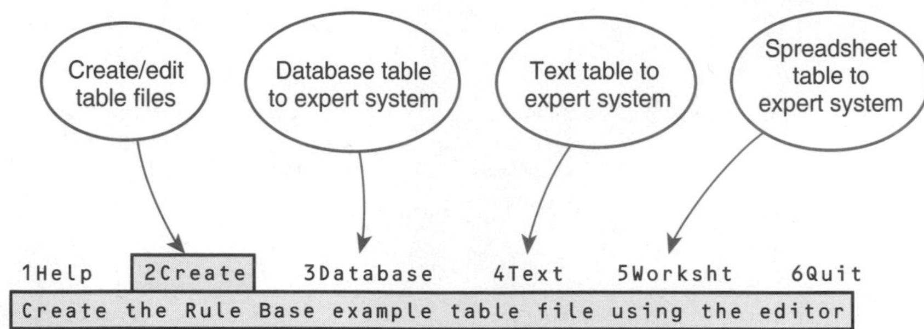

FIGURE M3.7 MENU FOR THE 2Induce OPTION

An expert system can be induced from an existing database table (3Database), a worksheet (5Worksht), or a text table (4Text). The 2Create option will allow us to create or change an existing table of text. The 2Create will invoke the regular text editor (word processor).

VP-Expert must have a file name with an extension of .tbl for a text table in order to create an expert system with a .kbs extension. We already have a table called wp1.tbl in the sample file directory. Let's take a look at this file by following these steps:

Step 1. Select 2Create from the Induce Menu. The message "Choose a text file" will appear, along with a list of files. The file names that appear are only files with a .tbl extension.

Step 2. Select wp1 either by entering wp1 or moving the light bar to this choice and pressing Enter. You are now in the editor for text tables; your screen should look like Figure M3.8.

Note that the message, "editing old file c:\VPX3ed\wp1.tbl" (or a portion thereof), is displayed in the upper-right corner. The table file is on the screen, and the menu bar is across the bottom. The only difference between this editor (2Create) and the editor from the Main Menu is that the editor loads only table files (.tbl). The first row of the table identifies the variables; the remainder of the rows represent potential rules. The last column is the conclusion, and the conditions connected by AND come from the first two columns.

Load the file wp1.tbl with the 2Create option simply to see the file. If we wished, we could change the table or create a new text table for a new expert system. For now, just print out the table by pressing F10, then the End key. (This procedure is identical to the printing option in the Main Menu editor.) To exit the table editor, press ALT-F8. Answer y to the question that appears on the screen. You should now be back to the 2Create option.

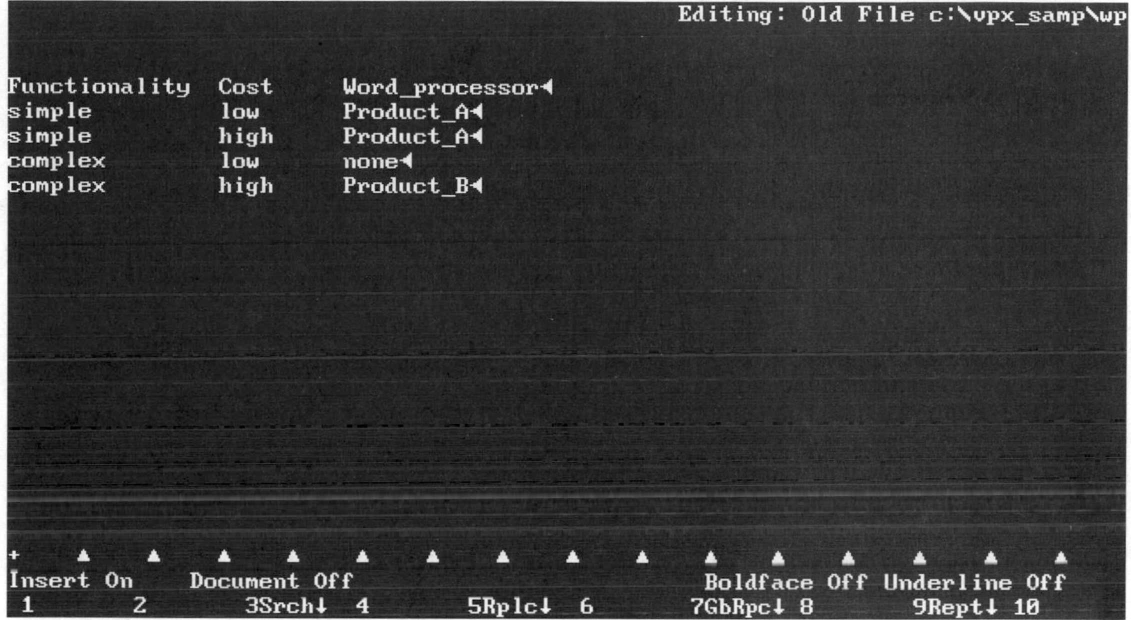
Editing: Old File c:\vpx_samp\wp

Functionality	Cost	Word_processor
simple	low	Product_A
simple	high	Product_A
complex	low	none
complex	high	Product_B

Insert On   Document Off   Boldface Off Underline Off
1   2   3Srch↓ 4   5Rplc↓ 6   7GbRpc↓ 8   9Rept↓ 10

FIGURE M3.8 THE TABLE FILE (wp1.tbl) IN THE INDUCE EDITOR (2Create)

We will use the table file wp1.tbl to induce an expert system or knowledge base file. We are now ready to generate, or induce, a new expert system program using our table file. (If you do not have at least 640K of memory, the induced file will not be created.) Exit the editor with ALT-F8 and select the 4Text option from the Induce Menu. The following screens, shown in Figure M3.9, will appear.

The first message asks for the table file (.tbl) that we want to use. Again, select wp1. The next screen message, "Choose a file:," asks for the name of the file (.kbs) to be induced (formed or created) from the selected table file. Because we already have a knowledge base called wp1.kbs, we will enter a different file name. Enter inwp1, which stands for "inducing from wp1.tbl," and press Enter. VP-Expert will now create a file called inwp1.kbs, which is a minimal expert system.

If you do not have enough conventional memory, VP-Expert displays the following message:

> A runtime error has occurred:
>
> Not enough memory to induce.
>
> (Press ENTER to go on)

If this happens in your environment, it means you cannot use the induce option, but you can examine an induced backup example by loading the sample file inwp1b.kbs at the Main Menu.

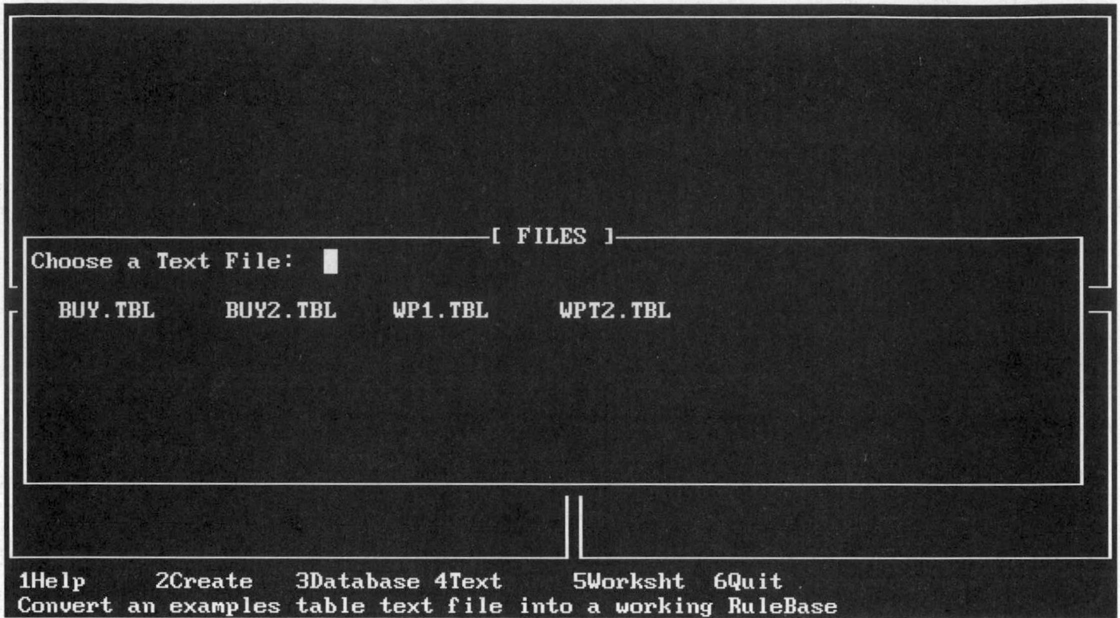

FIGURE M3.9 IDENTIFICATION OF THE TABLE FILE

With a table (.tbl) file, the system will generate a knowledge base. The new file will be saved as inwp1.kbs on the default directory. When the light bar returns to the 2Create option, select the 6Quit option of the Induce Menu and return to the Main Menu. To run the induced expert system, inwp1.kbs, we will follow the steps used to run a regular consultation:

Step 1. Select the 6FileName option and then enter inwp1.

Step 2. Select 4Consult to get to the Consult Menu. The familiar message that the file is being loaded will be displayed.

Step 3. Select the 2Go option.

You are now in the regular Consultation mode. In response to the questions on the screen, select Simple and then High. The consultation should respond with "The value of word processor is Product_A." At the end of the consultation, your screen should look like Figure M3.10. Your display may show an empty screen because VP-Expert displayed advice, which scrolled rapidly off the screen, and returned to the Consult menu. This consultation was generated from a table; no programming was required.

There are some differences, however, in the first wp1.kbs program and the one induced from a table in inwp1.kbs. The induced version has a rigid format. The questions to the user are phrased as, "What is the value of _____?" No introductory comments tell the user what the system is about or give instructions to start or exit the consultation. Yet, the expert system is usable and was created without any programming knowledge.

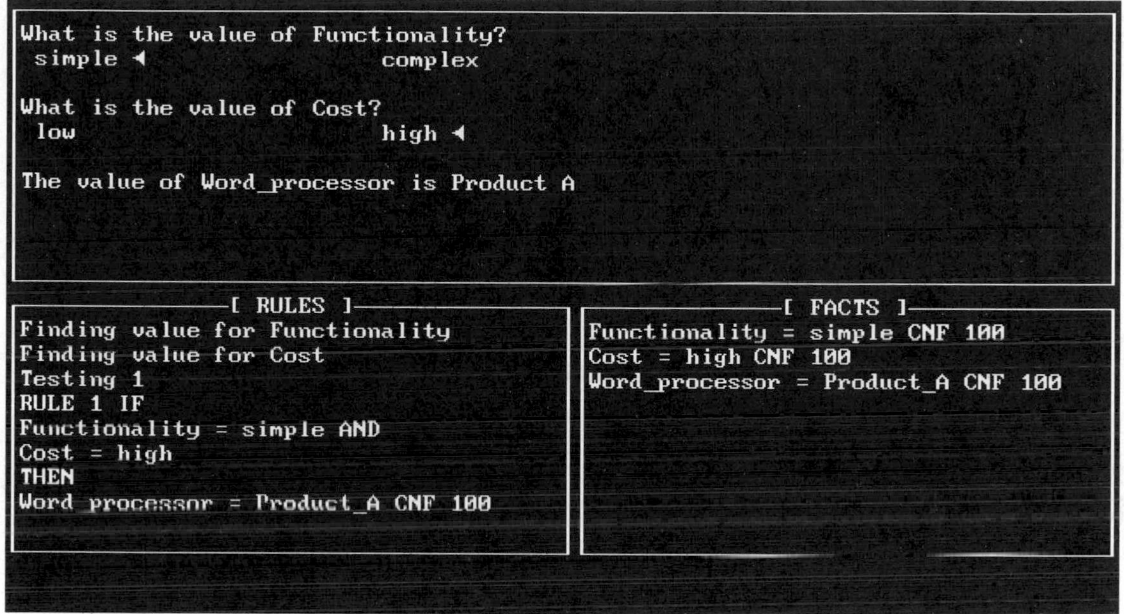

FIGURE M3.10 CONSULTATION SCREEN FOR INDUCED EXPERT SYSTEM (inwp1.kbs)

Let's take a look at the software for the induced expert system. To print out the file inwp1.kbs, we will follow the usual steps. Assuming that inwp1.kbs has already been selected under the 6FileName option, follow these steps to obtain a printout of the induced file (or any .kbs file):

Step 1. Select the 7Edit option of the Consult Menu.

Step 2. In the editor, press the F10 key, then the End key.

Step 3. Exit to the Consult Menu by pressing ALT-F8 and typing y.

Step 4. Return to the Main Menu with 8Quit.

The printout of inwp1.kbs, along with the table file, is shown in Figure M3.11. This is the code generated when you selected 4Text (from the Induce Menu). If you didn't create the induced file because of memory limitations, use the backup copy called inwp1b.kbs in your Sample Files directory.

Let's examine the **canned code** and compare the table file to the knowledge base file from your printouts. The conclusion heading (Word_processor) is taken as the variable for the FIND command (Comment 1 or C1 in Figure M3.11). The four rows under the headings from the table file give us the knowledge base (C2). The numbering of the rules, however, starts with zero. The value of the heading and the values in the first two columns are connected by AND. The two ASK (C3) statements are generated because of the two headings (columns) that appear in the conditions of the

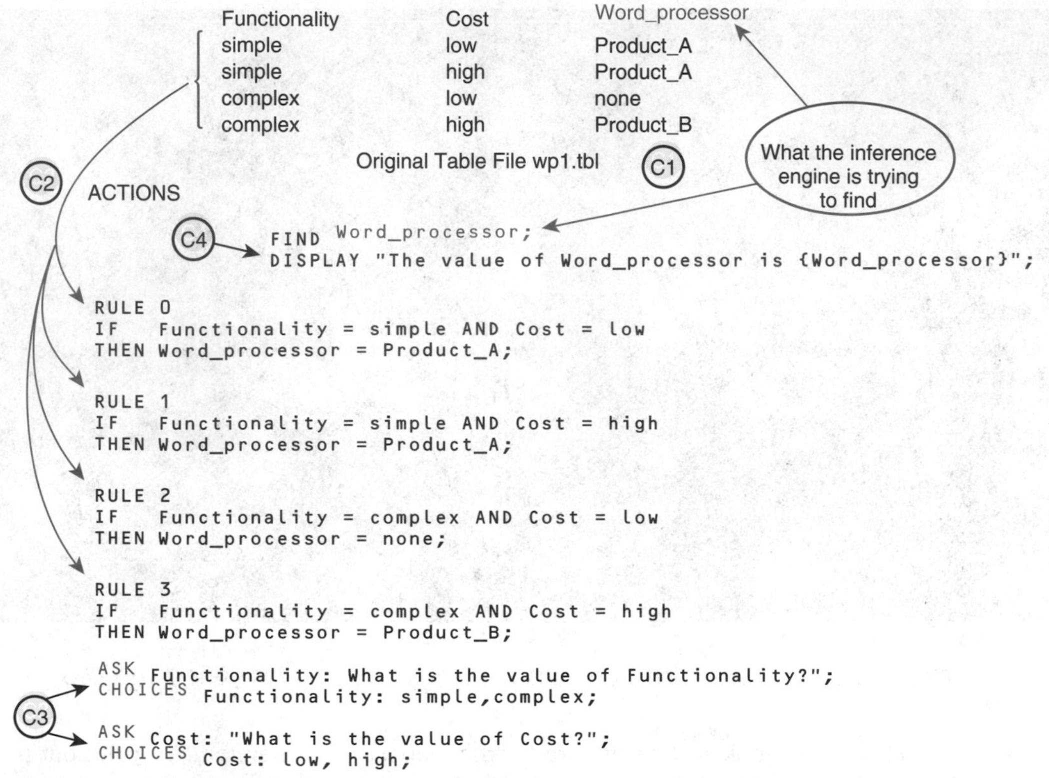

FIGURE M3.11 COMPARISON OF A TABLE FILE (wp1.tbl) AND THE INDUCED EXPERT SYSTEM FILE (inwp1.kbs)

rules. The values of the variable in the corresponding CHOICES command come from reading down the column with the appropriate heading. For example, reading down the Cost column, we find two distinct values, low and high; consequently, these values are placed in the CHOICES command after the ASK command for that variable, Cost. Because there are two headings for conditions, there are two ASK commands with the corresponding CHOICES commands. The DISPLAY command (C4) prints to the screen the value (Product_A) the inference engine finds. Your display may show an empty screen. This is because VP-Expert displayed advice, which scrolled rapidly off the sreen, and returned to the Consult menu. You can use the 4Variable menu of the 4Consult menu to display the value of Word_processor. The original wp1.kbs file, moreover, had a RUNTIME statement that did not allow the display of the Rules or Facts windows. An induced knowledge base, however, has the windows so that the activity of the inference engine is revealed.

What is the real power of the 2Induce option? Obviously, it creates a **form**, or **canned**, expert system without programming. Another advantage of using the 2Induce option is to generate the basics of the program and then make a few changes to meet specific needs. This option can save a lot of time spent in keying and debugging a program.

Using this option, however, has limitations. You must be sure that the table accurately represents the rules of the intended knowledge base. All conditions must be included for the 2Induce option to create a valid expert system. The old adage, GIGO (Garbage In, Garbage Out), is appropriate here. As we proceed through the modules, we will let you decide on the value of the 2Induce option of VP-Expert.

Let's summarize the logical concepts behind the use of the 2Induce option before proceeding. The 2Induce option (Induce Menu) allows you to create a knowledge-based system from a text, database, or spreadsheet table. For now we will use a text table. You can create your own table by using the 2Create option of the Induce Menu.

This option automatically produces a text file with a .tbl extension. An existing table file can also be modified or changed with this same option. Assuming the existence of a table file, you select the 4Text option from the same Induce Menu and indicate the table file you want and the intended name of the induced .kbs file. Once the induced file is created under the 4Text option (in the Induce Menu), the induced file can be run in a regular consultation session by selecting the 6FileName (.kbs) from the Main Menu and then by selecting 4Consult and 2Go. Note that table files, moreover, are created and changed in the Induce Menu, but .kbs files are manipulated within the 3Edit and 4Consult options of the Main Menu. The commands within the 2Create option of the Induce Menu and 3Edit option of the Main Menu, however, are the same editing or word processing commands.

The Complete Induction Cycle: Table to Expert System

We will now look at the mechanics of creating a table file to induce an expert system and then use the Main Menu editor to change and modify the induced program. The editor (2Create) of the Induce Menu will work only on text or table files, whereas the editor of the Main Menu will utilize only program (.kbs) files.

Creation of Table Files. Use the 2Create option of the Induce Menu to create the following table:

TUTORIAL	*TELEPHONE_HELP*	*BUY_PRODUCT_X*
yes	yes	yes
yes	no	maybe
no	yes	maybe
no	no	no

The information in this table is designed to tell you whether you should purchase a product called X. The logic is to buy a particular Product X if both a tutorial and telephone help are yes (available). If only one source of help is available, the advice to buy is maybe. If neither type of assistance is available, our expert system will advise you not to buy Product X.

To create this table file, follow these steps:

Step 1. Get to the VP-Expert working environment.

Step 2. Set the 7Path to the default Sample Files subdirectory (VPX_samp):.

Step 3. Select the 2Induce option from the Main Menu.

Step 4. Select the 2Create option from the Induce Menu.

Step 5. In reply to the message, "Choose a text File," enter the word buy. (Note that this file does not yet exist.)

Step 6. When you see a screen that is blank except for the message "Editing new file C:\VPX_samp\buy.tbl" in the upper-right corner and a menu bar at the bottom, you are in the text editor for tables. Now enter the rows and columns of text that are given in the table. To move from one column to another, you may use the Tab key (which may have two arrows pointing in opposite directions) or the Spacebar (for blanks). When you come to the end of a row, be sure to press the Enter key. You will see the familiar wedge mark when you do. If you make a mistake, type over it and use the DEL key to delete unwanted letters. Use the arrow keys to move around on the screen. When you enter the underscore as part of the variables (telephone_help and buy_product_x), you will have to use the Shift key. Once you have the table entered as described in the steps, press ALT-F5 to save the file but remain in the editor. To obtain a printout of the table, press the F10 key. When the next screen appears, press the End key. When the printing is complete, exit the table editor by pressing ALT-F6, and answer Y to the question, "Save as C:\VPX_samp\buy.tbl" (Y or N)?

The Induced Knowledge Base File. Now that you have created the buy.tbl text table and stored it on your default directory, you are ready to *induce* a knowledge base from this file. Select the 4Text option from the Induce Menu. In response to the question "Choose a file" enter buy (without an extension) and press Enter. In response to the next question "Choose a file" for the induced kbs file, type inbuy (without an extension) and press Enter. The light bar will return to 2Create when the induced file has been created. We are now finished with the options of the Induce Menu. Select the 6Quit option and return to the Main Menu.(If you have memory problems and can't induce inbuy.kbs, use the backup file in.buy2.kbs.)

Once in the Main Menu, we need to run the consultation. Follow these familiar steps:

Step 1. Select 6FileName and enter inbuy.

Step 2. Select 4Consult from the Main Menu.

Step 3. Select 2Go from the Consult Menu.

When the Consultation screen appears, select no to the first question and yes to the second. Remember, to make selections, press the Enter key, then the End key. When the consultation is over, the system advises maybe for Product_X. Your screen should be similar to Figure M3.12 after the consultation, depending on the "scrolling" of your display terminal.

```
What is the value of TUTORIAL?
 yes                        no ◄

What is the value of TELEPHONE_HELP?
 yes ◄                      no

The value of BUY_PRODUCT_X is maybe

[ RULES ]
BUY_PRODUCT_X = maybe CNF 100
Testing 2
RULE 2 IF
TUTORIAL = no AND
TELEPHONE_HELP = yes
THEN
BUY_PRODUCT_X = maybe CNF 100
Finding value for TELEPHONE_HELP

[ FACTS ]
TUTORIAL = no CNF 100
TELEPHONE_HELP = yes CNF 100
BUY_PRODUCT_X = maybe CNF 100
```

FIGURE M3.12 CONSULTATION SCREEN FOR THE SECOND INDUCED KNOWLEDGE BASE PROGRAM

Even though the messages to the user could be improved, the induced file runs just like a program written by a programmer. Rerun the consultation three times to verify that the advice is the same as that given in the original table file.

Now your task is to obtain a printout of the program file, inbuy.kbs, which was induced from the table file (buy.tbl) that we created. The familiar steps are as follows:

Step 1. Select the 7Edit option from the Consult Menu. (The selected file name is still inbuy.kbs.)

Step 2. When you are in the editor with the file loaded, press the F10 key, then the End key to obtain a printout. If you want to read the program file from the display screen, use the Page Up (PgUp) or Page Down (PgDn) keys.

The program file in your printout should be identical to the program file in Figure M3.13. You should be congratulated; you have just generated your first expert system program! In the next section, we will modify the induced file (inbuy.kbs) to make it run better for the user. Press ALT-F8 and enter Y to go to the Consult Menu. Return to the Main Menu with 8Quit.

Changes in the Induced Knowledge Base File. Now we know how to generate an expert system program without programming. To make it easy to use in the business environment, the program should be adjusted to be more understandable during a consultation run. Hence, we will make changes to the .kbs file with the editor in VP-Expert, just as we would make changes to a letter or other document with a word processor. Even though the emphasis in this module is on the knowledge base,

Tutorial	Telephone_help	Buy_product_x
yes	yes	yes
yes	no	maybe
no	yes	maybe
no	no	no

Table File buy.tbl

```
ACTIONS
     FIND Buy_product_x
     DISPLAY "The value of Buy_product_x is {Buy_product_x}";
RULE 0
IF   Tutorial = yes AND Telephone_help = yes
THEN Buy_product_x = yes;

RULE 1
IF   Tutorial = yes AND Telephone_help = no
THEN Buy_product_x = maybe;

RULE 2
IF   Tutorial = no AND Telephone_help = yes
THEN Buy_product_x = maybe;

RULE 3
IF   Tutorial = no AND Telephone_help = no
THEN Buy_product_x = no;

ASK Tutorial: "What is the value of tutorial?";
CHOICES Tutorial: yes, no;

ASK Telephone_help: "What is the value of telephone_help?";
CHOICES Telephone_help: yes, no;
```

FIGURE M3.13 INDUCED KNOWLEDGE BASE FILE (inbuy.kbs)

making the knowledge base usable is appropriate. Therefore, in this last section, we will briefly take you through the mechanics of basic changes to a program file. The complete set of commands for using the VP-Expert editor is available from the VP-Expert Help facility, and in the command summary in Appendix E.

Before we start, we offer this advice to experienced users of word processors. If you already know a word processor, we recommend that you may want to use it to change program files (.kbs) or to create table files (.tbl). There are only two points to consider. First, your word processor must let you translate a document file to a nondocument (ASCII) file. This means that the file will not have hidden and special control characters and symbols, which can cause errors if the file is read by different software. Second, table files must have a .tbl extension and programs must have a .kbs extension. If your word processor creates a different extension automatically, you can change the file extension with the DOS command, RENAME. To change the name of a file at the DOS prompt, enter the DOS command:

RENAME oldfile.name newfile.ext

This command assumes that the old file is on a disk in the default drive. Hence, if you can translate document files to nondocument files with the proper extension on a word processor that you already know, we recommend that you use that software package. We leave that decision to you. We will, however, give you the basics for using the VP-Expert editor. If you are familiar with word processing, skip over to Figure M3.15.

Our task is to take the program file inbuy.kbs and make changes to it, as illustrated in Figure M3.14. We will make six changes to the program file, and we will discuss the logic of the changes before their mechanics.

Tutorial	Telephone_help	Buy_product_x
yes	yes	yes
yes	no	maybe
no	yes	maybe
no	no	no

Table File buy.tbl

```
(C1)
(C2) ACTIONS
          FIND Buy_product-X
(C3)      DISPLAY "The value of Buy_product_X is {Buy_product_X}";

     RULE 0
     IF   Tutorial = yes AND Telephone_help = yes
     THEN Buy_product_X = yes;

     RULE 1
     IF   Tutorial = yes AND Telephone_help = no
     THEN Buy_product_X = maybe;

(C4) RULE 2
     IF   Tutorial = no AND Telephone_help = yes
     THEN Buy_product_X = maybe;

     RULE 3
     IF   Tutorial = no AND Telephone_help = no
     THEN Buy_product_X = no;

     ASK Tutorial: "What is the value of Tutorial?"; (C5)
     CHOICES Tutorial: yes, no;

     ASK Telephone_help: "What is the value of Telephone_help?"; (C6)
     CHOICES Telephone help: yes, no;
```

C1	Insert comments and blank line
C2	Insert DISPLAY message
C3	Insert DISPLAY message
C4	Rename rule labels
C5	Change question
C6	Change question

FIGURE M3.14 PROPOSED CHANGES TO INDUCED KNOWLEDGE BASE FILE (inbuy.kbs)

In Figure M3.14, the changes are represented by circles. Change 1 is to add three new lines—a comment line, a blank line, and the RUNTIME command, which will turn the Rules and Facts windows off during a consultation run. Change 2 will insert a DISPLAY command that gives an introductory message to the user. When we do this, we will have to delete the semicolon from the DISPLAY command. Change 3 will modify the DISPLAY statement so the advice from the expert system shows at the end of the consultation, along with an ending message.

For Change 4, we will modify the Rule labels from 0, 1, 2, and 3 to One, Two, Three, and Four. The last two, Changes 5 and 6, will reword the questions to the user. When we finish these program alterations, our induced file (inbuy.kbs) will be like the updated program in Figure M3.15. Once these changes have been made, with your own word processor or with the VP-Expert editor, we must test the program file to make sure it runs correctly.

```
!The file is inbuy.kbs.                                    (C1)
                    RUNTIME;

ACTIONS
                    DISPLAY "Welcome to WPHELP."            (C2)
                    DISPLAY " "
                    FIND Buy_product_X
                    DISPLAY "Concerning the purchase of Product X,   (C3)
                    we say {Buy_product_X}."
                    DISPLAY ""
                    DISPLAY "      End of consultation.";

RULE ONE
IF   Tutorial = yes AND Telephone_help = yes
THEN Buy_product_X = yes;

RULE TWO
IF   Tutorial = yes AND Telephone_help = no
THEN Buy_product_X = maybe;
                                                            (C4)
RULE THREE
IF   Tutorial = no AND Telephone_help = yes
THEN Buy_product_X = maybe;

RULE FOUR
IF   Tutorial = no AND Telephone_help = no
THEN Buy_product_X = no;
                                                            (C5)
ASK Tutorial: "Is a tutorial available for product X"?;
CHOICES tutorial: yes, no;
                                                            (C6)
ASK Telephone_help: "Is telephone help available from
the company that sells product X?";
CHOICES Telephone_help: yes, no;
```

FIGURE M3.15 UPDATED VERSION OF INDUCED KNOWLEDGE BASE FILE (inbuy.kbs)

We will now perform the changes using the VP-Expert editor. The basic instructions are as follows:

1. To move around the screen, use the arrow keys.
2. To insert a letter or word, use typeover.
3. To delete a letter or word, use the DEL key.
4. To add a blank line above, use the CTRL and Enter keys.
5. To delete a line, use the CTRL and Y keys.

There are many other useful commands, but this minimal set will let us make the desired changes. A summary of the VP-Expert editing commands is given in Appendix D. (The experienced microcomputer user is encourage to skip over the rest of this section on editing a text file.)

To get started, we assume that we are in the standard VP-Expert working environment. We are in the Main Menu; the 7Path is C:\VPX_samp. Select 6FileName and identify the needed file as inbuy by typing the file name or selecting it. Select 3Edit. We are now ready to make the first change, indicated as C1 in both Figure M3.14 and Figure 3.15. Follow these steps:

Step 1. Move the light bar to the first line, first column (on top of the A in Actions).

Step 2. Press the CTRL and Enter keys simultaneously to insert a blank line above the cursor. Press the same keys two more times.

Step 3. Move the light bar to the first row, first column and enter "!The file is inbuy.kbs." as a comment line. (The wedge mark on the screen indicates that the Enter key has been pressed.) The exclamation point (!) is used to indicate a comment line.

Step 4. Move the cursor to the first row below the comment line. After pressing the Spacebar eight times, enter RUNTIME; before pressing Enter.

To effect the changes for C2 and C3, follow these steps:

Step 1. Move the light bar to the first position of the line with the FIND command. Create two blank lines with the CTRL and Enter keys.

Step 2. Move the light bar to the first position of the blank line below ACTIONS. Press the Spacebar eight times and then enter the DISPLAY command,"Welcome to WPHELP."

Step 3. Move the light bar to the first position of the next blank line. Press the Spacebar eight times and enter the DISPLAY command, " " (so that a blank line will appear on the user's screen).

Step 4. Move the light bar to the line that has the DISPLAY instructions underneath FIND. Delete this line by pressing CTRL and Y. Press CTRL and Enter simultaneously four times to create four blank lines.

Step 5. Enter the three DISPLAY instructions (four lines) from Change 3 (C3) of Figure M3.15.

Be sure to include the semicolon after the "End of Consultation" phrase; it signals the end of the ACTIONS block to VP-Expert. Your program file should now include the changes indicated in C2 and C3 in Figure M3.15.

We next will make the changes to the Rule labels. Changing the Rule labels is not necessary, but we think zero is an awkward way to label rules—and we want to illustrate that Rule labels can be words, not merely numbers.

To make the changes for C4, follow these steps:

Step 1. Move the light bar to the 0 of RULE 0. Press the DEL key and then enter *ONE*.

Step 2. Repeat this same procedure to change 1 to *TWO*, 2 to *THREE*, and 3 to *FOUR* in the appropriate rule.

(Notice that we did not change the meaning of the IF-THEN rules at all.)

To complete the fifth change (C5), follow these steps:

Step 1. Move the light bar to the W of the question, "What is the value of tutorial?". Press the DEL key until all the text is deleted except the two pairs of quotation marks.

Step 2. Enter the new question, "Is a tutorial available for product X?".

To complete the last change (C6), follow these steps:

Step 1. Move the light bar to the W in, "What is the value of telephone_help?" Press the DEL key until only one quotation mark and the wedge mark are left.

Step 2. Enter the following partial new question: Is telephone help available from. A wedge mark will indicate that an Enter command is still at the end of this line.

Step 3. To put the rest of the question on the next line, move the light bar to the first C on the last CHOICES command. Put a blank line above this line by pressing CTRL and Enter simultaneously.

Step 4. Move the light bar to the first position of this blank line. Type the words: "the company that sells product X?";. Don't forget to include a semicolon after "X?". Now compare your program changes to the updated file in Figure M3.15. The placement of semicolons (;) at the end of certain commands is important. There is only one semicolon at the end of the ACTIONS block, but there is a semicolon at the end of each rule, and at the end of each ASK and CHOICES command, and at the end of the RUNTIME command.

Now that we think we have made all the changes correctly, we need to save the program and run it to double-check. To save and exit the editor, press ALT-F6 and reply Y to the question.

In the Main menu, select the 4Consult option for a test run. Each time a program file is loaded from the 4Consult option, the file is checked for spelling and syntax errors. This procedure is one reason for the lengthy delay in loading. If there are no typing errors, you should see a screen indicating that the file has been loaded. Select the 2Go option. In response to the two questions posed, select No and then Yes. Your Consultation screen should be the same as or similar to the screen in Figure M3.16.

The screen's messages make the expert system easier to use. In Figure M3.16, Change 1 does not appear on the Consultation screen because it is a comment line used for information only. The RUNTIME instruction closes the Rules and Facts windows. Change 2 is a welcome statement. Changes 5 and 6 are the rephrasing of the induced questions. The system's advice resulted from Change 3.

You may not have obtained the advice shown in Figure M3.16. Instead, you may have received a message that looks like the one displayed in Figure M3.17, which says that something is wrong with line 7. Hence, the consultation cannot be run.

```
Welcome to WPHELP.

Is a tutorial available for product X ?
 yes                         no ◄

Is telephone help available from the
company that sels product X ?
 yes ◄                       no

Concerning the purchase of PRODUCT X,
        We say maybe.

      End of Consultation.
```

FIGURE M3.16 CONSULTATION SCREEN FOR UPDATED FILE (inbuy.kbs)

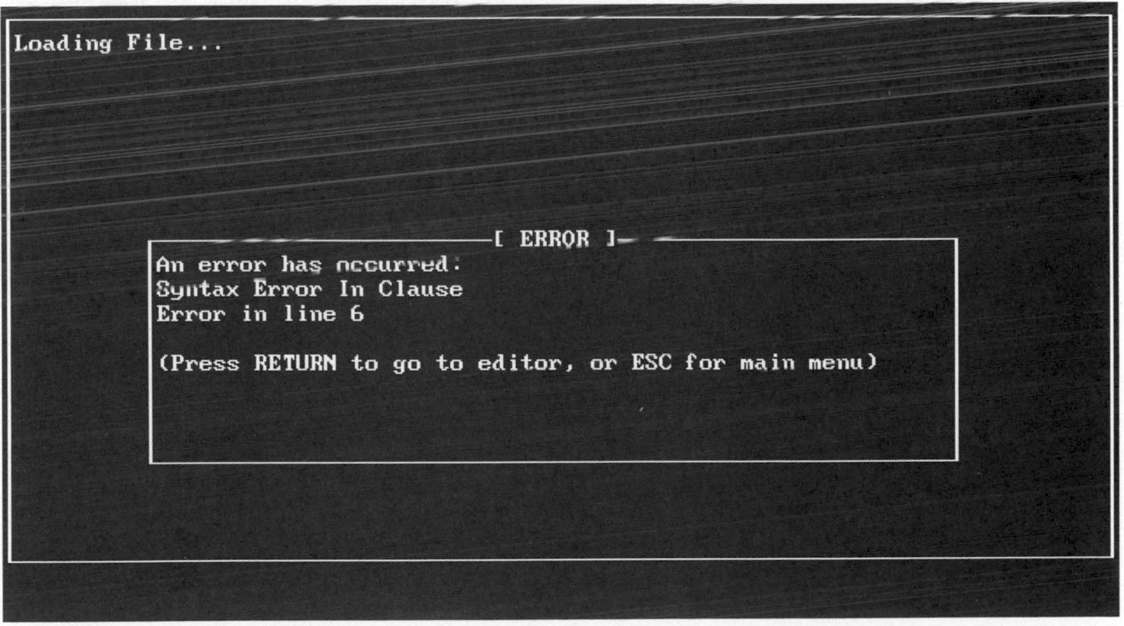

FIGURE M3.17 POSSIBLE ERROR MESSAGE FROM VP-EXPERT FOR UPDATED FILE (inbuy.kbs)

Press the Enter key to return to the editor, where you will correct the mistake. Actually, the error may be above line 7, on line 6, or even on line 5. Identifying the error is now the challenge. Compare your program to the one in Figure M3.15 to try to find your mistake. If you are a good typist, you probably didn't make a mistake. If you can't get your program to run, you will find a copy of the original and induced files under the file names buy2.tbl, inbuy2.kbs, and inbuy3.kbs on the Sample File directory.

We have changed an induced program to make it more usable. The larger the knowledge base, the more time the 2Induce option will save you. We do recommend, however, that you modify the program to make the knowledge base more usable. In this section, we concentrated on the knowledge base—how to form rules, create table files, induce a knowledge base, and update induced program files. In the next module, we will discuss the inference engine and how it works.

Our final task for this module is to turn off the system and deactivate the printer. If you are not at the DOS prompt, choose 8Quit from the Main Menu and exit to DOS.

Key Points

- VP-Expert's knowledge base is structured by production rules—not semantic nets, frames, or predicate calculus.
- The production rules take the form of IF-THEN, with ELSE and BECAUSE as optional parts.
- The IF part of a rule can include one or more conditions. The THEN part contains the conclusion(s). The optional ELSE part also contains alternative conclusions if the condition is false. The BECAUSE option does not affect the logic of a rule, but adds an explanation if the user uses the 2How? or 3Why? options during a consultation run.
- The IF part of a rule may be a simple condition or a compound condition using AND or OR. Up to 20 conditions are allowed in the IF part of a rule.
- The IF part of an IF-THEN rule is evaluated to be either true or false. When the IF part is true, the conclusion fires or is true. When the IF part is false, the rule does not fire.
- IF-THEN-ELSE rules always fire. When the IF statement is true, the condition(s) between the THEN and ELSE will be true. When the IF statement is false, the condition(s) after the ELSE are true.
- When the IF part contains a compound statement connected by ANDs, the conclusion is true only if all parts of the compound statement are true.
- When the IF part is composed of compound statements connected by ORs, the condition is false only when all parts of the compound conditions are false.
- If the condition of an IF-THEN rule contains both AND and OR, the OR will be evaluated first, regardless of its placement in the condition.
- The IF part of a rule is composed of a variable, a relational operator, and a constant, number, or other variable. If an additional variable is used, it is enclosed in parentheses.

- VP-Expert variables are composed of up to forty letters, numbers, or special characters, which are $ % ^ ¦.
- The relational operators for the IF part of a rule are:
  = equals
  < less than
  < = less than or equal to
  > greater than
  > = greater than or equal to
  <> not equal to
- The OR connective is not allowed in conclusions, but compound conclusions equivalent to ANDs are implied.
- Arithmetic calculations can be a part of a condition or a conclusion. The four arithmetic operations are:
  + Addition
  - Subtraction
  * Multiplication
  / Division
- Special mathematical functions, @ABS(X), @EXP(X), @LOG(X), and @SQRT(X), can be used in VP-Expert programs.
- Trigonometric functions are also available.
- When using mathematical calculations, they must always be to the right of the relational operator and inside parentheses.
- An assignment statement means that the value on the right side of the equal sign (=) is moved to the variable on the left side of the equal sign. Assignment statements are found in conclusions
- VP-Expert uses indirect and double-indirect addressing.
- For a variable to have more than one value during a consultation run, it must be declared as PLURAL. A program can have more than one PLURAL variable.
- A canned expert system program can be generated from a database, spreadsheet, or text table.
- The 2Induce option from the Main Menu contains the Induce Menu for creating/editing text tables (2Create) and generating (4Text) a knowledge base file (.kbs).
- The Induce Menu is as follows:
  1Help 2Create 3Database 4Text 5Worksht 6Quit
- To induce an expert system from a table file, first create the table (2Create) and then generate the knowledge base file (4Text) of the Induce Menu.
- The 3Edit option of the Main Menu may be used to make changes or modifications to any induced .kbs file.
- Induced files need modifications to make them more usable to the user community.
- The mechanics of using the VP-Expert editor are given in the VP-Expert's Help Facility.
- If you already know a word processor, you may use it to create and change table or knowledge base files. Be sure that the word processor will translate your files to nondocument files and that the extension name is .tbl for table files and .kbs for knowledge base files (for use with VP-Expert).

Exercises

1. Indicate which of the following variables are valid (V) and which are invalid (I):
   a. COST
   b. cost
   c. COST
   d. cOST/
   e. $COST
   f. 7COST
   g. COST7
   h. COST_7
   i. COST-7
2. Which of the following statements are valid (allowed) in VP-Expert? Mark V (valid) or I (invalid) by each.
   a. x=3
   b. x?3
   c. x=(y−3)
   d. x = abs(y)
   e. x = @abs(y)
   f. x = (@abs(y))
   g. x−3=y
   h. x−3*
   i. (y−3)=x
   j. abs(y) = x
   k. @abs(y) = x
   l. (@abs(y)) = x
3. Write a BECAUSE statement appropriate for each of the rules shown in Figure M3.13.
4. Consider the following rule:
   IF Savings > 5000 OR
   Gross_pay_per_month > 3000
   THEN qualify = yes.
   a. Indicate (Y or N) whether the rule will "fire" for the following values of savings and gross_pay_per_month.

Savings	Gross pay	Qualify?	Rule fires?
2000	40000		
2500	2500		
6000	4000		
7000	2700		

   b. Change OR to AND in the rule. Does the rule fire for the same pairs of values?
5. Obtain a printout of wp3.kbs.
   a. Circle all variables (on the paper).
   b. Draw a box around all relational operators.

c. Underline all words that the user sees on the screen.
d. What are condition 2 and conclusion 2 in Rule 1?
e. What is the advice given if all questions are answered No?

6. Repeat Exercise 5 with wp4.kbs, wp5.kbs, and inbuy.kbs.

7. Use the following table to induce an expert system and then answer the questions that follow it:

Name	Type	Model	Advice
A	B	C	D
E	F	G	H
I	J	K	L

a. How many conditions would be in each rule?
b. What column represents the conclusion?
c. How many rules would be induced?
d. How many ASK statements would be induced?
e. What is the last rule?

8. Use Exercise 7 to answer the following.
a. What ASK statements would be induced?
b. What CHOICE statements would be induced?

9. Given the rules below, write out the table file that could induce it.

```
RULE 0
IF Cost = 400 and Warranty = Yes
THEN Advice = OK;

RULE 1
IF Cost = 400 and Warranty = No
THEN Advice = No;

RULE 2
IF Cost = 800 and Warranty = Yes
THEN Advice = Maybe;

RULE 3
IF Cost = 800 and Warranty = No
Then Advice = No;
```

10. Use the VP-Expert Help Facility to obtain a printout for each of the following topics:
a. IF THEN
b. AND
c. OR
d. RULE
e. ELSE
f. BECAUSE
g. INDUCE MENU

h. CURSOR COMMANDS
i. ALTERNATE KEYS
j. FUNCTION KEYS
k. EDITOR COMMANDS

11. If necessary, correct the following syntax in mathematical calculations.
    a. Amount = 3 × 4
    b. Amount = (3 * Increment
    c. Amount = tan(x)
    d. Amount = @(sqrt(x)
    e. Amount = Amount+3/4

12. If necessary (or possible), correct the syntax of the following rules:
    a. IF Count >Min AND
       <Max
       THEN Level is High;
    b. IF Count = 2*Delta
       THEN Level is Low;
    c. IF Count > Min AND
       Delta < Mid OR
       Error = None
       THEN Level = Low or
       Level = Medium
    d. IF Count = Rule
       THEN Find = OK;
    e. IF Count = OK
       THEN FIND{Amount};

Assignments

Your Sample File directory is C:\vpx_samp.

1. Run the expert system invest1.kbs, which is on your Sample File directory. Answer the following questions:
   a. What is the purpose of the expert system?
   b. How many rules are in the knowledge base? What kind of rule(s)?
   c. How many variables are in the program (.kbs) file?
   d. Write a text table that could be used to induce invest1.kbs before modifications.

2. Run the expert system invest2.kbs from your Sample File directory. Answer the following questions:
   a. What is the purpose of the expert system?
   b. How many rules are in the knowledge base? What kind of rule(s)?

c. Compare the rules from invest1.kbs to those of invest2.kbs.
d. Could a text file induce this knowledge base?

3. Run the expert system taxhelp1.kbs from your Sample File directory. Answer the following questions:
   a. What is the purpose of the system?
   b. How many rules are in the knowledge base? What kind of rule(s)?
   c. How many variables are in the program file?
   d. Run the program and answer No to all questions. What happens?

4. Run taxhelp2.kbs from the Sample File directory. Answer the following questions:
   a. What is the purpose of the expert system?
   b. How many rules are in the knowledge base? What kind of rule(s)?
   c. How many variables are in the program?
   d. Answer No to all the questions. Is the advice valid?
   e. Why is a PLURAL needed in the program?

5. Run taxhelp3.kbs from your Sample File directory. Answer the following questions:
   a. What is the purpose of the program?
   b. How many rules are in the knowledge base? What kind of rules?
   c. How many variables are there in the program?
   d. Compare taxhelp2.kbs and taxhelp3.kbs.

6. Use the 2Create option of the Induce Menu and do the following:
   a. Create a table file from the table in Exercise 7. Call it example.tbl.
   b. Use the 4Text option of the Induce Menu to induce example.kbs.
   c. Run a consultation with example.kbs.
   d. Obtain a printout of example.kbs.
   e. Divide the printout into the three parts of a VP-Expert program: ACTIONS, RULES, and STATEMENTS.

7. Using the table below, perform the following operations:

Backups	Spell_Checker	Advice
Yes	Yes	Buy
Yes	No	No
No	Yes	No
No	No	No

   a. Create a table file called example2.tbl using the 2Create option of the Induce Menu.
   b. Induce example2.kbs from example2.tbl by using 4Text of the Induce Menu.
   c. Run a consultation with example2.kbs.
   d. Obtain a printout.
   e. Divide the printout into three parts: ACTIONS, RULES, and STATEMENTS.

8. Assume that Assignment 7 (example2.kbs) is advising you not to buy a particular word processor unless the package creates automatic file backups and comes with a spelling checker.
   a. On the printout, write new questions and messages to the user.
   b. Using the 3Edit option of the Main Menu, implement your changes to improve example2.kbs.
   c. Obtain a printout.
   d. Run a consultation with the changed version of example2.kbs.
9. Using the results of Assignment 8, do the following:
   a. Convert the rules of example2.kbs to one IF-THEN-ELSE rule.
   b. Using the 3Edit option of the Main Menu, delete the four rules (using CTRL-Y) and replace with one IF-THEN-ELSE rule.
   c. Run a consultation with your new example2.kbs.
   d. Does the consultation offer the same advice as in Assignment 8?
10. Find an article that compares the features of word processing packages. Use the summaries of the features of the packages to do the following:
    a. Select a subset of information and induce a potential expert system.
    b. Create a table with 2Create of the Induce Menu and call it features.tbl.
    c. Induce an expert system features.kbs using 4Text from the Induce Menu.
    d. Run a consultation with features.kbs.
    e. Obtain a printout of features.tbl and features.kbs.
    f. Does your expert system need a PLURAL variable to give accurate and complete advice from the table?
11. In using the Induce Menu, we created new files on the Sample File directory. At the C:\VPX_samp prompt, find the following by using DOS commands:
    a. How many free bytes are left on the disk?
    b. The 2Create option uses VPXI.EXE. How large is this file?
    c. How large is wp1.tbl?
    d. How large is inwp1.kbs?
12. The file wprelate in your Sample File directory is a variation of WPHELP using compound AND and OR conditions with < and >= operators. The COST variable has been arbitrarily split at $50 and $250 intervals.
    a. Run the consultation for 25, 50, 100, and 250 for COST.
    b. Insert parentheses around the OR condition of RULE 5. Does the consultation work?
    c. How could you rewrite RULE 5 to make it equivalent to a simpler rule using only one AND?
13. The file wpmath on your Sample File directory contains commands involving mathematical calculations.
    a. Run the consultation.
    b. How do you display math calculations to the screen?
    c. What happens when you delete the @ symbols from the sample file?

14. The table file wpt2.tbl on your Sample File directory contains a column of relational operators (<, >=).
    a. Load wpt2.tbl using the 2Create option of the Induce Menu. Observe the columns.
    b. Predict the rules that you would expect the induction method to produce.
    c. Induce the expert system file with the 4Text option of the Induce Menu.
    d. Run a consultation with the induced file wpt2.kbs. What happened?
    e. Use the Edit option to make any necessary changes and to improve the user interface.

Module 4

The VP-Expert Inference Engine

In this module, we will examine how the inference engine works to solve a problem. The inference engine tries to move from an initial state to a goal state. Two ways for achieving the solution are backward chaining and forward chaining. We will use the TRACE command and 5Tree from the Main Menu to trace the behavior of the inference engine. These results can be viewed as a Text tree or a Graphics tree. VP-Expert uses backward chaining techniques, but forward chaining can be implemented with the WHENEVER command. We will look also at the placement of goals and rules within the structure of a program or .kbs file. This module will end with a discussion of how the inference engine deals with certainty and uncertainty.

Inference Engine Logic

The purpose of problem solving is to move from an initial state to a goal state, possibly by going through intermediate states. In an expert system, the inference engine has to search through the rules to find the goal state or to reach a conclusion. VP-Expert uses **deductive logic** for the operation of its inference engine, not induction or abduction (see Chapter 4). We assume that the rules in the knowledge base are valid; hence, the rules fire when the conditions are true.

Backward Chaining

The inference engine operates or reasons with rules by using forward and backward chaining. VP-Expert uses **backward chaining**; this means that the goal state, or conclusion, is assumed and the conditions are checked to see if that assumed goal is correct. Consider the rules given in Figure M4.1.

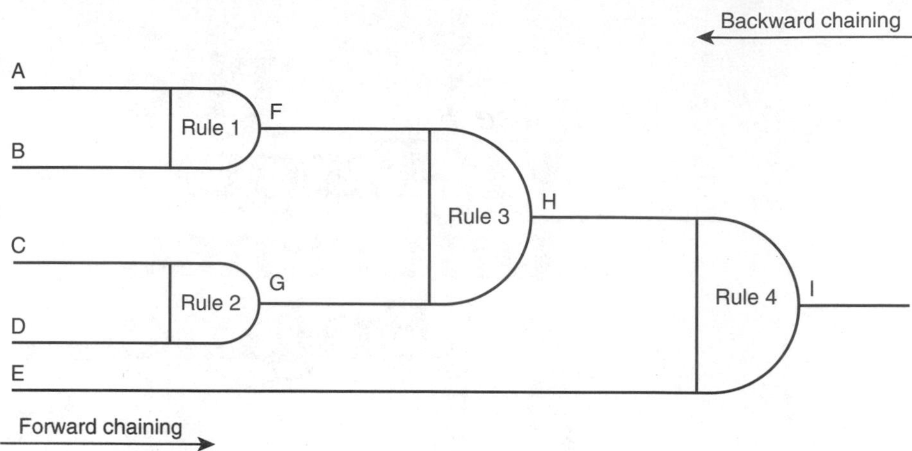

FIGURE M4.1 FORWARD AND BACKWARD CHAINING

Here the goal state is named I; a VP-Expert program file would state this as FIND I. If the goal state I is not known, then FIND I turns control over to the inference engine. The inference engine would work backward and look at the conditions of rules where the conclusion was I, the goal state. The only rule with I as the conclusion is Rule 4. The inference engine would examine Rule 4 and try to find the truth value of its conditions, H and E. If the truth values are known, the search stops and a conclusion is reached if both H and E are true. If H is unknown, the inference engine will look for rules where H is the conclusion and try to determine its value. If its value is not known, then VP-Expert will ask or query the user. This process continues until VP-Expert finds the truth values of the initial states, A, B, C, D, and E, and makes a determination on a value for the goal state I.

Forward Chaining

Forward chaining, on the other hand, would start at the conditions (initial states) and work toward the conclusions (goal states). You may hit many conclusions, none of which you may want. VP-Expert's WHENEVER command, placed in front of a rule, tests the rule whenever the value of a variable in the condition changes value. This command partially implements forward chaining.

Confidence Factors (CNF)

Before we proceed to examples of backward and forward chaining in VP-Expert, we should consider **confidence (certainty) factors** or **CNF**. Confidence factors are whole numbers ranging from 1 to 100. If a rule has a CNF 100, we are 100% sure that we are correct. A CNF 0 indicates that we are absolutely sure that the conclusion is false (invalid). In general, the larger the CNF value the better. If a rule has a CNF 75, we are 75% confident that the rule is true. Both the IF and the THEN parts of a rule may have individual confidence factors. We will discuss how these values are assigned in detail in Module 6; here we will encounter CNFs as we trace the behavior of the inference engine.

Trace of the Inference Engine: Backward Chaining

Let's turn our attention to tracing the activity of the inference engine when an expert system is active. Assuming that the 7Path (C:\VPX_samp), 6FileName, and 4Consult options have been selected, we will use the 6Set option from the Consult Menu to specify that we want to trace the activity of the inference engine. The trace must be set before you run a consultation (2Go) with the knowledge-based file. After the consultation is over, you can see the results by exiting the Consult Menu and using the 5Tree option of the Main Menu. The Tree Menu will let us view both a *textual trace* and a *graphical trace* of the consultation activity. Once you exit the Consult Menu, the trace is no longer active. The next time you enter the Consult Menu, you have to reset the Trace option. Figure M4.2 shows the logic of this procedure. The trace is set from the Consult Menu but viewed from the 5Tree option of the Main Menu.

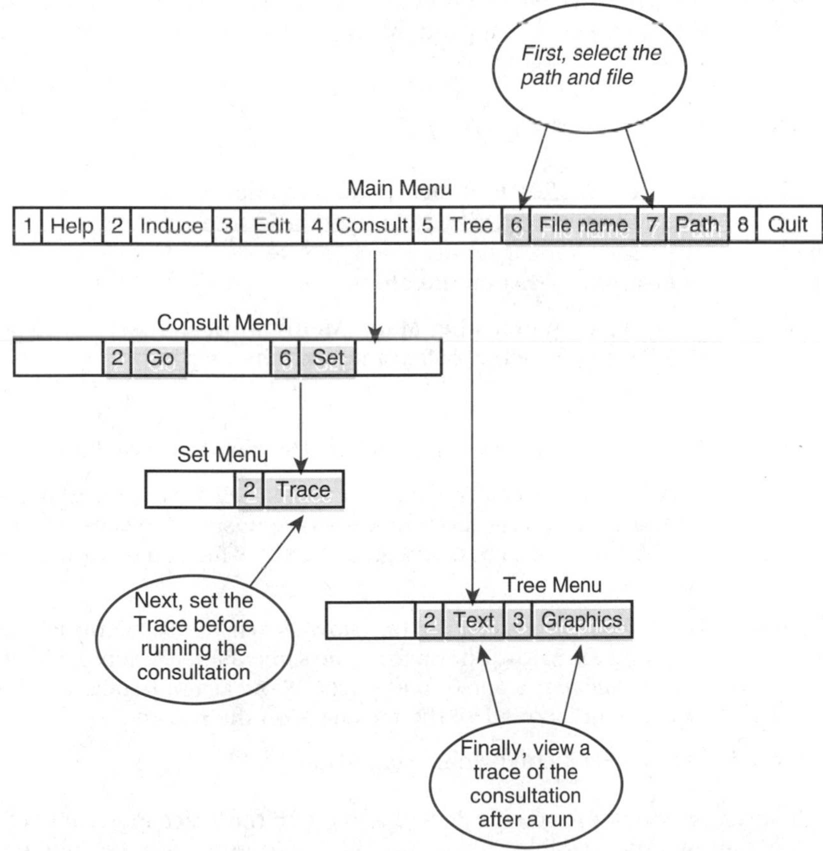

FIGURE M4.2 PROCEDURE FOR USING THE TRACE OPTION

We will use the file wp2.kbs as our example. The knowledge base consists of the following four rules:

```
RULE 1
IF      Functionality = simple AND Cost = low
THEN Word_processor = Product_A;

RULE 2
IF      Functionality = simple AND Cost = high
THEN Word_processor = Product_A;

RULE 3
IF      Functionality = complex AND Cost = low
THEN Word_Processor = none;

RULE 4
IF      Functionality = complex AND Cost = high
THEN Word_processor = Product_B;
```

All of these rules have Word_processor as a conclusion, our goal state. The command "FIND Word_processor" appears in the .kbs file and the value of the variable Word_processor is not known.

Using the TRACE Command

We will now set the TRACE command and slow down the scrolling of the messages in the Rules and Facts windows by following these steps.

Step 1. Get to your VP-Expert directory.

Step 2. Type vpx. When the Main Menu appears, select 7Path and enter C:\VPX_samp:. Select 6FileName, and then wp2.kbs.

Step 3. Select 4Consult.

Step 4. At the Consult Menu, choose 6Set. The Set Menu will have six choices.

Step 5. Now select 2Trace. This command will tell VP-Expert to save the messages from the Rules and Facts windows as a consultation runs. The saved messages will be put in both text and graphics trees for you to view after the consultation.

Step 6. To view the messages better, slow down the scrolling in the windows by selecting the 3Slow option from the same menu. When the light bar returns to 2Trace, move the cursor and select 3Slow again. Repeat this four more times so that you can read all the messages on the screen.

Step 7. Exit the Set Menu by choosing 6Quit.

You are now ready to run the consultation with the Trace mode active. The following table shows the possible values that we could enter for the four possible runs of wp2.kbs.

Run	Functionality	Cost	Word processor
1	simple	low	Product_A
2	simple	high	Product_A
3	complex	low	none
4	complex	high	Product_B

The values for the conditions are in the Functionality and Cost columns. The advice is the value of the variable Word_processor in the last column. Select 2Go and trace Run 1 by selecting Simple and Low during the consultation. As you run the consultation, the messages in the Rules window will be similar to these:

Testing (Rule) 1

RULE1 IF

Functionality = simple AND

Cost = low

THEN

Word processor = Product_A CNF 100

Finding Functionality

Finding Cost

The messages in the Facts window will be similar to these:

Functionality = simple CNF 100

Cost = low CNF 100

Word_processor = Product_A CNF 100

These messages document the activity of the inference engine as it searches for a value for Word_processor. The CNF is the confidence factor, which has defaulted to 100. As the inference engine tests a rule, the entire rule is displayed in the Rules window.

Viewing a Text Tree

After we have completed the consultation with Run 1, we can view the trace. This process will take us back to the Main Menu to the 5Tree option. Follow these steps:

Step 1. Exit the Consult Menu with 8Quit.

Step 2. Choose the 5Tree option from the Main Menu. A menu of 1Help 2Text 3Graphics 4Quit will appear.

Step 3. Select 2Text for a text trace of our consultation with wp2.kbs.

A text file should appear on your screen, as shown in Figure M4.3.

The familiar menu bar of the editor appears at the bottom. (Text under the Tree menu is different from Text under the Induce menu.) The message "Editing Old File c:\vpx_samp\wp" appears in the upper-right corner. The file with extension .trc was created during our consultation run when the Trace mode was active.

Let's interpret this text file. Three different messages could appear in the lines of the **Text tree.** If the line containing "Testing" is followed by a Rule label, then that rule is being tested. If the line contains the name of a variable (Word_processor, Functionality, Cost), then the inference engine is trying to find a value for that variable. If the line contains a parenthesis, then the value inside has been assigned (found) for the indicated variable.

Let's discuss the Text tree (trace) shown in Figure M4.3. Line 1 tells us that we are tracing file c:\vpx_samp\wp2.kbs. Line 2 contains useless information so we will ignore it. Line 3 indicates that the goal is to find a value for the variable Word_processor. Line 4 tells us that the first rule in the knowledge base that contains Word_processor as a conclusion is Rule 1. Hence, the inference engine is examining the conditions in Rule 1. Line 5 tells us that Functionality is a variable in the condition of Rule 1. Line 6 identifies that Simple is a value for the variable Functionality. (This fact came from the user input.) Line 7 indicates that Rule 1 has another variable called Cost as part of the conditions. Then line 8 tells us that Cost has been assigned a value

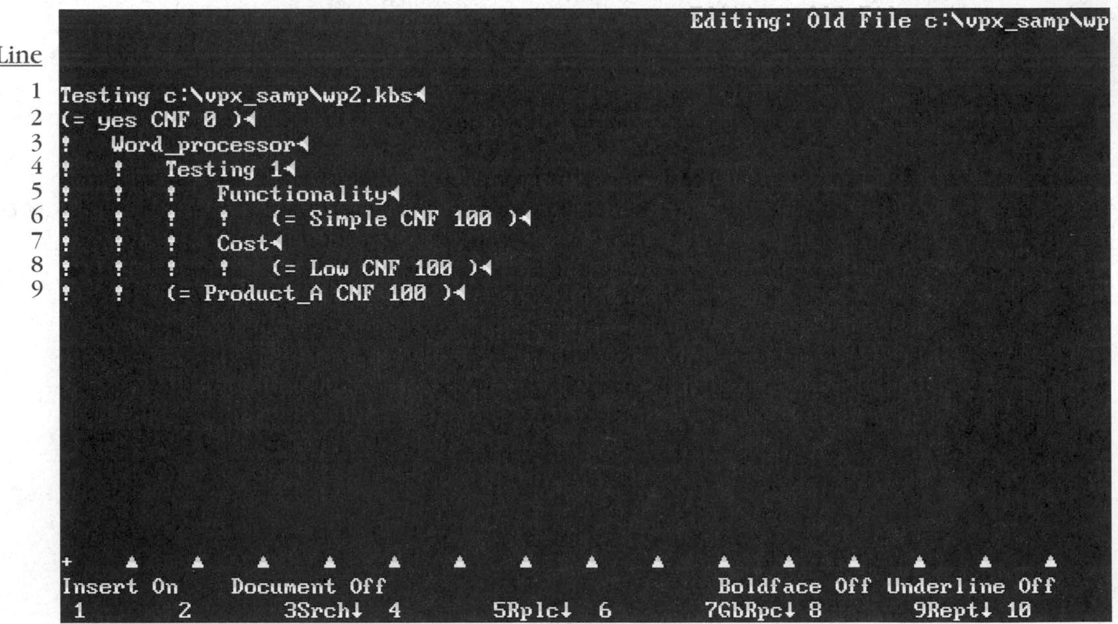

```
                                          Editing: Old File c:\vpx_samp\wp
Line
1  Testing c:\vpx_samp\wp2.kbs◄
2  (= yes CNF 0 )◄
3  !   Word_processor◄
4  !   !   Testing 1◄
5  !   !   !   Functionality◄
6  !   !   !   !   (= Simple CNF 100 )◄
7  !   !   !   Cost◄
8  !   !   !   !   (= Low CNF 100 )◄
9  !   !   (= Product_A CNF 100 )◄

Insert On    Document Off                          Boldface Off Underline Off
 1      2       3Srch↓  4       5Rplc↓  6       7GbRpc↓ 8       9Rept↓ 10
```

FIGURE M4.3 TEXT TREE FOR RUN 1 OF wp2.kbs

of Low. The last line of the trace, line 9, indicates that the value of Product_A has been assigned to the variable Word_processor. Again, this was part of Rule 1. We match the variable to the value by moving left on the value line and up until we hit a variable name (not a rule name).

To obtain a printout of the Text tree, use the familiar editor commands, F10 followed by End. To exit the editor without making changes to the trace file contents or name, press ALT and F6 and reply Y when asked to save the file. Obtain a printout and then exit the editor (text trace). You will be at the Tree Menu. If you have a graphics card in your system, you can view the Graphics tree of our past run. If you don't have a graphics card, exit to the Main Menu with 4Quit.

Viewing a Graphics Tree

To view a **graphical tree structure** of the trace, select 3Graphics from the Tree Menu. Your screen should appear similar to Figure M4.4.

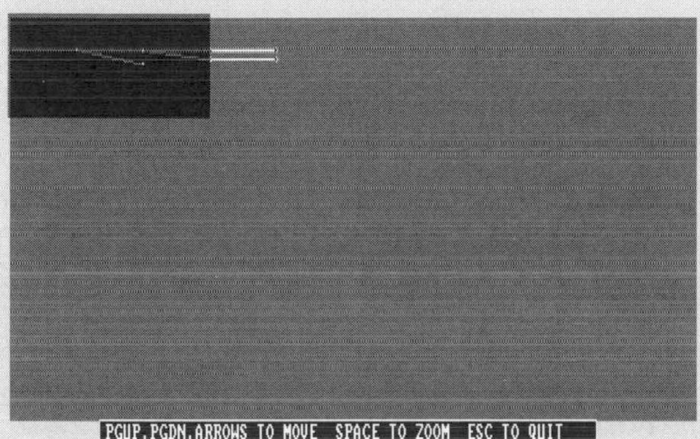

FIGURE M4.4 OPENING SCREEN FOR DISPLAYING A GRAPHICS TREE

A highlighted box appears in the upper-left corner of the display. Pressing the Right Arrow key moves the lighted box to another part of the Graphics tree. If you press the Spacebar, the contents of the lighted box will be enlarged. The movable lighted box and the zoom (enlarge) features are necessary to examine a large Graphics tree. Press the Right Arrow key and then the Spacebar. Your Graphics tree should now resemble Figure M4.5. The enlargement shows text and arrows on the tree.

We will interpret the tree reading left to right (the tree is sideways). The top of the tree (root) identifies that the goal state is a value for Word_processor. This leads to a test of Rule 1, then a search for Functionality, and finally an assignment of Simple to Functionality. Rule 1 also requires a value for Cost; a search for Cost is made and a value Low is found. With these two known conditions, Rule 1 fires, giving a conclusion of Product_A. The Text tree and the Graphics tree give the same information, but in different formats.

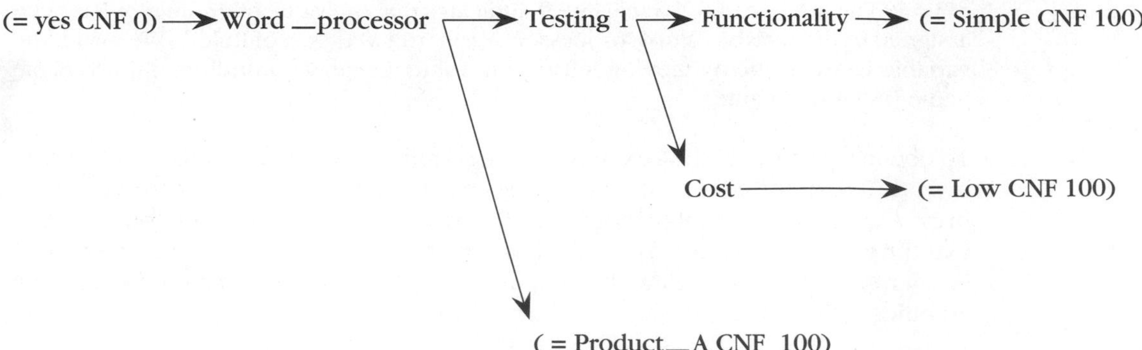

FIGURE M4.5 GRAPHICS TREE FOR RUN 1 OF wp2.kbs

If you have a color monitor, you can see the trace in color. An additional feature of the Graphics tree is an "animated effect" to show the flow of logic. To activate this feature in the Graphical tree mode, press the Home key. In response to the message on the screen, enter the name of a variable you want to trace. The screen will show an animated line to the rules involved in finding a value for the given variable. Let's try it. Press the Home key and enter the variable Word_processor. Can you see the logic flow?

There are no instructions here on how to obtain a printout of the Graphics tree. If you press Shift and Print Screen, the DOS operating system will print out the words, but you will have draw the lines by hand. To exit the Graphics tree section, you have to return to Normal mode from Enlarged mode. Press the Spacebar to return to the normal view of the Graphics tree. Press Escape to exit the Graphics tree. When you return to the Trace Menu, exit to the Main Menu with 4Quit. We will now obtain traces of the other runs of wp2.kbs.

Tracing Additional Consultations

The first run (Run 1) of wp2.kbs was simple because it involved only the first rule in the knowledge base. We will now run wp2.kbs with different values. In each case, set the trace before each run of the consultation. Then exit to the Main Menu and enter the Tree Menu to view the Text and Graphics trees. Obtain printouts of each tree.

To start, select 6FileName (wp2.kbs), 4Consult, 6Set, 2Trace, and 6Quit. At the Consult Menu, run wp2.kbs (2Go), but enter the values Simple and High (Run 2). Your Text tree and Graphics tree should resemble Figure M4.6.

Note on line 8 of the Text trace that Rule 1 did not fire; hence, the inference engine went to the next rule, in this case Rule 2, that contained Word_processor. The same information is displayed with the Graphics tree.

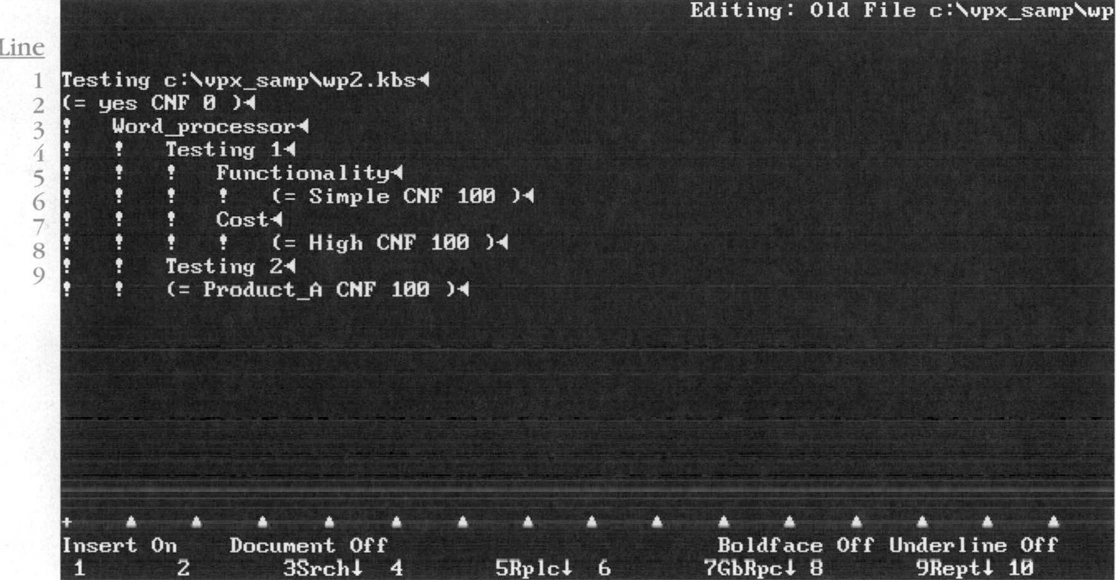

FIGURE M4.6 TEXT AND GRAPHICS TREES FOR RUN 2 OF wp2.kbs

Run wp2.kbs again, entering values Complex and Low (Run 3). The resulting traces are given in Figure M4.7.

Observe line 6 where Functionality = Complex made the compound condition false so that the rule did not fire; hence, the inference engine went to the next rule, Rule 2. Because Rule 2 also has a false compound condition, it too will not fire. The search is now on Rule 3, where the inference engine is hunting a value for Cost (line 9).

Again, activate a consultation with wp2.kbs, but enter values of Complex and High (Run 4). As you can observe in Figure M4.8, the inference engine has to test all four rules (Lines 4, 7, 8, and 11) before it can reach a conclusion. Because the inference engine searches sequentially (top to bottom), the placement of the rules can affect the length of the search, but not the final results. Let's move on to another example with the 2Trace option.

Tracing a More Complex Example

Each of the rules in our example wp2.kbs had the goal variable Word_processor as the conclusion. What does the inference engine do if the rules are not as simple? Let's consider an example in which the conclusion in some rules is the condition in other rules. We will modify our knowledge base so that complex or simple "Functionality" is somewhat more precise. Eliminating *fuzzy logic* is desirable (see Chapter 4).

```
                                            Editing: Old File c:\vpx_samp\wp

Testing c:\vpx_samp\wp2.kbs◄
(= yes CNF 0 )◄
!   Word_processor◄
!   !   Testing 1◄
!   !   !   Functionality◄
!   !   !   !   (= Complex CNF 100 )◄
!   !   Testing 2◄
!   !   Testing 3◄
!   !   !   Cost◄
!   !   !   !   (= Low CNF 100 )◄
!   !   (= none CNF 100 )◄

+    ▲    ▲    ▲    ▲    ▲    ▲    ▲    ▲    ▲    ▲    ▲    ▲    ▲    ▲    ▲
Insert On    Document Off                     Boldface Off Underline Off
 1       2       3Srch↓  4       5Rplc↓  6       7GbRpc↓ 8       9Rept↓ 10
```

FIGURE M4.7 TEXT TREE FOR RUN 3 OF wp2.kbs

```
                                            Editing: Old File c:\vpx_samp\wp

Testing c:\vpx_samp\wp2.kbs◄
(= yes CNF 0 )◄
!   Word_processor◄
!   !   Testing 1◄
!   !   !   Functionality◄
!   !   !   !   (= Complex CNF 100 )◄
!   !   Testing 2◄
!   !   Testing 3◄
!   !   !   Cost◄
!   !   !   !   (= High CNF 100 )◄
!   !   Testing 4◄
!   !   (= Product_B CNF 100 )◄

+    ▲    ▲    ▲    ▲    ▲    ▲    ▲    ▲    ▲    ▲    ▲    ▲    ▲    ▲    ▲
Insert On    Document Off                     Boldface Off Underline Off
 1       2       3Srch↓  4       5Rplc↓  6       7GbRpc↓ 8       9Rept↓ 10
```

FIGURE M4.8 TEXT TREE FOR RUN 4 OF wp2.kbs

We will assume that different types of documents can be classified according to their degree of complexity. Writing a short memo is quite different from generating a two-column newsletter with graphics. If we know the kind of document, we can classify it as one of four types, called type1, type2, type3, or type4. The higher the type number, the more complex the word processor required. If we know the type of document, then we can classify the functionality as simple or complex. In other words, functionality is the conclusion in one rule, but the condition in other rules. Our new program, wp6.kbs, is given in Figure M4.9.

Note that Functionality is part of the conditions in Rules 1, 2, 3, and 4, but Functionality is a conclusion in Rules 5 and 6. Also, a new ASK command identifies the document type for the variable Document. The ASK command for the Functionality variable has been eliminated.

We will now create a Text tree and Graphic tree trace of wp6.kbs. Assuming that we are now in the Main Menu with C:\VPX_samp as the default drive, follow these abbreviated steps to run the trace:

```
6FileName (wp6)
4Consult
    6Set
        2Trace
        3Slow (Five times)
        6Quit
    2Go (Select Type3 and High)
    8Quit
5Tree
    2Text (Get a printout)
    3Graphics (Get a printout)
    4Quit
(Back to Main Menu)
```

The indentation of a command in the above list indicates the menu depth of the command. All commands aligned in a column under a heading belong to the same menu.

After you have completed these steps, your Text tree should be the same as shown in Figure M4.10.

The inference engine tests Rule 1 because it is the first rule that contains Word_processor as a conclusion. Hence, the inference engine is now trying to find Functionality (the first condition in the compound IF part). The first rule to contain Functionality as a conclusion is Rule 5, but Rule 5 will fire only if the variable Document is known. In line 8 the inference engine finds the value of Document, Type3, from the user input. In line 10, Rule 6 is tested and fired to conclude that Functionality = complex. The inference engine now searches for rules (after Rule 1) with Functionality = complex as the condition; hence, Rule 2 and Rule 3 are tested. On line 13, after finding Functionality = complex as part of the compound condition,

```
!The file wp6.kbs has six rules. Conclusions of some rules are
conditions in other rules.

ACTIONS
     DISPLAY "This expert system WPHELP will advise you on
selecting a word processor for the IBM compatible
microcomputer in your business.

Press the Spacebar to start the consultation."
     CLS
     FIND Word_processor
     DISPLAY "WPHELP advises {Word_processor}."
     DISPLAY " "
     DISPLAY "   Press the Spacebar to exit WPHELP."
     CLS;

RULE 1
IF   Functionality = simple AND cost = low
THEN Word_processor = Product_A;

RULE 2
IF   Functionality = simple AND cost = high
THEN Word_processor = Product_A;

RULE 3
IF   Functionality = complex AND cost = low
THEN Word_processor = none;

RULE 4
IF   Functionality = complex AND cost = high
THEN Word_processor = Product_B;

RULE 5
IF   Document = type1 OR Document = type2
THEN Functionality = simple;

RULE 6
IF   Document = type3 OR Document = type4
THEN Functionality = complex;

ASK Document : "Select the column of the most complex documents that you
need to produce in your business
       Type1          Type2          Type3          Type4
       Letter         Report         Resume         Newsletter
       Memo           Term Paper     Form Letter    Brochure
";
CHOICES Document : Type1, Type2, Type3, Type4;

ASK Cost: "Estimate the amount of money that you can afford.";

CHOICES Cost : Low, High;
```

FIGURE M4.9 PROGRAM CODE FOR wp6.kbs

the inference engine now tries to find a value for Cost. From the user input, the Cost is High. Since Rule 3 does not fire, Rule 4 is tested next and fires; the conclusion is Word_processor = Product_B. Again, note the inference engine's sequential search down the list of rules that have Word_processor as the conclusion. If Rule 4 had been

```
                                          Editing: Old File c:\vpx_samp\wp

Testing c:\vpx_samp\wp6.kbs◄
(= yes CNF 0 )◄
!   Word_processor◄
!   !   Testing 1◄
!   !   !   Functionality◄
!   !   !   !   Testing 5◄
!   !   !   !   !   Document◄
!   !   !   !   !   !   (= Type3 CNF 100 )◄
!   !   !   !   Testing 6◄
!   !   !   !   (= complex CNF 100 )◄
!   !   Testing 2◄
!   !   Testing 3◄
!   !   !   Cost◄
!   !   !   !   (= High CNF 100 )◄
!   !   Testing 4◄
!   !   (= Product_B CNF 100 )◄


Insert On    Document Off                    Boldface Off Underline Off
 1      2        3Srch↓  4        5Rplc↓  6        7GbRpc↓ 8        9Rept↓ 10
```

FIGURE M4.10 TEXT TREE FOR wp6.kbs

placed as the first rule in the list, the search would have been shorter, but the same conclusion would have been reached. The graphics tree of wp6.kbs is given in Figure M4.11.

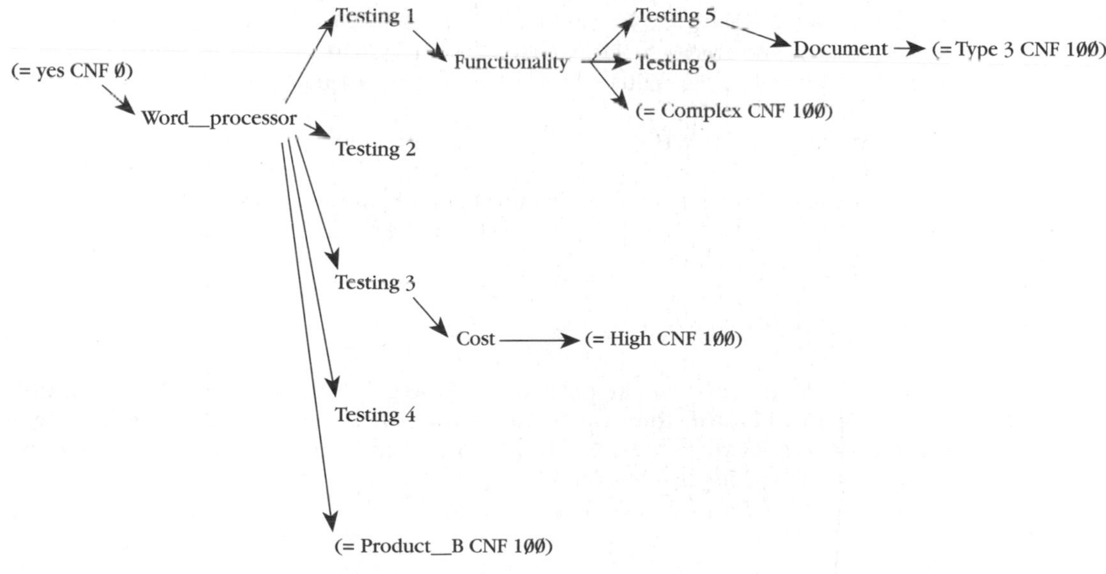

FIGURE M4.11 GRAPHICS TREE FOR wp6.kbs

In fact, this figure combines two enlarged sections of the graphics tree. The same trace information is given, but in a left-to-right and up-to-down orientation. For now, exit VP-Expert and return to the DOS prompt.

All of the examples tracing wp2.kbs and wp6.kbs show the backward chaining activity of the inference engine. The search starts from conclusions and goes back to conditions. In the next section, we will discuss the opposite control strategy, forward chaining, moving from condition to conclusion.

Use of WHENEVER: Forward Chaining

Even though backward chaining is VP-Expert's control strategy, **forward chaining** can be achieved. The use of the WHENEVER command (as the beginning of a rule) causes the rule to be tested whenever the value of a variable contained in the condition changes value. When a WHENEVER rule is found to have a true condition, the THEN part is executed. In other words, when there is a change in the variable in the condition, the rule is automatically checked for firing, regardless of the strategy of the VP-Expert inference engine.

Consider the following example:

```
WHENEVER  Cost_too_high
IF        Cost >= 500
THEN      DISPLAY "Consider volume discounts.";
```

Keywords are WHENEVER, IF, THEN, and DISPLAY. The phrase Cost_too_high serves as a Rule label and abides by Rule label restrictions. The rest of the syntax is identical to IF-THEN rules. If the value of the variable Cost changes, the WHENEVER clause checks the rule for firing. The WHENEVER format does not have an ELSE or BECAUSE part. If more than one WHENEVER rule names the same changed variable (Cost), then the WHENEVER rules are checked in the order that they occur in the knowledge base. The WHENEVER rule is known in AI programming as a "demon" because it watches the variables referenced by the IF condition and executes each time the condition is true.

FIND as the Goal Variable

It is appropriate to discuss the placement of the goal variable, the **FIND command**, in a program. Because this command turns control over to the inference engine, understanding any rules regarding its placement and relation to the rest of an expert system program (.kbs) is important.

Instruction Types and Order

VP-Expert has two types of commands: statements and clauses. **Statements** may be used (almost) anywhere in the program, but **clauses** can be used only in the ACTIONS block and in the conclusion of rules. The commands we have discussed and their classifications as to instruction (command type) are as follows:

Statements	Clauses
ACTIONS	CLS
ASK	DISPLAY
CHOICES	FIND
PLURAL	

RULES

As you can observe, the FIND instruction is a clause, not a statement. It may be more meaningful to say that FIND is a word that starts a clause. The VP-Expert Help Facility will always indicate the instruction type (statement or clause) in its explanation of the command.

The three parts of a VP-Expert program, ACTIONS, RULES, and STATEMENTS, behave in different ways. The commands listed in the ACTIONS block are executed in the order they appear. When a FIND command is encountered, the inference engine takes over. When the inference engine is finished, control passes back to the next line in the ACTIONS block. The inference engine searches the RULES block in the order that we have discussed and uses statements (ASK, CHOICES) from the STATEMENTS block as needed. The order of the statements in the STATEMENTS block has no effect on the inference engine. (CHOICES statements should come after the appropriate ASK statements, of course.) We can summarize this information concerning instruction type, parts of a program, and effect of order in a table as follows:

PART	CONTAINS	ORDER
ACTIONS	Statements, clauses	Order dependent
RULES	Statements, clauses in conclusions only	Partially order dependent
STATEMENTS	Statements	Not order dependent

The ACTIONS block contains statements and clauses that are executed in the order they are listed. The RULES block is composed of statements and clauses in the conclusions of rules. These rules are partially "order dependent" if they contain the goal variable. The STATEMENTS block contains only statements and is not order dependent. When the inference engine needs to know the value of a variable (which it cannot find in the rules), then the appropriate ASK statement will be invoked, regardless of its order in the last part of the program. The inference engine takes no action, however, if a FIND variable has a known value.

Placement of the FIND Clause

Because the **FIND clause** identifies the goal variable, it is restricted to the ACTIONS Block and the conclusion of rules. We have already seen many examples of the FIND command in this position.

The FIND clause can be part of the conclusion of a rule, but not the condition. For example, the rule below is correct in VP-Expert:

```
RULE Networking
IF   Environment = Networked
THEN Network = Yes
     FIND Networked_word_processor;
```

In this rule, if our environment has networked computers, then two (AND) conclusions are reached. The variable Network is assigned the value Yes, and the FIND instruction turns control over to the inference engine to find a value for Networked_word_processor. If this rule fired when the inference engine used another FIND instruction, then work would cease on that FIND until a value of Networked_word_processor was found. In this example, placing a FIND command in the IF part would be incorrect.

Because the WHENEVER clause indicates a rule, you might conclude that the FIND command could be in the corresponding conclusion. Unfortunately, VP-Expert does not allow this. The FIND command cannot be placed in the conclusion of a WHENEVER rule. In other words, the following example is incorrect:

```
WHENEVER Networking
IF       Environment = Networked
THEN     Network = Yes
FIND     Networked_word_processor;
```

The FIND command must be left out of WHENEVER rules.

Can the FIND command occur more than once in the STATEMENTS block or in appropriate rule conclusions? The answer is yes. If multiple FIND instructions occur, then the inference engine completes its search for the first FIND command before it starts a search for the second. Hence, it is possible to have multiple goal variables (or states) in VP-Expert applications.

Uncertainty and the Inference Engine

The inference engine not only has to search through the rules for values, but has to deal with certainty, or confidence, levels of the results. We will explore how the inference engine handles various confidence factors (CNF) for rules. We will also take a look at how the inference engine handles "I don't know" or "?" as user input during a consultation.

Confidence Factors

So far, the **confidence factors** or **CNF** that we have seen in the Rules window, the Facts window, or Tree traces have been CNF 100. If no confidence factors are programmed into the rules, then the value defaults to 100, complete certainty. Obviously, not all decisions are black and white, but somewhere in between. A well-developed expert system's confidence factors will come from the *experts* in the domain field.

Confidence factors in VP-Expert are whole numbers between 0 and 100. Complete uncertainty is indicated by CNF 0, complete confidence by CNF 100. If the CNF is 50, then we are only 50% sure of our premise, conclusion, or rule. In general, an expert system is more reliable as the CNF value assigned to the final conclusion or goal variable increases.

Even though it is possible to program CNF into the rules, there is another way users can explore how the inference handles and calculates CNF values. When a user runs a consultation, he or she can enter a confidence factor before making a selection. These are the logical steps:

Step 1. When a question with various selections appears on the Consultation screen, move the cursor to your selection.

Step 2. Press the Home key, and a small box will appear by your selection.

Step 3. Enter a number between 0 and 100. The number will appear by your selection. Press Enter, followed by End to make your regular selection.

As the consultation runs, the CNF numbers will appear by the appropriate rules and variables in the bottom windows. When you view a trace, either Text tree or Graphics tree, the CNF numbers will be listed. If you make a mistake entering a number during a consultation, there are two ways to correct it. One way is to press Home again and enter the correct number. The second way is to press the DEL key to erase the selection and the CNF number.

We will set the Trace function, run a consultation, enter CNF numbers, and then observe the results from the Rules and Facts windows and the Text tree of the trace.

We will use wp6.kbs, the knowledge base with the six rules. During the consultation we will enter 90 for the CNF of Type3 (Document) and 80 for the CNF of Cost. Return to the VP-Expert Menu with your Sample Files directory in drive B and follow these abbreviated steps:

7Path (C:\VPX_samp)

6FileName (wp6)

4Consult
- 6set
  - 2Trace
  - 3Slow (Five times)
  - 6Quit
- 2Go

  For Type, move the cursor to Type3, press the Home key, enter 90, and make the selection with the Enter and End keys.

  For Cost, position the cursor on High, press the Home key, enter 80, and make the selection with the Enter and End keys.
- 8Quit

5Tree
- 2Text
- 4Quit

(End at Main Menu)

If everything went smoothly, your last Consultation screen with wp6.kbs should look like Figure M4.12.

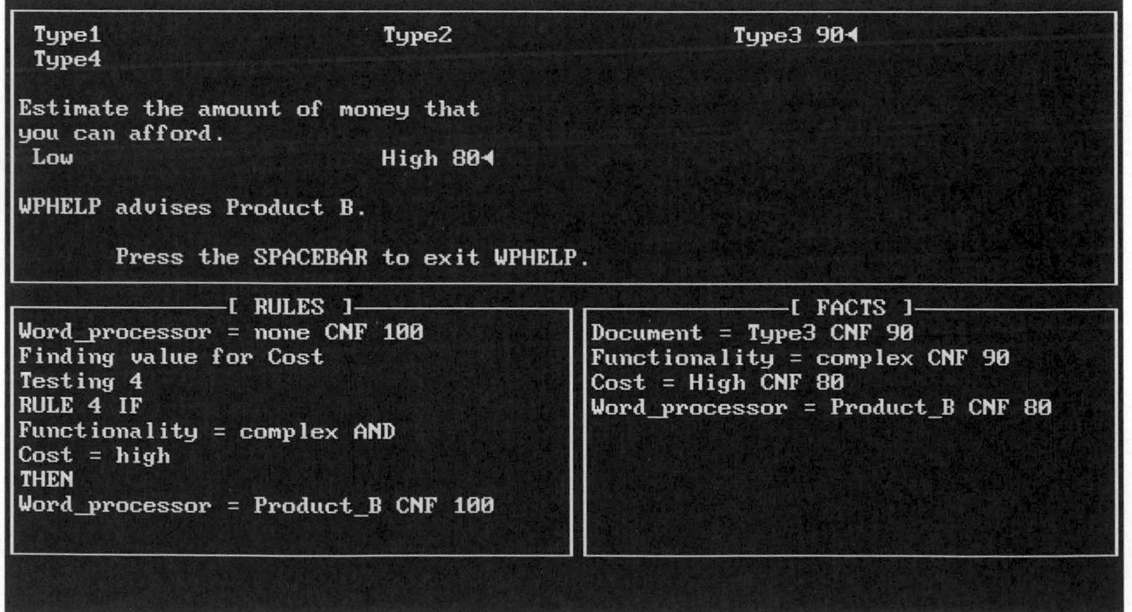

FIGURE M4.12 LAST CONSULTATION SCREEN WITH CNF NUMBERS FOR wp6.kbs

Note that the entered CNF value of 80 shows on the Consultation window. The default values of 100 appear in the Rules window, and the input values (CNF 90 for

Type3 and CNF 80 for Cost) appear in the Facts window. The inference engine reports that the *intermediate variable*, Functionality, has been assigned CNF 90 and that the goal state, Word_processor = Product_B, has CNF 80. These calculations are done by the inference engine. The same information is given with the Text tree (trace) of wp6.kbs in Figure M4.13.

```
                                        Editing: Old File c:\vpx_samp\wp


Testing c:\vpx_samp\wp6.kbs◄
(= yes CNF 0 )◄
!   Word_processor◄
!   !   Testing 1◄
!   !   !   Functionality◄
!   !   !   !   Testing 5◄
!   !   !   !   !   Document◄
!   !   !   !   !   !   (= Type3 CNF 90 )◄
!   !   !   !   Testing 6◄
!   !   !   !   (= complex CNF 90 )◄
!   !   Testing 2◄
!   !   Testing 3◄
!   !   !   Cost◄
!   !   !   !   (= High CNF 80 )◄
!   !   Testing 4◄
!   !   (= Product_B CNF 80 )◄


+   ▲   ▲   ▲   ▲   ▲   ▲   ▲   ▲   ▲   ▲   ▲   ▲   ▲   ▲   ▲
Insert On    Document Off                    Boldface Off Underline Off
 1      2       3Srch↓  4        5Rplc↓  6        7GbRpc↓ 8        9Rept↓ 10
```

FIGURE M4.13 TEXT TREE TRACE FOR wp6.kbs

Let's see how VP-Expert handles CNF for compound rules. Select wp3.kbs, the version that helps a user decide whether to buy now or wait. The rule base of wp3.kbs is one compound AND rule. When you run the consultation, enter the following values:

First question, (Afford), 90, Yes

Second question, (Help), 80, Yes

Third question, (Time), 70, Yes

Fourth question, (Need), 60, Yes

Use the same instructions given for running wp6.kbs, but change the file name and input during the consultation.

After running your consultation, your last Consultation screen should resemble Figure M4.14 and your Text tree should look like Figure M4.15.

```
 Yes 70◄                    No

Do you need a word processor?
 Yes 60◄                    No

Concerning the purchase of a word processor,
WPHELP advises you to proceed.


        Press the SPACEBAR to exit WPHELP.

──────────────[ RULES ]──────────          ──────────────[ FACTS ]──────────
Need = Yes                                 Afford = Yes CNF 90
THEN                                       Help = Yes CNF 80
Advise = proceed CNF 90                    Time = Yes CNF 70
ELSE Advise = wait CNF 70                  Need = Yes CNF 60
Finding value for Afford                   advise = proceed CNF 54
Finding value for Help
Finding value for Time
Finding value for Need
```

FIGURE M4.14 LAST CONSULTATION SCREEN WITH CNF NUMBERS FOR wp3.kbs

```
                                          Editing: Old File c:\vpx_samp\wp

Testing c:\vpx_samp\wp3.kbs◄
(= yes CNF 0 )◄
!   advise◄
!   !   Testing 1◄
!   !   !   Afford◄
!   !   !   !   (= Yes CNF 90 )◄
!   !   !   Help◄
!   !   !   !   (= Yes CNF 80 )◄
!   !   !   Time◄
!   !   !   !   (= Yes CNF 70 )◄
!   !   !   Need◄
!   !   !   !   (= Yes CNF 60 )◄
!   !   (= proceed CNF 54 )◄






+   ▲   ▲   ▲   ▲   ▲   ▲   ▲   ▲   ▲   ▲   ▲   ▲   ▲   ▲   ▲
Insert On    Document Off                         Boldface Off Underline Off
 1       2       3Srch↓  4       5Rplc↓  6       7GbRpc↓ 8       9Rept↓ 10
```

FIGURE M4.15 TEXT TREE WITH CNF NUMBERS FOR wp3.kbs

Observe that when the CNF values are 90, 80, 70, and 60 for the four AND conditions, the inference engine gives the conclusion a CNF value of 54.

Let's try an OR rule. Use wp4.kbs and enter the following values for the indicated selections:

> First question, (Letter), Yes, 90
>
> Second question, (Resume), Yes, 80
>
> Third question, (Notice), Yes, 70
>
> Fourth question, (Paper), Yes, 60

Use the same instructions for running wp6.kbs but change the file name and the user inputs. Your last Consultation screen and Text tree (trace) should appear as shown in Figures M4.16 and M4.17, respectively.

Interestingly enough, the inference engine assigned a CNF Value of 99 when the user entered 90, 80, 70, and 60 as the CNF values for the four OR conditions. The calculated CNF is higher than the input values. (We will discuss this calculation further in Module 6.)

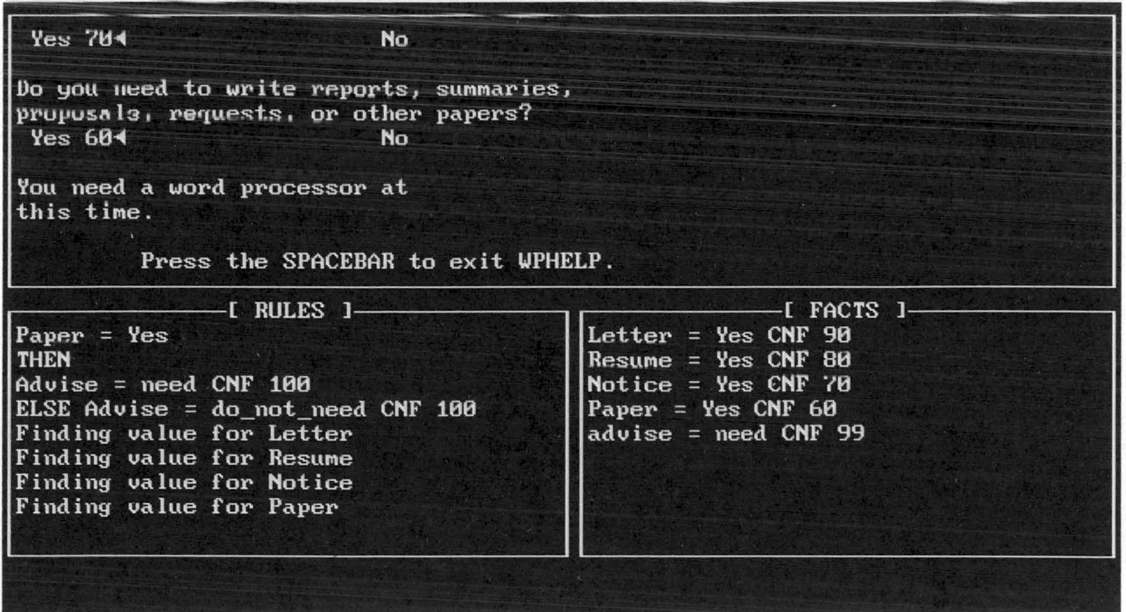

FIGURE M4.16 LAST CONSULTATION SCREEN WITH CNF NUMBERS FOR wp4.kbs

```
                                          Editing: Old File c:\vpx_samp\wp

Testing c:\vpx_samp\wp4.kbs◄
(= yes CNF 0 )◄
!   advise◄
!   !   Testing 1◄
!   !   !   Letter◄
!   !   !   !   (= Yes CNF 90 )◄
!   !   !   Resume◄
!   !   !   !   (= Yes CNF 80 )◄
!   !   !   Notice◄
!   !   !   !   (= Yes CNF 70 )◄
!   !   !   Paper◄
!   !   !   !   (= Yes CNF 60 )◄
!   !   (= need CNF 99 )◄

+   ▲   ▲   ▲   ▲   ▲   ▲   ▲   ▲   ▲   ▲   ▲   ▲   ▲   ▲   ▲
Insert On    Document Off                       Boldface Off Underline Off
 1       2       3Srch↓  4        5Rplc↓  6        7GbRpc↓ 8        9Rept↓ 10
```

FIGURE M4.17 TEXT TREE WITH CNF NUMBERS FOR wp4.kbs

Unknown Values

The inference engine must deal with another aspect of uncertainty, unknown values. What happens during a consultation if the inference engine can't find the value of a variable? Usually the programmer has included an ASK statement to query the user for a needed value. What if the programmer left out the ASK statement? VP-Expert has a built-in command, AUTOQUERY, to assist in this situation. If this command is active, VP-Expert will automatically generate a preformatted question. For example, if the inference engine can't find a value for Functionality through ASK statements or rule examination, the AUTOQUERY feature generates the following:

"What is the value of Functionality?"

The user then enters a value for Functionality (but no possible values for Functionality are displayed). To activate this option, just include

AUTOQUERY;

at the beginning of a .kbs file.

What happens if the user does not know or understand a question during the consultation? VP-Expert allows the user to enter the question mark (?) to indicate *I don't know*. The inference engine can handle this entry if the designer or programmer included rules dealing with the appropriate variable and a key word,

UNKNOWN. When the user enters a ? at the screen, the inference engine searches for a rule with variable = UNKNOWN. Including rules to handle direct user input of ? can make an expert system more usable. (If the question mark is given as an option to select from a list, then the inference engine does not translate to the key word UNKNOWN.)

Let's consider an example using UNKNOWN as part of the condition in rules. We have added two rules to wp6.kbs and changed the user input to direct screen entry. Our new file is wp7.kbs. The two new rules are:

```
RULE 7
IF      Document = UNKNOWN
THEN Word_processor = wait_for_now
        DISPLAY "Examine your business functions
        for word processing before buying.";

RULE 8
IF      Cost = UNKNOWN
THEN Word_processor = none_for_now
        DISPLAY "Consult your manager or boss to
        determine your budget.";
```

Each consultation has two parts, an assignment statement and a message to the user. If the user enters a ? for type of document, the system will advise him or her to review the business functions before buying. If a ? is entered for cost, then the system recommends that your manager or boss be consulted to determine your budget.

From the Main Menu, select wp7 as your file, then set the Trace option. When running the consultation, enter ? as your reply to type of document. The last Consultation screen for this run with wp7.kbs is shown in Figure M4.18 and the Text tree in Figure M4.19.

In both figures, Rule 7 was tested for Document = UNKNOWN.

After completing this consultation, rerun wp7.kbs, but this time enter Type3 as the answer to the documents question and a ? for cost. The last Consultation screen and Text tree for this run appear in Figures M4.20 and M4.21, respectively.

In both figures, Rule 8 was fired for the condition Cost = UNKNOWN. Remember that the question mark (?) is used only for typing replies, not for selecting from lists. In our examples, the CNF numbers defaulted to 100 because we did not program them into our knowledge base (rules).

After you have finished your consultations with wp7.kbs, exit VP-Expert and return to the DOS prompt to end your session with VP-Expert for Module 4.

In this section, we have explored how the inference engine searches through a rule base (backward chaining), how to create a Text or Graphics tree of a consultation, how to identify the different instruction types, how to place the FIND command, how

```
          Memo         Term Paper  Form Letter    Brochure

 ?

Examine your business functions for
word processing before buying.

WPHELP advises wait for now.

        Press the SPACEBAR to exit WPHELP.
```

```
[ RULES ]
Cost = high
THEN
Word_processor = Product_B CNF 100
Testing 7
RULE 7 IF
Document = Unknown
THEN
Word_processor = wait_for_now CNF 100
```

```
[ FACTS ]
Word_processor = wait_for_now CNF 100
```

FIGURE M4.18 LAST CONSULTATION SCREEN WITH DOCUMENT UNKNOWN FOR wp7.kbs

```
                                            Editing: Old File c:\vpx_samp\wp


Testing c:\vpx_samp\wp7.kbs◄
(= yes CNF 0 )◄
!   Word_processor◄
!   !   Testing 1◄
!   !   !   Functionality◄
!   !   !   !   Testing 5◄
!   !   !   !   !   Document◄
!   !   !   !   Testing 6◄
!   !   Testing 2◄
!   !   Testing 3◄
!   !   Testing 4◄
!   !   Testing 7◄
!   !   (= wait_for_now CNF 100 )◄









+    ▲    ▲    ▲    ▲    ▲    ▲    ▲    ▲    ▲    ▲    ▲    ▲    ▲    ▲    ▲
Insert On    Document Off                       Boldface Off Underline Off
 1      2       3Srch↓  4       5Rplc↓  6       7GbRpc↓ 8       9Rept↓ 10
```

FIGURE M4.19 TEXT TREE WITH DOCUMENT UNKNOWN FOR wp7.kbs

```
Estimate the amount of money that
you can afford?  Enter low, high, or ?.
 ?

Consult your manager or boss to
determine your budget.

WPHELP advises None for now.

        Press the SPACEBAR to exit WPHELP.

[ RULES ]
Document = Unknown
THEN
Word_processor = wait_for_now CNF 100
Testing 8
RULE 8 IF
Cost = Unknown
THEN
Word_processor = None_for_now CNF 100

[ FACTS ]
Document = type3 CNF 100
Functionality = complex CNF 100
Word_processor = None_for_now CNF 100
```

FIGURE M4.20 LAST CONSULTATION SCREEN WITH COST UNKNOWN FOR *wp7.kbs*

```
                                        Editing: Old File c:\vpx_samp\wp

Testing c:\vpx_samp\wp7.kbs◄
(= yes CNF 0 )◄
!   Word_processor◄
!   !   Testing 1◄
!   !   !   Functionality◄
!   !   !   !   Testing 5◄
!   !   !   !   !   Document◄
!   !   !   !   !   !   (= type3 CNF 100)◄
!   !   !   !   Testing 6◄
!   !   !   !   (= complex CNF 100 )◄
!   !   Testing 2◄
!   !   Testing 3◄
!   !   !   Cost◄
!   !   Testing 4◄
!   !   Testing 7◄
!   !   Testing 8◄
!   !   (= None_for_now CNF 100 )◄

+   ▲   ▲   ▲   ▲   ▲   ▲   ▲   ▲   ▲   ▲   ▲   ▲   ▲   ▲   ▲
Insert On    Document Off                   Boldface Off Underline Off
 1       2       3Srch↓  4       5Rplc↓  6       7GbRpc↓ 8       9Rept↓ 10
```

FIGURE M4.21 TEXT TREE WITH COST UNKNOWN FOR *wp7.kbs*

to enter CNF or confidence factors, and how to use the question mark as input. In the next module, we will add more information about developing the knowledge base.

Key Points

- The inference engine is the reasoning mechanism of an expert system.
- VP-Expert uses deductive logic, not induction or abduction.
- The two types of control strategies for inference engines are backward chaining and forward chaining. Backward chaining moves from the conclusion to condition, while forward chaining moves from condition to conclusion.
- VP-Expert uses backward chaining. Forward chaining can be implemented with WHENEVER rules.
- The goal state is the value of the variable named by the FIND command.
- The behavior of the inference engine can be traced with a Text tree or a Graphics tree. The Trace option must be selected before the consultation is run. The steps are as follows:

```
7Path
6FileName
4Consult
    6Set
        2Trace
        6Quit
    2Go
    8Quit
5Tree
    2Text
    3Graphics
    4Quit
```

- The Text tree is displayed as text from the editor using a trc file. Use F10 and End to get a printout. Exit with ALT and F8.
- The Graphics tree appears as a sideways tree. Individual sections can be enlarged with the zoom feature. To obtain a partial printout, use the Shift and Print Screen keys and draw the arrows (lines). To exit, get out of Zoom (Spacebar) and then press Escape.
- There are two types of commands or instructions for VP-Expert programs: statements and clauses. Statements can occur almost anywhere, but clauses are restricted to the Actions block and the conclusions of rules.
- FIND is a clause. There can be multiple FIND commands in the ACTIONS block or in conclusions. The FIND command, however, cannot appear in WHENEVER rules.
- CNF numbers are whole numbers between 0 and 100 that represent the degree of certainty or confidence in a condition, conclusion, or rule. Zero is complete uncertainty; 100 is complete confidence.
- Users can input CNF values during a consultation with direct data entry. When the cursor is on the appropriate key, press the Home key and enter the number. Then make the selection with Enter and End.

- The question mark (?) can be allowed as user input during direct entry in a consultation if rules contain the appropriate variable and the key word UNKNOWN.
- The AUTOQUERY statement will automatically generate a preformatted question for the user if the inference engine cannot determine a value for a given variable.

Exercises

1. Which of the following options belong to the Set Menu?
   a. Trace
   b. Track
   c. View
   d. Fast
   e. Go
2. Which of the following options belong to the Tree Menu?
   a. View
   b. Trace
   c. Text
   d. Graphics
   e. Go
3. Arrange the following options into a logical sequence to produce a Text tree of a consultation:
   a. Go
   b. Consult
   c. Trace
   d. Path
   e. FileName
   f. Set
   g. Text
   h. Tree
4. Assume that these rules are in order from 1 to 6 and backward chaining is used:

   RULE 1
   IF A
   THEN F
   RULE 2
   IF B
   THEN G
   RULE 3
   IF F
   THEN G
   RULE 4
   IF D
   THEN G
   RULE 5
   IF E
   THEN F

RULE 6
IF F AND G
THEN H

a. Which rule will the inference engine test first for FIND F?
b. Which rule will the inference engine test first for FIND G?
c. Which rule will the inference engine test first for FIND H when H has a known value?

5. Given the trace in Figure M4.22, match the variables to the values.

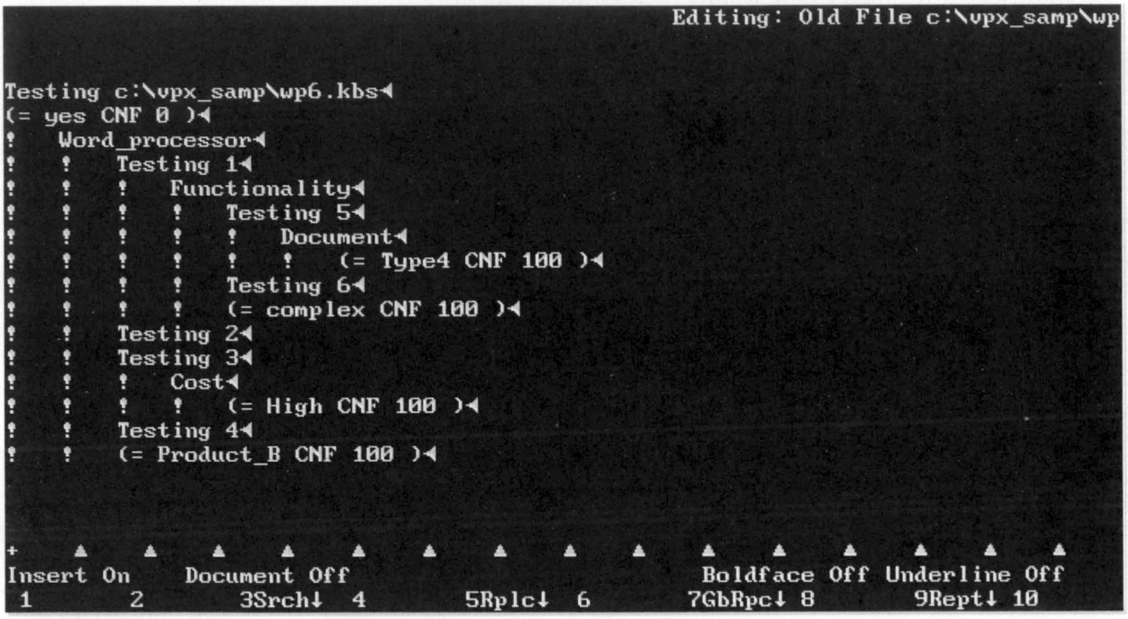

```
                                              Editing: Old File c:\vpx_samp\wp

Testing c:\vpx_samp\wp6.kbs
(= yes CNF 0 )
!   Word_processor
!   !   Testing 1
!   !   !   Functionality
!   !   !   !   Testing 5
!   !   !   !   !   Document
!   !   !   !   !   !   (= Type4 CNF 100 )
!   !   !   !   Testing 6
!   !   !   !   (= complex CNF 100 )
!   !   Testing 2
!   !   Testing 3
!   !   !   Cost
!   !   !   !   (= High CNF 100 )
!   !   Testing 4
!   !   (= Product_B CNF 100 )

Insert On     Document Off                        Boldface Off Underline Off
 1        2       3Srch   4        5Rplc   6        7GbRpc  8       9Rept  10
```

FIGURE M4.22 TEXT TREE

6. Tell whether order is important or not important for each of the following:
   a. Two ASK instructions
   b. Two FIND instructions
   c. Two DISPLAY instructions
   d. Two RULES
7. Identify each of the following instructions as a clause (C) or statement (S):
   a. RUNTIME
   b. ASK
   c. CHOICES
   d. FIND
   e. DISPLAY

8. Identify each of the following instructions as valid (V) or invalid (I):
   a. FIND X FIND Y
   b. IF FIND X = FIND Y
      THEN Z = M;
   c. IF X = Y
      THEN FIND Z;
   d. WHENEVER Low
      IF X = Y
      THEN FIND Z;
   e. IF X = Unknown
      THEN FIND Y;
9. CNF numbers in VP-Expert are numbers between
   a. 0 and 1
   b. -1 and 1
   c. 0 and 100
   d. -100 and 100
   e. 1 and 100
10. The user may enter CNF numbers during a consultation
    a. when the questions have selections.
    b. when the user types in a reply.
    c. when a trace is on.
    d. when the Slow option is active.
    e. all of the above.
11. Identify each of the following as true (T) or false (F):
    a. Unlike standard rules, WHENEVER does not require a conclusion.
    b. The FIND command can be used in the conclusion of a WHENEVER rule.
    c. WHENEVER rules implement forward chaining.
    d. One WHENEVER rule can cause the execution of another WHENEVER rule.
    e. The WHENEVER rule has the ELSE option.
12. Correct the syntax errors in the following:
    a. WHENEVER Sunshine:
       IF Sunshine = Yes
       THEN DISPLAY "Today is sunny."
       ELSE DISPLAY "Today is not sunny."
    b. What should be added to the above WHENEVER rule?
13. Identify each of the following as true (T) or false (F):
    a. UNKNOWN is a keyword used to identify variables whose values are not known.
    b. Rules using UNKNOWN can be induced in VP-Expert.
    c. The keyword UNKNOWN and the ? are the same.
    d. FIND Cost UNKNOWN is a valid instruction or command.
    e. The user can enter the ? symbol to indicate "unknown."
14. What will VP-Expert do if the AUTOQUERY is active and the inference engine requires a value for a certain variable Cost (for which there is no rule or ASK statement)?

Assignments

1. Using the VP-Expert Help Facility, obtain a printout of each of the following topics:
   SET MENU
   TREE MENU
   CLAUSE
   STATEMENT
   CNF
   WHENEVER
   FIND
   BACKWARD CHAINING
   FORWARD CHAINING
   UNKNOWN
   AUTOQUERY
2. With the Trace option on, rerun file wp4.kbs from your Sample File directory. During the consultation, enter No to the first question and Yes to the rest. Obtain both a Text tree and a Graphics tree of the resulting consultation.
3. With the Trace option on, run program wp3.kbs from your Sample File directory. Answer No to the first question. Obtain both a Text tree and a Graphics tree of the resulting consultation.
4. With the Trace option on, run taxguide.kbs from the Sample File directory. During the consultation, enter confidence factors (CNF) above 50 and make any selections you like. Obtain both a Text tree and Graphics tree of the resulting consultation.
5. Let's change the order in which rules are listed and compare the path of the inference engine. Do the following:
   a. Select wp6.kbs from your Sample File directory.
   b. Get into VP-Expert and use the editor to change the knowledge base so that the rules appear as:
      RULE 4
      RULE 5
      RULE 1
      RULE 2
      RULE 3
      RULE 6
      Be sure to save these changes.
   c. Obtain a Text tree trace of a consultation with the changed wp6.kbs.
   d. Compare your Text tree to Figure M4.22.
   e. Can the order of rules affect the work of the inference engine?
6. Let's trace the activity of the inference engine when the FIND variable has a value.
   a. Select wp2.kbs from the Sample File directory.
   b. Using the VP-Expert editor insert the statement

      Word_processor = Product_X

   above the FIND instruction. Be sure to save this change.

c. Obtain a Text tree trace of a consultation with the changed wp2.kbs.
d. Explain the results of the consultation and Text tree trace of the changed version of wp2.kbs.

7. For a challenge, take the knowledge base (rules) on investing from Assignment 2 in Chapter 2 and do the following:
   a. Write the necessary commands and rules to create a .kbs file.
   b. Using the editor from the Main Menu enter your code.
   c. Save the file as INVEST.
   d. Run the expert system entering different CNF numbers.
   e. Obtain a Text tree and a Graphics tree of one consultation.
8. When you run a consultation with the Trace option active, VP-Expert creates a file with a .trc extension. Use the DOS CD command (change directory) to get to your Sample File directory.
   a. Type DIR *.trc.
      How many trace files were created when you used Module 4?
   b. Delete the trace files with the DOS command ERASE *.trc.
   c. How many free bytes are left on the Sample Files directory?
9. Using wp2.kbs from your Sample Files directory, delete all the ASK and CHOICES commands at the bottom of the program.
   a. Run a consultation with the new wp2.kbs. What happens?
   b. Insert the statement AUTOQUERY at the top of the program.
   c. What happens when you run the AUTOQUERY version of wp2.kbs?
10. Rewrite invest .kbs to handle user input of "I don't know."
11. Rewrite the original wp2.kbs using forward chaining instead of backward chaining. The new program should give the same advice as the original wp2.kbs.
12. Using the forward chaining .kbs file from Assignment 11, obtain a printout of a Text tree and view its Graphics tree.

Module 5

Using Databases and Spreadsheets with the Knowledge Base

Business and scientific communities over the past decade have stored tremendous amounts of information in database files and worksheet files. These files may be on mainframes or microcomputers. Acquiring the knowledge and expertise for an expert system, however, is a difficult process. Specialists, called knowledge engineers, are responsible for extracting such information from experts and then translating it into usable expert systems. One of the exclusive features of VP-Expert is its interface to selected database and spreadsheet (or worksheet) files. The information contained in these files is the backbone of most enterprises; its incorporation into expert systems could save both the knowledge engineer and the company time and money.

VP-Expert provides an interface to data in existing database files that were created with dBASE III, dBASE III PLUS, or dBASE IV. The worksheet linkage is to spreadsheets created with Lotus 1-2-3. The readme.vpx system file contains the most recent dBASE and Lotus 1-2-3 version numbers that are accessible through VP-Expert. Hence, if your enterprise has any applications using these tools, the existing files can become the front end for an expert systems application. This ability is a tremendous time-saver for the knowledge engineer. Another advantage is that the data has been used repeatedly and so is (probably) valid and its procedures verified.

We will start in this module with database access and finish with worksheet access.

Using a Database

We will look at how to retrieve records from a database file by using a **loop structure** called WHILEKNOWN-END. We will use the MENU command to help us generate a list of choices for a user to select. We will explore ways to search for specific records and a method to restrict our records before searching. A procedure for letting the user repeat the searching process will also be explained.

Displaying a Record

Let's tackle the problem of **displaying** a database record through VP-Expert. Figure M5.1 shows a dBASE III PLUS file of five records.

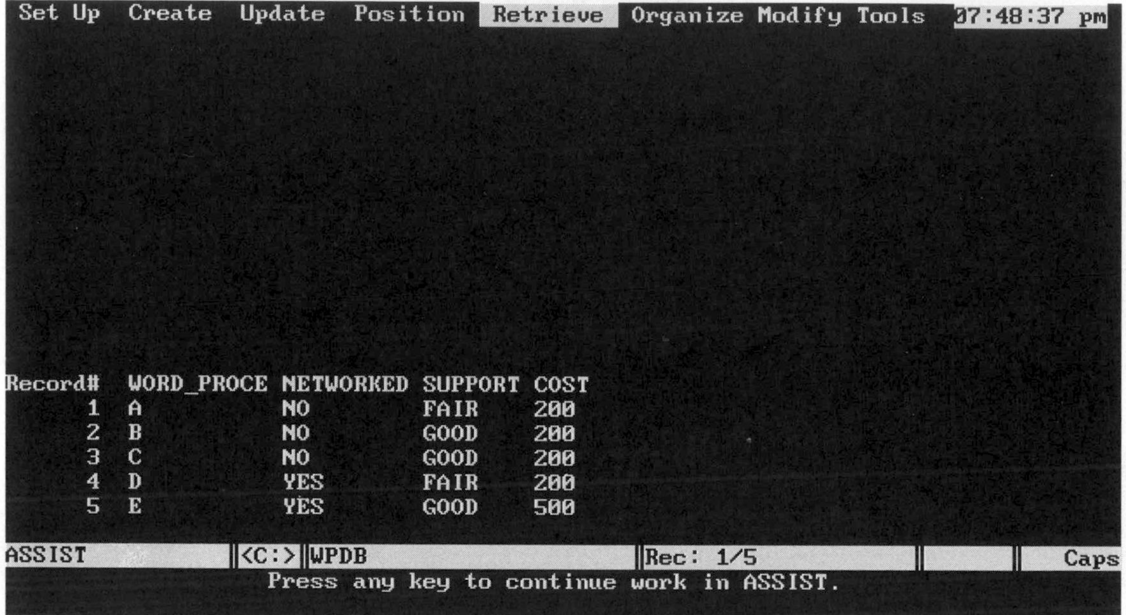

FIGURE M5.1 DATABASE FILE (wpdb.dbt) CREATED WITH dBASE III PLUS

The database file or table has four columns, entitled Word_proce, Networked, Support, and Cost. (Record number is not a column in the database; it is generated by the dBASE print facility.) The column headings are called field names or attributes. The field name of Word_proce uniquely identifies each record or row in the database. Each row or record is information about a word processor. For example, record 4 tells us that word processor D has a network version, support is rated as fair, and the price for one licensed copy is $200. From the nature of the data, you would conclude that all columns are character data except the Cost column, which is numeric.

To read or access this information, you must know the name of the database file, the exact name of the database fields, and the meaning of the information in the table. All accessible database files must have an extension of .dbt. The field names as shown in Figure M5.1 are not case-sensitive; they may be uppercase or lowercase.

We will now run a consultation that will use this database file with VP-Expert. Our example will ask us what kind of support is required for our word processing. The database file in Figure M5.1 shows that there are two values for Support, Good or Fair. If the user needs fair (or modest) support, then word processor A or D could be used. If good support is required, then word processors B, C, and E are possibilities. We will now use the database file and search for the needed information.

The database file from Figure M5.1, wpdb.dbt, is on your Sample File directory. The example expert system that reads this file, wp8.kbs, is also on the Sample File directory. The database file and the expert system file must be in the same directory (or on the same disk) when you are running a consultation. Get to the VP-Expert directory and follow these abbreviated steps to search the database file for word processors with good support:

Type vpx

7Path C:\VPX_samp

6FileName (wp8)

4Consult

2Go (select Good)

The first question from the consultation asks you to select your level of required support. Our selection is good. This screen should look like Figure M5.2.

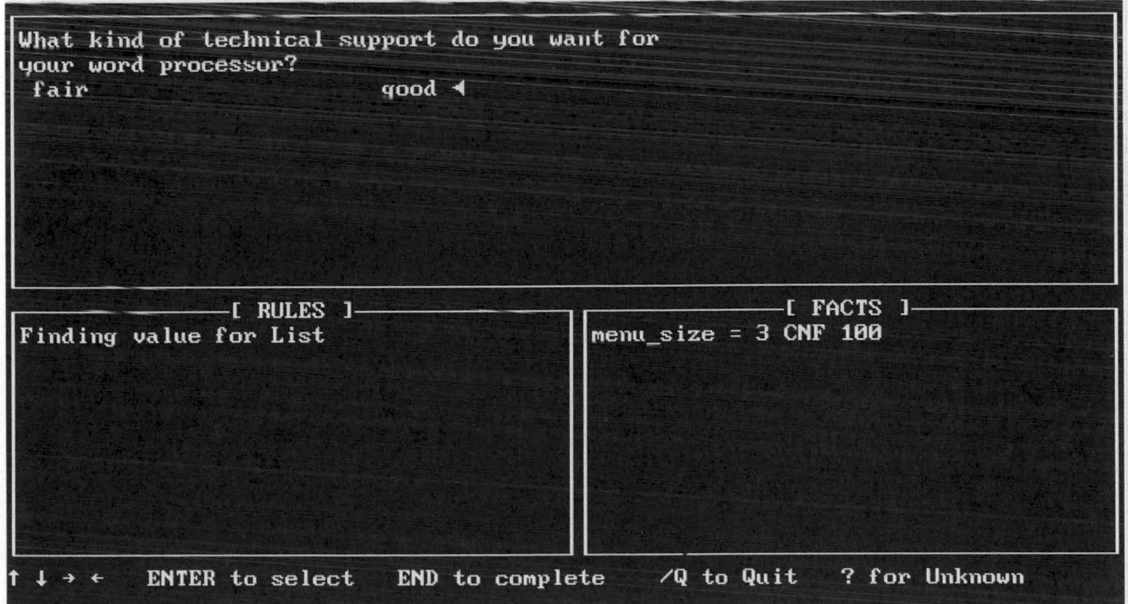

FIGURE M5.2 FIRST QUESTION FROM wp8.kbs

Somehow the expert system knows the possible values of Support (Good, Fair) listed in the database file. At the end of the consultation, your screen should look like Figure M5.3.

The expert system has replied that word processor B is rated as good for technical support, that the cost is $200, and that Networked is no.

There is a problem here. If you examine the database file from Figure M5.1, you will find two additional records, numbers 3 and 5, that satisfy the search. Both word processor C (record 3) and word processor E (record 5) are rated as good for support. If you examine Figure M5.3, you will see that the Rules window did not display any rules. The Facts window, however, does show the values of the field names for record 2:

record_num = 2

WORD_PROCE = B

NETWORKED = no

SUPPORT = good

COST = 200

```
What kind of technical support do you want for
your word processor?
 fair                       good ◄

Word processor B is rated as
good for technical support.

The cost is $200 and networked is no.

          Press the SPACEBAR to exit WPHELP.

--------------[ RULES ]--------------      --------------[ FACTS ]--------------
Finding value for List                     menu_size = 3 CNF 100
                                           List = good CNF 100
                                           record_num = 2 CNF 100
                                           WORD_PROCE = B CNF 100
                                           NETWORKED = no CNF 100
                                           SUPPORT = good CNF 100
                                           COST = 200 CNF 100
```

FIGURE M5.3 LAST CONSULTATION SCREEN FOR wp8.kbs

We will now examine the program file wp8.kbs to determine how the expert system created the first question from the database values and accessed one correct record. We will then have to correct that program to make it give all the answers available from the database file.

The immediate task is to obtain a printout of the program code. Make sure the printer is on, then follow these steps for printing:

7Edit
ALT-F10, End (To obtain a printout)
ALT-F8 (To exit without saving)
8Quit (Return to the Main Menu)

Your printout should be the same as Figure M5.4.

```
!The file wp8.kbs will access a record
!               from the database file called wpdb.dbt.
!               The field names in the database file are:
!               Word_proce, Networked, Support, Cost.

ACTIONS
                DISPLAY "This expert system WPHELP will display
information from a database file.

Press the Spacebar to start the consultation.~"
                CLS
     (C1)-----> MENU List, ALL, wpdb, Support
                FIND List
     (C2)-----> GET List=Support,wpdb,ALL
                DISPLAY "Word processor {Word_proce} is rated as"
                DISPLAY "{Support} for technical support."
                DISPLAY ""
                DISPLAY "The cost is ${Cost} and networked is
{Networked}

                Press the Spacebar to exit WPHELP.~"
                CLS;
RULE 1
                IF Support <> UNKNOWN
                THEN DISPLAY "The technical support for word
processor"
                     DISPLAY "{Word_processor} is stated as
{Support}."
                ELSE DISPLAY "We cannot find the information in the"
                     DISPLAY " database file.";

ASK List: "What kind of technical support do you want for your
word processor?";
```

FIGURE M5.4 PROGRAM CODE FOR wp8.kbs

Two new clauses, MENU (Comment 1) and GET (Comment 2), allow us to get to the database records. We will need to understand these in detail. The MENU instruction (along with an ASK) generates the results shown in Figure M5.2. The MENU instruction is a clause that has four distinct parts (parameters) after the keyword MENU. The first parameter identifies a VP-Expert variable; the second parameter tells us that we want to search all the records from the database file (the third parameter without its extension). The fourth parameter tells us what database column the menu choices will come from. MENU follows this format:

MENU

VP-Expert variable,

what part of the database is to be searched,

name of the database file,

what database column do we search and place as choices on a menu

In our example, List is the VP-Expert variable, ALL of the database records are to be searched, the name of the database file is wpdb, and the column values under the database field Support are to be made available to a VP-Expert-generated ASK question via the variable List. The MENU clause generates the *choices* from the database column. A FIND clause naming the same VP-Expert variable list activates the inference engine's search for a value of the variable. If no rules apply, the inference engine will search for an ASK statement for that variable. In other words, MENU is like *choices* with an ASK statement. VP-Expert stores the user's choice from the menu in the variable List.

Let's look at more examples of the MENU clause.

Example 1

The clause

MENU Price, ALL, dwpdb, Cost

with an ASK statement would generate a list of two choices, 200 and 500. When the user selects one, its value is stored in the variable called Price. Price is a VP-Expert variable, and Cost is a database variable. Don't use the same name for a VP-Expert variable and a database field name (variable). The clause

MENU Cost, ALL, wpdb, Cost

would cause unpredictable results because the VP-Expert variable and the database variable have the same name. Moreover, the physical name of the database file must be wpdb.dbf, but only the first part, wpdb, is used. The extension must exist, but it is excluded from the clause.

Example 2

The clause

MENU Package, ALL, wpdb, Word_proce

would generate the following choices for the user:

A B C
D E

When the user makes his or her selection for a word processor, its value will be stored in the VP-Expert variable Package. Note that the proper spelling of that field name is Word_proce because that is how it is known to the database. Of course, the proper question must be identified with the variable Package in an ASK statement.

Now let's take a look at another new clause, the GET instruction in Figure M5.4. The GET clause searches the database to retrieve the designated records. The clause in wp8.kbs

GET List = Support, wpdb, ALL

tells VP-Expert to find a database record in wpdb.dbf where List = Support and to make all the fields of that record available to VP-Expert. In Figure M5.4, note that the DISPLAY commands reflect both VP-Expert values and database values. In the Facts window of Figure M5.3, we saw that the search found all values of fields for record 2, but that VP-Expert did not locate all of the proper records.

The GET clause finds the *first* record that meets the condition List = Support. We have to force VP-Expert to continue or repeat its search through the database. In the next section, we will examine a WHILEKNOWN instruction that will solve the problem.

Using Loops to Display Records

A repetitive or **loop structure** will solve the problem of retrieving all records with *Support = good* from the database. The GET clause retrieves only the next record in the database that meets the condition. The List = Support parameter does not force it to repeat the search after it has found a record. Envision a pointer or marker at the top of the database file. When a GET clause is executed, the imaginary marker or pointer moves to the next record in the database that meets the condition. The pointer stays at this record until another GET clause is encountered or until it is pointing to the *end of file*. We need a mechanism to force repetitive use of the GET clause. One answer is the WHILEKNOWN-END structure.

Let's run an example that will display all the correct records from the database. Because the Consultation window is small, we have inserted commands in our new example to send DISPLAY statements to the printer as well as the screen. Make sure the printer is on and use wp9.kbs to retrieve and display all the records. Both wp9.kbs and wpdb.dbf are in the Sample File directory. If you have access to a printer, use the VP-Expert editor to insert the PRINTON and PRINTOFF commands in wp9.kbs, as shown in Figure M5.11 on page 305 (C1). At the Main Menu, follow these steps to run our example with printer output (and with the Trace option in effect):

6FileName (wp9)

4Consult

    6Set

        2Trace

        6Quit

    2Go (select Good)

(Remain at the Consult Menu.)

After pressing the Spacebar to start the consultation, select Good as the level of needed support. Figure M5.5 shows the Consultation screen. The Consultation window scrolled through all the responses even though only two now show on the screen.

```
    has good support
    costs $200
    networked = no

Word processor E
    has good support
    costs $500
    networked = yes

   Press the SPACEBAR to exit WPHELP.

-[ RULES ]-                              -[ FACTS ]-
Finding value for Screen                 Screen = Yes CNF 100
Testing 1                                record_num = 5 CNF 100
RULE 1 IF                                WORD_PROCE = E CNF 100
Support <> UNKNOWN                       NETWORKED = yes CNF 100
THEN                                     SUPPORT = good CNF 100
Screen = Yes CNF 100                     COST = 500 CNF 100
ELSE Screen = No CNF 100                 Screen = Yes CNF 100
Finding value for Support                Screen = No CNF 100
```

FIGURE M5.5 LAST CONSULTATION SCREEN FOR wp9.kbs (WITH SUPPORT = GOOD)

The Rules window reveals that it was searching for Screen and testing Rule 1. The Facts window scrolled through many messages. The ending messages reveal record_num = 5 and the values of the fields for record 5. The last message reveals that Screen = No.

The results of the consultation are better surveyed from the printout, as shown in Figure M5.6, than from the consultation itself.

Word processor B
has good support
costs $200
networked = no

Word processor C
has good support
costs $200
networked = no

Word processor E
has good support
costs $500
networked = yes

FIGURE M5.6 PRINTER OUTPUT FROM wp9.kbs (WITH SUPPORT = GOOD)

Three word processors, B, C, and E, have good support ratings. If you scan the database file in Figure M5.1, you will conclude that the output is correct. Record numbers 2, 3, and 5 represent the correct search. Before we examine the Text tree, let's check the value of the variables after the consultation with wp9.kbs. At the 2Go of the Consult Menu, select 4Variable. Your screen should now resemble Figure M5.7.

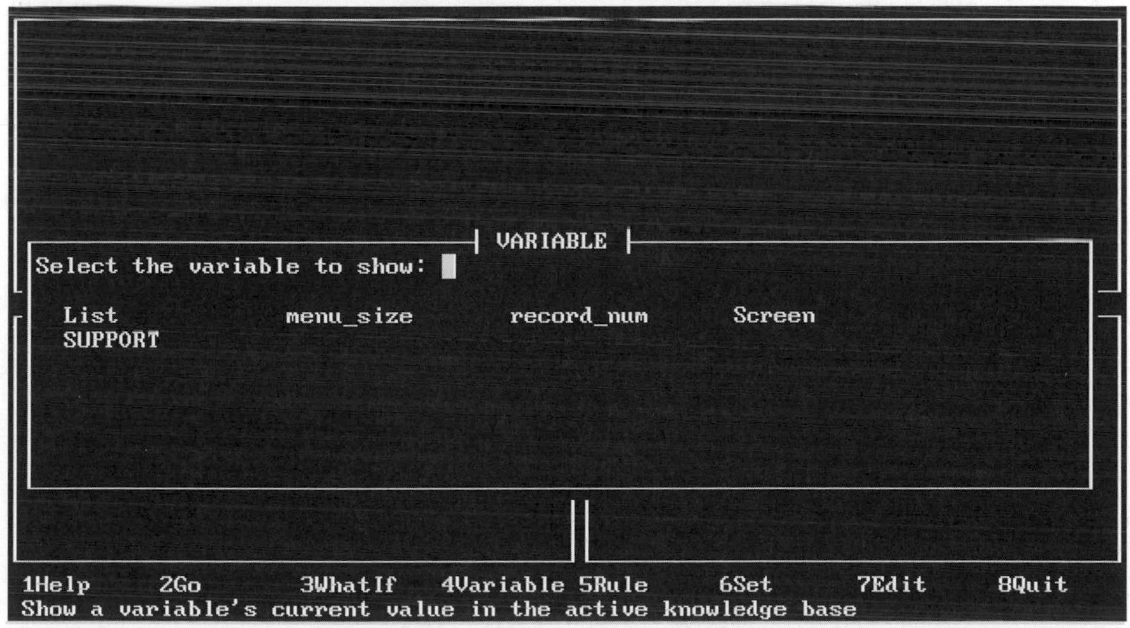

FIGURE M5.7 VARIABLE OPTION AFTER CONSULTATION WITH wp9.kbs (WITH SUPPORT = GOOD)

Note that the database variables (field names) are in capital letters and the VP-Expert variables are in lowercase. Check the value of each variable by moving the cursor to the desired variable and pressing Enter. The value of the selected variable will appear on the screen. To see another variable, select the 4Variable option again.

Your efforts should reveal some interesting characteristics of the values of the variables. You should have found the following:

List = Good

menu_size = 3

NETWORKED = undefined

SUPPORT = undefined

COST = undefined

record_num = 5

WORD_PROCE = undefined

Screen = no

Another word for undefined is UNKNOWN; in other words, there is no known value assigned to this variable at the end of the consultation. List has the value Good because we selected that value. Obviously, VP-Expert has an error because the number of menu options (Good, Fair) is 2, not 3. Record num = 5 means that the last record found by the search was number 5. (These values will help us explain how the WHILEKNOWN-END loop structure works when we analyze the program code.)

Let's examine the Text tree (trace) of the wp9.kbs consultation run with Support = Good. Follow these steps to obtain a printout of the associated Text tree file:

8Quit

5Tree

2Text

F10, End (For a printout)

ALT, F6, y (To save and exit)

4Quit

(End at the Main Menu.)

The resulting printout of the Text tree should match Figure M5.8.

Note on line 4 of the Text tree that a variable List has value Good. Lines 6 to 15 indicate that Rule 1 is being repeatedly tested for a value for Screen. In three tests the value for Screen is found to be Yes. In the last test of Rule 1, Screen has the value no (Line 18). Line 17 reports a variable Support (database field name) and indicates no value (undefined or UNKNOWN).

```
                                                Editing: Old File c:\vpx_samp\wp
Line
 1  Testing c:\vpx_samp\wp9.kbs◄
 2  (= yes CNF 0 )  #1wpdb.dbf◄
 3  !   List◄
 4  !   !   (= good CNF 100 )◄
 5  #1wpdb.dbf◄
 6  !   Screen◄
 7  !   !   Testing 1◄
 8  !   !   (= Yes CNF 100 )◄
 9  !   Screen◄
10  !   !   Testing 1◄
11  !   !   (= Yes CNF 100 )◄
12  !   Screen◄
13  !   !   Testing 1◄
14  !   !   (= Yes CNF 100 )◄
15  !   Screen◄
16  !   !   Testing 1◄
17  !   !   !   Support◄
18  !   !   (= No CNF 100 )◄

+   ▲   ▲   ▲   ▲   ▲   ▲   ▲   ▲   ▲   ▲   ▲   ▲   ▲   ▲   ▲
Insert On    Document Off                          Boldface Off Underline Off
 1       2       3Srch↓  4        5Rplc↓  6        7GbRpc↓ 8        9Rept↓ 10
```

FIGURE M5.8 TEXT TREE FOR wp9.kbs (WITH SUPPORT = GOOD)

Let's repeat this exercise but select Fair as our level of support. Repeat the abbreviated steps we outlined earlier. If you examine Figure M5.1, you can conclude that the results should be word processors A and D because they are the only records with Support = Fair. The last Consultation screen for Support = Fair is shown as Figure M5.9. The printer output is the same as the Consultation screen.

When you obtain your printout of the Text tree (trace), it should look like Figure M5.10.

Lines 4 and 5 indicate that the VP-Expert variable List has the value Fair. The next lines indicate a search for the variable Screen and subsequent testing of Rule 1. Rule 1 is tested twice and concludes that Screen = Yes. The last time the same rule is tested, the result is Screen = No, but the consultation ends. Line 15 lists Support but with no value (undefined or UNKNOWN). We have somehow corrected our problem: the program is reading the database file, and a repetitive process is displaying the correct answers (Products A and D) to the condition Support = Fair. Let's examine the program code to find out how this repetitive structure works.

Obtain a printout of the program code for wp9.kbs by following these abbreviated steps starting from the Main Menu with the printer on:

```
    has fair support
    costs $200
    networked = no

Word processor D
    has fair support
    costs $200
    networked = yes

   Press the SPACEBAR to exit WPHELP.

[ RULES ]
Finding value for Screen
Testing 1
RULE 1 IF
Support <> UNKNOWN
THEN
Screen = Yes CNF 100
ELSE Screen = No CNF 100
Finding value for Support

[ FACTS ]
Screen = Yes CNF 100
record_num = 4 CNF 100
WORD_PROCE = D CNF 100
NETWORKED = yes CNF 100
SUPPORT = fair CNF 100
COST = 200 CNF 100
Screen = Yes CNF 100
Screen = No CNF 100
```

FIGURE M5.9 LAST CONSULTATION SCREEN FOR wp9.kbs (WITH SUPPORT=FAIR)

```
                                        Editing: Old File c:\vpx_samp\wp

Testing c:\vpx_samp\wp9.kbs◄
(= yes CNF 0 )◄
#1wpdb.dbf◄
!   List◄
!   !   (= fair CNF 100 )◄
#1wpdb.dbf◄
!   Screen◄
!   !   Testing 1◄
!   !   (= Yes CNF 100 )◄
!   Screen◄
!   !   Testing 1◄
!   !   (= Yes CNF 100 )◄
!   Screen◄
!   !   Testing 1◄
!   !   !   Support◄
!   !   (= No CNF 100 )◄

Insert On    Document Off                  Boldface Off Underline Off
 1       2       3Srch↓  4       5Rplc↓  6       7GbRpc↓ 8       9Rept↓ 10
```

FIGURE M5.10 TEXT TREE FOR wp9.kbs (WITH SUPPORT = FAIR)

3Edit

F10, End (For a printout)

ALT F8, y (To exit without saving)

Your printout of wp9.kbs is shown in Figure M5.11. Let's take a detailed look at this program.

```
!The file wp9.kbs will access information
!                  from a database file called wpdb.dbf.
!                  The field names in the database file are:
!                  Word_proce, Networked, Support, Cost.
!                  A WHILEKNOWN-END loop will be used.
!                  Print classes will be used.

ACTIONS
                   DISPLAY "WPHELP will send information from a
database file to the screen and printer.
Make sure that your printer is on.

Press the Spacebar to start the consultation.~"
          CLS
          PRINTON  <-- (C1)
 (C2) --> MENU List, ALL, wpdb, Support
          FIND List
        [ WHILEKNOWN Support
        [      Get List = Support, wpdb, ALL
   (C3) {      RESET Screen
        [      FIND Screen
        [ END
          PRINTOFF  <-- (C1)
          DISPLAY " "
          DISPLAY "Press the Spacebar to exit WPHELP.~"
          CLS;

RULE 1
          IF SUPPORT <> UNKNOWN
          THEN Screen = Yes
               DISPLAY " "
               DISPLAY "Word processor {Word_proce}"
               DISPLAY "      has {support} support"
               DISPLAY "      costs ${Cost}"
               DISPLAY "      networked = {Networked}
          ELSE Screen = No;

(C2) ASK List: "What kind of technical support do you want for your
word processor?";
```

FIGURE M5.11 PROGRAM CODE FOR wp9.kbs

The printout during the consultation was caused by two commands, PRINTON and PRINTOFF (Comment 1 or C1). After the PRINTON command is executed, every DISPLAY statement will be sent to both the screen and the printer. Because the ASK statement is not a DISPLAY statement, the question "What kind of technical support do you want for your word processor?" was not sent to the printer. Every PRINTON command should be followed (somewhere) by a PRINTOFF.

After you pressed the Spacebar to start the consultation, the MENU clause with the ASK statement (Comment 2) generated the question with the two choices for Support: Fair and Good. In other words, the size of the menu is 2. These values came from the database column of Support values. When the user selected a value, VP-Expert stored the value (Fair or Good) in the variable called List. The GET command then initiated the inference engine's search to find a database record whose support value is the same as the user's selection.

The looping for the correct records is performed with the WHILEKNOWN-END structure. (Comment 3 on Figure M5.11). Whenever VP-Expert encounters a variable name, its value or contents is either known or UNKNOWN (undefined). If the variable is known, then it has a value and there is no reason to start a search (FIND) for that variable. If the variable has no value assigned to it, VP-Expert gives it the value UNKNOWN. When you tell the system to FIND a variable with value UNKNOWN, the inference engine takes over to conduct the search. The goal is to change the value of the variable from UNKNOWN to a known value.

This concept of ***known or unknown*** in a WHILEKNOWN-END structure is used to cause a set of instructions to repeat. The start of the loop is marked with the key word WHILEKNOWN followed by a variable. In our case, WHILEKNOWN is followed by Support, a database field name. The end of the WHILEKNOWN loop is designated by END. All the instructions between the top and bottom may or may not be repeated; the decision depends on the value of the variable (Support) named after WHILEKNOWN. When the system encounters a WHILEKNOWN loop, it will execute all instructions between the WHILEKNOWN command and the END one time. At that point (the END), VP-Expert will test for the value of the variable that appears after the WHILEKNOWN key word.

If that variable has a value, the instructions inside the loop will be executed again. If the value of the variable is UNKNOWN, then VP-Expert will not repeat the instructions, but will continue with the next instruction after the END. If there is no instruction after the END, the consultation is over. In summary, the loop is executed once regardless of the value of the WHILEKNOWN variable. The loop test is made at the end of the loop to determine if repetition is to occur.

In our case, the test variable is Support. The system executes a GET clause to locate a record in the database wpdb.dbt that has List = Support. List is where the user's MENU choice is stored. The possible values for List are Good or Fair. Two more instructions are executed before the loop test, in this case, the RESET and FIND clauses. The RESET instruction changes the value of a variable to UNKNOWN, regardless of it having or not having an assigned value. When the FIND command encounters a variable name with a value of UNKNOWN, control passes to the inference engine to find the value. At this point, the inference engine searches the Rule base. Rule 1 has a conclusion involving the variable Screen, so that rule is tested.

Let's examine Rule 1. The condition on this rule is expressed as

```
IF Support <> UNKNOWN
```

This reads as "If the database variable Support is not UNKNOWN." It is equivalent to the condition, "If Support is known." Because VP-Expert does not recognize the key word or system value known, the condition must be expressed with is *not UNKNOWN* or <> *UNKNOWN*. In essence, as long as we can obtain a value for support from the database file, the rule will fire the THEN part and print out the field values. If the Support variable becomes UNKNOWN, then the ELSE part of the rule fires and no display to the screen is made. Because the database variable Support has a known value (either Fair or Good from the database), the rule fires and the subsequent DISPLAY statements transfer the information to the screen. Control passes back to the loop and the END clause.

The test to loop or not to loop is now made. The system looks at the value for Support. Because this value is known from the database, the test of WHILEKNOWN Support is true, so the GET command moves to the next record in the database. The Screen value is changed to UNKNOWN by the RESET command, which tests Rule 1 again. The rule fires, and the new record information is sent to the screen. Back in the loop, the Support variable is examined. If it has a value (Good or Fair), then the process is repeated. At some point, a GET clause moves from the last (specified) record and can't find any more. The GET clause will not go back to the top of the database file. When the GET is past the last record or at the bottom of the file, the system assigns UNKNOWN to the values of the variables in the database, as illustrated in Figure M5.12.

List = Support?

List = Good

	Record #	Word_Proce	Networked	Support	Cost
Get	1	A	no	Fair	200
	2	B	no	Good	200
Get	3	C	no	Good	200
	4	D	no	Fair	200
Get	5	E	yes	Good	500
Get	6	unknown	unknown	unknown	unknown

(List = Good)
Get List = Support, wpdb, ALL

FIGURE M5.12 MOVING THROUGH THE DATABASE WITH GET INSIDE A WHILEKNOWN LOOP (SUPPORT = GOOD)

The VP-Expert variable List has the value Good, and the GET statement moves the pointer to record 2, then record 3, and then record 5. The next GET moves the pointer to the end of file; at this point, VP-Expert assigns UNKNOWN to Support. Hence, the loop test fails and no more repetitive actions of the loop are allowed. Note that the value of List has not changed; it is still assigned the value Good.

In summary, the loop is repeated four times. The first three times cause Rule 1 to fire the THEN part because Support is known; hence, the three records are displayed on the screen. The last GET clause goes past the end of the database, and Support is now UNKNOWN; hence, the ELSE part of the rule fires and no printing occurs. The RESET command activated the search because the variable Screen was reset to UNKNOWN to activate a rule search.

If the user had selected Fair from the menu choices shown on the Consultation screen, then the WHILEKNOWN Support loop would have been executed three times, as illustrated in Figure M5.13.

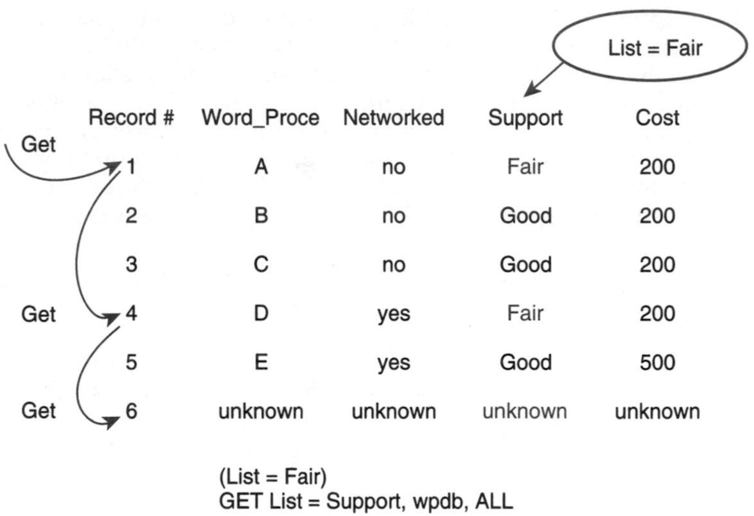

FIGURE M5.13 MOVING THROUGH THE DATABASE WITH GET INSIDE A WHILEKNOWN LOOP (SUPPORT = FAIR)

Rule 1 would be true the first two times it was tested, but would be false the third time. Hence, two records would be printed out, and no record would be printed of the last test.

In summary, we have solved our problem. When the user identifies either Good or Fair for support, we can get VP-Expert to access the database to display the desired records. To do this, we use several new instructions:

MENU	Creates a menu of choices from a column of the database.
GET	Gets a record from the database that matches what we want.
RESET	Changes the value of a variable to UNKNOWN.
WHILEKNOWN-END	Tests at the bottom of the loop to determine if the loop is to be repeated.

More examples will help, so let's run another one using databases. The file wp10.kbs will send two questions to the screen with two MENU clauses. The first menu will let you select a rating for Support, and the second a value for Cost. Because there are two values for each variable, there are four possible combinations of runs. We will select Fair and 200. If you survey Figure M5.1, you can see that two records (1 and 4) meet this condition (Support = Fair and Cost = 200). From the Main Menu, load this file, set the trace, run the consultation, and examine the consultation results, Text tree, and program code. The abbreviated steps are as follows:

6FileName (wp10)

4Consult

    6Set

        2Trace

        6Quit

    2Go (Fair, 200)

    7Edit

        F10, End (for a printout)

        ALT-F8, y

    8Quit

5Tree

    2Text

        F10, End (For a printout)

        ALT-F8, y

    4Quit

    (End at the Main Menu)

As you run your consultation with wp10, your screen should look like Figure M5.14 when both menus are displayed. (You have selected Fair and the second question appears.)

Note that the Rules window is finding List and Budget. In the Facts window, List = Fair. The selection for Budget (200) has not been made yet. Interestingly, the inference engine reports that menu_size = 3 for both questions. Obviously, it is off by one.

Your final consultation screen with wp10 will look like Figure M5.15.

The screen shows that two word processors, A and D, have fair support at a $200 cost. Notice also that record 4 and its values are displayed in the Facts window. The printer

```
What kind of technical support do you want for
your word processor?
 fair ◄                      good

At what cost?
 200 ◄                       500

─────────[ RULES ]──────────          ─────────[ FACTS ]──────────
Finding value for List                menu_size = 3 CNF 100
Finding value for Budget              List = fair CNF 100
                                      menu_size = 3 CNF 100

↑ ↓ → ←   ENTER to select   END to complete   /Q to Quit   ? for Unknown
```

FIGURE M5.14 TWO MENU-GENERATED QUESTIONS FROM CONSULTATION WITH wp10.kbs (WITH SUPPORT = FAIR)

```
    has fair support
    costs $200
    networked = no

Word processor D
    has fair support
    costs $200
    networked = yes

Press the SPACEBAR to exit WPHELP.

─────────[ RULES ]──────────          ─────────[ FACTS ]──────────
Finding value for Screen              Screen = Yes CNF 100
Testing 1                             record_num = 4 CNF 100
RULE 1 IF                             WORD_PROCE = D CNF 100
Support <> UNKNOWN                    NETWORKED = yes CNF 100
THEN                                  SUPPORT = fair CNF 100
Screen = Yes CNF 100                  COST = 200 CNF 100
ELSE Screen = No CNF 100              Screen = Yes CNF 100
Finding value for Support             Screen = No CNF 100
```

FIGURE M5.15 LAST CONSULTATION SCREEN FOR wp10.kbs (WITH SUPPORT = FAIR AND COST = 200)

output in Figure M5.16 reports the same results. This output was activated by setting PRINTON in the program code (and PRINTOFF at the end of the consultation).

Word processor A
has fair support
costs $200
networked = no

Word processor D
has fair support
costs $200
networked = yes

FIGURE M5.16 PRINTER OUTPUT FROM wp10.kbs

The Text tree (trace) of the consultation with wp10 in Figure M5.17 reveals the activity of the program code. Lines 3 through 6 indicate that List = Fair and Budget = 200. The next lines show a search for the variable Screen. Two times Screen = Yes from Rule 1, but the third time, Screen = No. Support is shown as having no value (UNKNOWN) on Line 15. The program code is given in Figure M5.18.

```
                                              Editing: Old File c:\vpx_samp\wp
Line
1  Testing c:\vpx_samp\wp10.kbs◄
2  (= yes CNF 0 )  #1wpdb.dbf◄
3  !   List◄
4  !   !   (= fair CNF 100 )◄
5  #1wpdb.dbf◄
6  !   Budget◄
7  !   !   (= 200 CNF 100 )◄
8  #1wpdb.dbf◄
9  !   Screen◄
10 !   !   Testing 1◄
11 !   !   (= Yes CNF 100 )◄
12 !   Screen◄
13 !   !   Testing 1◄
14 !   !   (= Yes CNF 100 )◄
15 !   Screen◄
16 !   !   Testing 1◄
17 !   !   !   Support◄
18 !   !   (= No CNF 100 )◄

+   ▲   ▲   ▲   ▲   ▲   ▲   ▲   ▲   ▲   ▲   ▲   ▲   ▲   ▲   ▲
Insert On     Document Off                    Boldface Off Underline Off
 1        2       3Srch↓  4       5Rplc↓  6       7GbRpc↓ 8       9Rept↓ 10
```

FIGURE M5.17 TEXT TREE FOR wp10.kbs (WITH LIST = FAIR AND BUDGET = 200)

Let's highlight three points about this code. The two menus were created from the two MENU and two ASK clauses (Comment 1 or C1). Support and Cost are database names, whereas List and Budget are the VP-Expert variables where the values Fair and 200 are stored. (Hence, List and Budget are known values.) Next, examine the GET

```
!The file wp10.kbs will access information
!                   from a database file called wpdb.dbf.
!                   The field names in the database file are:
!                   Word_proce, Networked, Support, Cost.
!                   Two menus will be generated.
!                   Print statements are used.

ACTIONS
                    DISPLAY "This expert system WPHELP will
display information from a database file.

Press the Spacebar to start the consultation.~"
            CLS
            PRINTON
      (C1)  MENU List, ALL, wpdb, Support
            FIND List
            MENU Budget, ALL, wpdb, Cost
            FIND Budget

            WHILEKNOWN Support   <-(C3)
                    GET List=Support AND Budget=Cost,wpdb,ALL (C2)
                    RESET Screen
                    FIND Screen
            END
            PRINTOFF
            DISPLAY " "
            DISPLAY "Press the Spacebar to exit WPHELP.~"
            CLS;
                                          (C3)
RULE 1
            IF Support <> UNKNOWN   <-
            THEN Screen = Yes
                 DISPLAY " "
                 DISPLAY "Word processor {Word_proce}"
                 DISPLAY "has {support} support"
                 DISPLAY "costs ${Cost}"
                 DISPLAY "networked = {Networked}"
            ELSE Screen = No;

ASK List:  "What kind of technical support do you want for your
word processor?";
(C1)
ASK Budget: "At what cost?";
```

FIGURE M5.18 PROGRAM CODE FOR wp10.kbs

clause from Comment 2. The two questions imposed the two restrictions, as reflected by the AND statement in the GET command.

List = Support
Budget = Cost

List has the value Fair and Budget has the value 200. The variables on the left of the equal sign are VP-Expert names; the ones on the right are database names. A relational operator (=) is between them. This is known as a **database rule** in VP-Expert. Database rules in the GET clause search the database for the next record that meets the database rule. Be careful to observe the format of

VP-Expert-variable   relational operator   field name

Your results may be unreliable if you do not use this format and if you confuse field names and VP-Expert names. Our final comment on wp10 concerns the use of UNKNOWN on Comment 3. The phrases WHILEKNOWN Support, and IF Support <> UNKNOWN, are both used as control techniques. The first phrase uses Support as the test variable to determine if the loop will be repeated. The second phrase is used as the condition to Rule 1. Note that UNKNOWN pertains to a database value, not a VP-Expert variable. Cost would have worked as the control variable as well as Support in this example. In the next section, we will restrict the user's access to a subset of the database.

Before continuing with the next section, we should point out other loops besides WHILEKNOWN: WHILETRUE and FOR. The syntax of the WHILETRUE loop is

```
WHILETRUE  condition THEN
                :
                :
                :
END
```

The condition must evaluate to TRUE before the loop is executed. Thus, if the condition is FALSE, the commands between WHILETRUE and END will be skipped. This loop is unlike the WHILEKNOWN loop, which will always execute once even if the variable is known.

Another looping structure in VP-Expert is the FOR loop. The syntax for this structure is

```
FOR variable = start_value TO stop_value
                :
                :
                :
END
```

This loop condition only works with numeric values. When the FOR clause is executed, the start_value (number) is automatically assigned to the variable name after the FOR command. With each iteration of the loop, the variable is automatically increased by 1; thus, the loop continues to repeat until the variable is greater than the stop_value. For example, the following loop displays the numbers 1 to 10:

```
FOR count = 1 to 10
    DISPLAY "The number is {count}"
END
```

If the numbers 1 and 10 are stored in variables Start and Stop, respectively, these values can be indirectly specified by preceding the variable name with the @symbol. Thus, the loop above could be rewritten as

```
Start = 1
Stop = 10
FOR count = Start to Stop
   DISPLAY "The number is {count}"
END
```

In VP-Expert it is possible (though not advisable) to increment or decrement the loop variable within the loop using direct assignment (count = (count + 1)). The loop above could be rewritten with a WHILETRUE structure as shown below:

```
count = 1
WHILETRUE count <= 10 THEN
   DISPLAY "The number is {count}"
   count = (count + 1)
END
```

This same loop could be rewritten (rather laboriously) with additional variables as shown below:

```
count = 1
WHILEKNOWN  flag
   DISPLAY "The number is {count}"
   count = (count + 1)
   FIND flag
END
      :
      :
      :
RULE flag
IF flag > 10
   THEN RESET flag
ELSE
   flag = ON;
```

Remember that the WHILEKNOWN loop executes the first time, regardless of whether the variable is known or UNKNOWN.

In general, the FOR loop is used when the count or number of repeats is known or when the count can be obtained from the user. The WHILETRUE loop must have the loop condition changed; otherwise, an infinite loop occurs. The WHILEKNOWN loop is easy to use while "reading" information from a database, spreadsheet, or other file.

We will now turn our attention to restricting database input. This concept of restricting users to certain data subsets or views is a common and needed procedure in the real world.

Restricting the Database Records

Expert systems deal with specific domains. For example, in a medical application, we may have a large database of medical information, but our application is an expert system for cancer treatment and diagnosis. We need a way to restrict the database so that the application's user sees only information about cancer records. In the business area, we may have a large database of census information, but our expert system application for marketing needs information only on customers under 25 years of age. Restricting the database to a narrower domain is useful in expert system applications. VP-Expert can achieve this by forcing **conditions**, or **restrictions**, in the MENU and GET clauses of the program code.

Let's modify our WPHELP system again. We will restrict our original database to only those records where the Support rating is Good. To our application, the database appears as only three records, B, C, and E, as shown in Figure M5.19. When we search the database by Cost values, records concerning A and D will be excluded. To see the results, run a consultation with wp11.kbs and examine the printer output, a Text tree, and the program code. When the price is to be selected, enter 200. The results should be records B and C because they are the only records in the restricted database that have a cost of $200. If you need the instructions, refer to the previous abbreviated steps. The printer output from wp11.kbs is shown in Figure M5.20. Only two records, B and C, have a $200 cost in our restricted database. The Text tree trace shown in Figure M5.21 reveals records restricted to Good (Line 3), that Budget = 200 (Lines 4 and 5), and that Screen was evaluated twice as Yes and once as No. The Cost variable has no value (Line 14).

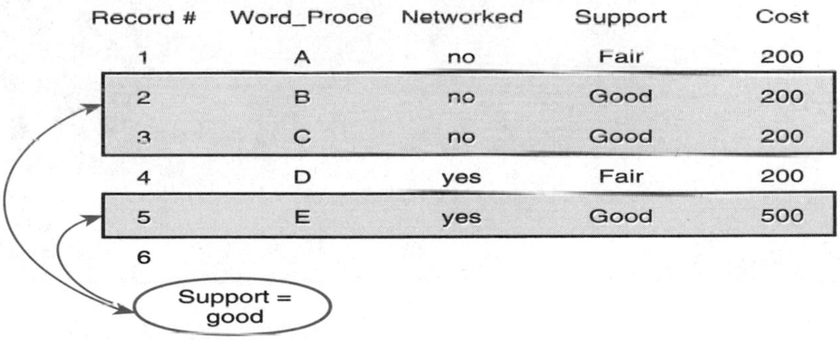

Record #	Word_Proce	Networked	Support	Cost
1	A	no	Fair	200
2	B	no	Good	200
3	C	no	Good	200
4	D	yes	Fair	200
5	E	yes	Good	500
6				

Support = good

FIGURE M5.19 RESTRICTING THE DATABASE RECORDS

Figure M5.22 gives the program code for this consultation. The value Good has been assigned to a VP-Expert variable Kbsupport (Comment 1). In both the MENU (C2) and GET (C3) clauses, we use Kbsupport = Support to tell VP-Expert to select the menu choices and to get only those records with Support = Good. We must use this method to fit the database rule format (variable relational_operator field name) for the parameter. You would probably call it a trick, but this method does restrict the database records as described. In the WHILEKNOWN-END construct, note that the

Word processor B
has good support
costs $200
networked = no

Word processor C
has good support
costs $200
networked = no

FIGURE M5.20 PRINTER OUTPUT FROM CONSULTATION WITH wp11.kbs (WITH RECORDS RESTRICTED TO SUPPORT = GOOD)

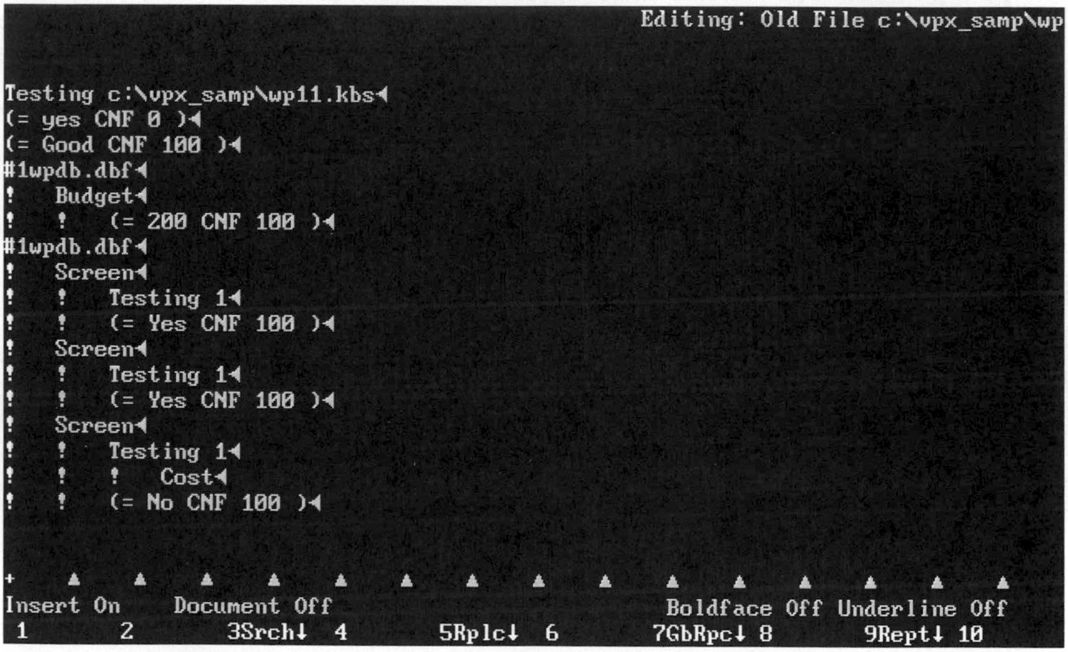

```
                                          Editing: Old File c:\vpx_samp\wp
Line
Testing c:\vpx_samp\wp11.kbs◄
(= yes CNF 0 )◄
(= Good CNF 100 )◄
#1wpdb.dbf◄
!   Budget◄
!   !   (= 200 CNF 100 )◄
#1wpdb.dbf◄
!   Screen◄
!   !   Testing 1◄
!   !   (= Yes CNF 100 )◄
!   Screen◄
!   !   Testing 1◄
!   !   (= Yes CNF 100 )◄
!   Screen◄
!   !   Testing 1◄
!   !   !   Cost◄
!   !   (= No CNF 100 )◄

+   ▲   ▲   ▲   ▲   ▲   ▲   ▲   ▲   ▲   ▲   ▲   ▲   ▲   ▲   ▲
Insert On    Document Off                 Boldface Off Underline Off
 1       2       3Srch↓  4       5Rplc↓  6       7GbRpc↓ 8       9Rept↓ 10
```

FIGURE M5.21 TEXT TREE FOR wp11.kbs (WITH RECORDS RESTRICTED TO SUPPORT = GOOD)

Cost field name is the control variable. This application, however, has only one ASK because there is only one MENU clause. Our next example will let you browse through all the database records.

Repeated Displaying of Database Records

Now we will let the user see any records that he or she wants to see. To do this, we need to display a list of the word processors and let the user select the record for

```
!The file wp11.kbs will access and restrict information
!         from a database file called wpdb.dbf.
!         The field names in the database file are:
!         Word_proce, Networked, Support, Cost.

ACTIONS
          DISPLAY "This expert system WPHELP will display
information from a restricted database file.
Turn on the printer.

Press the Spacebar to start the consultation.~"
          CLS
  (C1)--> Kbsupport = Good
          MENU Budget, Kbsupport = Support, wpdb, Cost <--(C2)
          FIND Budget
          PRINTON

          WHILEKNOWN Cost
               GET Budget = Cost AND Kbsupport = Support, wpdb,
     (C3)-->   ALL
               RESET Screen
               FIND Screen
          END
          PRINTOFF
          DISPLAY " "
          DISPLAY "Press the Spacebar to exit WPHELP.~"
          CLS;

RULE 1
          IF Cost <> UNKNOWN
                    THEN Screen = Yes
                    DISPLAY " "
                    DISPLAY "Word processor {Word_proce}"
                    DISPLAY "has {support} support"
                    DISPLAY "costs ${Cost}"
                    DISPLAY "networked = {Networked}"
          ELSE Screen = No;

ASK Budget: "What cost for a word processor is in your budget?";
```

FIGURE M5.22 PROGRAM CODE FOR wp11.kbs

display. This time, however, we will restrict the information going to the screen, but not the number of records. The user controls the length of the search. Program wp12.kbs will do this. Now run the consultation and obtain a printout of the Text tree and the program code. When you run the consultation, select C the first time, then A, then E, and then enter a ? to end the consultation. Refer to the abbreviated steps if you need to refresh your memory.

The results of this consultation did not go to the printer. The Text tree in Figure M5.23 shows that the variable Word_processor was assigned the value C, then the value A, then E, and finally UNKNOWN.

The program code in Figure M5.24 shows how we achieved this interactive consultation with wp12.kbs.

```
                                        Editing: Old File c:\vpx_samp\wp

Testing c:\vpx_samp\wp12.kbs◄
(= yes CNF 0 )◄
#1wpdb.dbf◄
!   Word_processor◄
!   !   (= C CNF 100 )◄
#1wpdb.dbf◄
#1wpdb.dbf◄
!   Word_processor◄
!   !   (=  CNF 100 )◄
!   Go◄
!   !   Testing 1◄
!   !   (= Yes CNF 100 )◄
#1wpdb.dbf◄
#1wpdb.dbf◄
!   Word_processor◄
!   !   (= A CNF 100 )◄
!   Go◄
!   !   Testing 1◄

+   ▲   ▲   ▲   ▲   ▲   ▲   ▲   ▲   ▲   ▲   ▲   ▲   ▲   ▲   ▲
Insert On    Document Off                   Boldface Off Underline Off
 1      2        3Srch↓  4        5Rplc↓  6        7GbRpc↓ 8        9Rept↓ 10
```

FIGURE M5.23 TEXT TREE FOR wp12.kbs FOR SELECTING C, A, E, AND ?

We have five basic points to consider. The menu of five choices (A, B, C, D, and E) was generated by the MENU clause (Comment 1). We used the ALL parameter for the scope of the choices from the column. The control variable for the WHILEKNOWN loop is Go (Comment 2), a VP-Expert variable, not a field name from the database. Hence, for the loop to stop repeating, its value must become UNKNOWN.

Rule 1, however, has a condition involving another variable, Word_processor and UNKNOWN. The GET clause (Comment 3) shows us that we will retrieve the word processor record the user identifies. The last parameter, Support, indicates that we are only interested in this field, not all fields. After the word processor record is found and displayed, the variable is reset. The DISPLAY statement (Comment 4) tells the user to enter a ? over a menu choice to quit the session. The ? is translated as an UNKNOWN. If the variable were not reset, then only one record would be found in the database.

A new instruction, MRESET (Comment 5), is used after a MENU clause to release the internal memory required for the menu choices. If the menus are large, memory problems could occur; we release the memory for the menu to avoid this problem.

So far, we have seen how to access records in a database with VP-Expert. We can search for any combination of conditions or we can restrict our search to a subset of the database. New instructions WHILEKNOWN, GET, MENU, RESET, and MRESET

```
!The file wp12.kbs will access information
!         from a database file called wpdb.dbf.
!         The field names in the database file are:
!         Word_proce, Networked, Support, Cost.
!         The user will loop until he or she enters a ?.

ACTIONS

          DISPLAY "This expert system WPHELP will display
information from a database file.

Press the Spacebar to start the consultation.~"
  (C1)──► CLS
          MENU Word_processor, ALL, wpdb, Word_proce
        ─►WHILEKNOWN Go
  (C2)         FIND Word_processor
             ─►GET Word_processor = Word_proce, wpdb, Support
  (C3)         CLOSE wpdb
               DISPLAY "Word processor {Word_processor} has
{Support} support."

               DISPLAY " "
               DISPLAY "Press any key to continue."
               CLS
  (C4)         RESET Word_processor
             ─►DISPLAY "Type a question mark ? over a menu choice
     if you want to quit.
               "
               MENU Word_processor, All, wpdb, Word_proce
               RESET Go
  (C5)         FIND Go
             ─►MRESET Word_processor

     END
     DISPLAY "Press the Spacebar to exit WPHELP.~"
     CLS;
RULE 1
     IF Word_processor <> UNKNOWN
               THEN Go = Yes;

ASK Word_processor: "Select the name of the word processor for
which you want support information.";
```

FIGURE M5.24 PROGRAM CODE FOR wp12.kbs

help us to accomplish this. Once the database information is accessed, it can be used in the knowledge base.

Changing Records in a Database

Accessing database information is important in an application, but **changing** or **updating** records is vital to the integrity (accuracy) of the database information. VP-Expert allows you to change the contents of a database. Program wp13.kbs will allow you to change the Cost field in the database. This is an appropriate modification in our word processing application because prices change quickly. We will use a copy of our database file (see Figure M5.1) called wpdbl.dbl because our program will change the values in this database through a command called PUT.

From the Main Menu, select wp13.kbs and run the change cost program. Select word processor D and change its cost to 500. As you run the consultation, the update screen will appear as in Figure M5.25, showing the old cost of 200 with a new cost of 500.

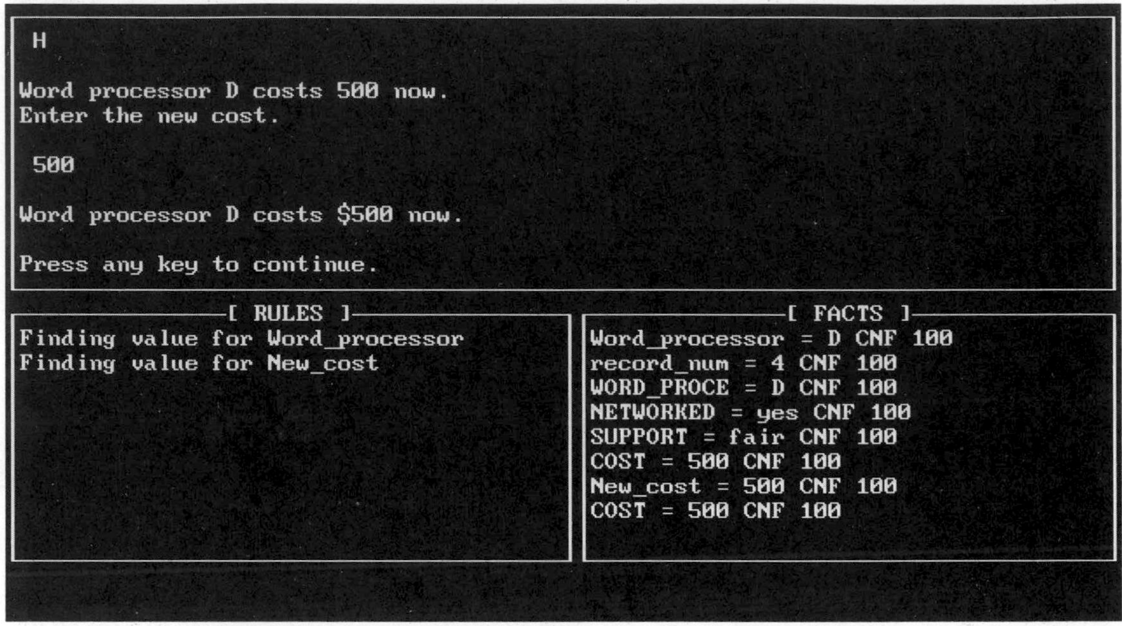

FIGURE M5.25 SAMPLE CONSULTATION SCREEN FOR CHANGING THE COST OF A WORD PROCESSOR (wp13.kbs)

The user supplies the new cost information, and the program writes the new cost to the proper record (in our case, word processor D). The user can then choose to change another price or exit the consultation. This program code is displayed in Figure M5.26.

The first task is to find the correct record, then obtain the new price and store it in the database. From Comment 1 on Figure M5.26, we use the MENU clause to let the user select the record and then retrieve that record. From Comment 2, the FIND clause causes the ASK New_cost statement to be displayed. The reply is stored in the variable New_cost. To transfer the contents of New_cost to the variable Cost, we have to enclose New_cost in parentheses (indirect addressing). If we omitted the parentheses, the Cost would equal the words, New_cost. The command that transfers the new value of Cost to the selected database record is the PUT clause (Comment 3). The format of this instruction is PUT database file. Notice that we had to close the database (Comment 3) so that when we repeat the search for another record, the search would start at the top of the file instead of on the record for word processor D.

```
!The file wp13.kbs will change or update
!         the Cost field in wpdb1.dbt.
!         The field names in the database file are:
!         Word_proce, Networked, Support, Cost.
!         The user will loop until he or she enters a ?.

ACTIONS
          DISPLAY "This expert system WPHELP will let you change
or update the cost of a word processor.

Press the Spacebar to start the consultation.~"
          CLS
   (C1)──► MENU Word_processor, ALL, wpdb1, Word_proce
          WHILEKNOWN Go
               FIND Word_processor
   (C2)        GET Word_processor = Word_proce, wpdb1, ALL
               RESET New_cost
            ──►FIND New_cost
               COST = (New_cost)
               DISPLAY "Word processor {Word_proce} costs ${Cost}
now."

               DISPLAY " "
   (C3)─────► PUT wpdbl
        ─────► CLOSE wpdbl
               DISPLAY "Press any key to continue.~"
               CLS
               RESET Word_processor
               DISPLAY "Type a question mark ? over a menu choice
if you want to quit.
               "
               MENU Word_processor, ALL, wpdbl, Word_proce
               RESET Go
               FIND Go
               MRESET Word_processor
          END
          DISPLAY "Press the Spacebar to exit WPHELP."
          CLS;
RULE 1
          IF Word_processor <> UNKNOWN
               THEN Go = Yes;

ASK Word_processor: "Select the name of the word processor whose
cost has changed.";

ASK New_cost: "Word processor {Word_proce} costs {Cost} now.
Enter the new cost.
";
```

FIGURE M5.26 PROGRAM CODE FOR UPDATE PROGRAM (wp13.kbs)

Another way to update a record is to use the SETPOS clause. This instruction allows you to tell the database what "record number" you want to mark for a subsequent GET instruction. The format for SETPOS is

SETPOS database_file_name, record_number.

A subsequent GET command would retrieve the record from the database at the specified record number, which is a valid number between 1 and the maximum value of the existing database file. This clause can be useful if the user knows the record number in advance. To use this command, the reader must know that RECORD_NUM is a system variable in VP-Expert. While a database consultation is running, the "system" is keeping track of the last database record number retrieved in the system variable RECORD_NUM. If you assign a number to this value, for example, RECORD_NUM = 23, the assignment does not affect the position of the pointer to the last database record retrieved (read, updated, etc.) until a subsequent GET clause is executed. The SETPOS clause can have a value of RECORD_NUM, but the correct syntax is

```
SETPOS wpdb1, @RECORD_NUM
```

because the value is a system variable. Correct syntax for incrementing this system variable is

```
RECORD_NUM = (RECORD_NUM + 1)
```

because the numeric variable is contained in a mathematical operation. Four system variables of interest (SETPOS, MENU_SIZE, TIME, and DATE) will be discussed later.

Adding Records to a Database

New products come out all the time, so we need a mechanism in VP-Expert that allows us to add new records continually to the database. The APPEND command does this. From the Main Menu, select file wp14.kbs. When you run a consultation with this example, add a new record with the following information:

Word processor	: H
Cost	: 599
Support	: Good
Networked	: Yes

The database is wpdb1.dbf, a copy of wpdb.dbf. The data entry screen will appear as shown in Figure M5.27. It is important to verify the information before it is written to the database. You may also add more records if you like.

The program code for this example is illustrated in Figure M5.28. To accomplish our addition, we had to find the new information from the user and then **add** or **append** it to the database. Comment 1 on Figure M5.28 shows a series of FIND clauses that activate the inference engine to send the appropriate questions to the screen. By embedding Update in the conclusion of Rule 1 and More in the condition of Rule 2, the FIND Reply clause will cause us to verify the screen information before firing the rule that appends (C2) the new record information to the database. The RESET ALL command of Comment 3 is needed so that the loop and FIND commands will be reactivated.

The important point here is that database information can be incorporated in an expert system. The database names can be used in the rule base. The results of a

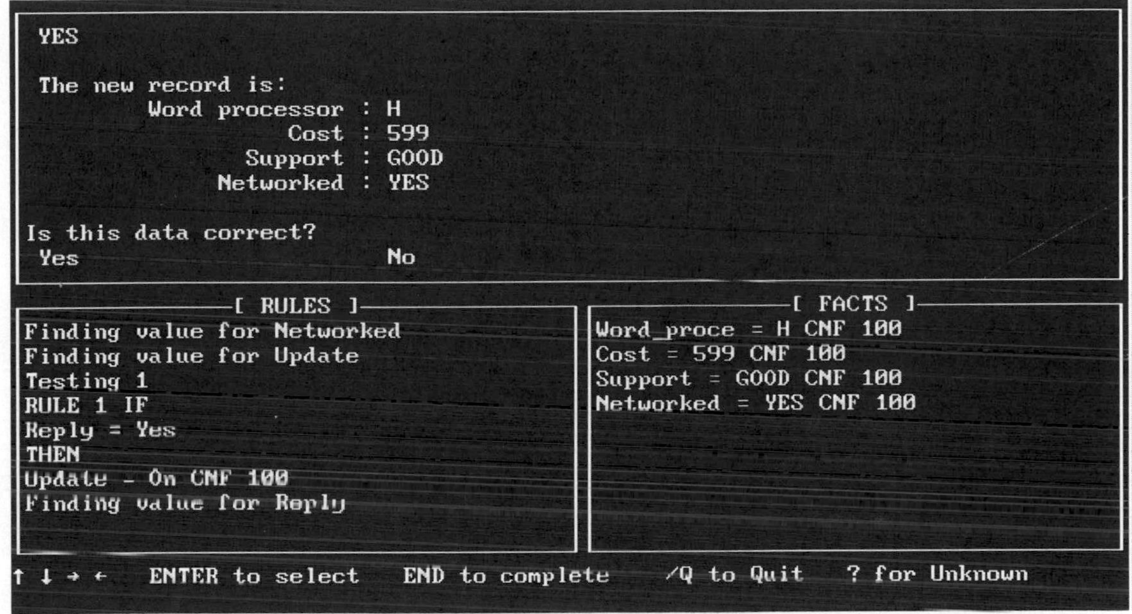

FIGURE M5.27 CONSULTATION SCREEN FOR ADDING A DATABASE RECORD (wp14.kbs)

consultation can then be written back to the database. For example, as the market grows toward increased networking, we could incorporate budget or maintenance estimates in a rule. For example,

IF Support = Good AND Network = Yes
THEN Anew_Maintenance = (0.3 * Cost)

Networked environments carry a 30% annual overhead for maintenance. The database information can be part of the expert system.

Thus, we have seen how VP-Expert can be the front end to a database application. We can access information with the MENU and GET commands, repeat our actions with the WHILEKNOWN structure, update records with the PUT clause, and add new records with the APPEND command. Let's now turn our attention to using worksheets (electronic spreadsheets) for the front end to our expert system.

Using a Spreadsheet

Just as a database can hold large amounts of information, so can **worksheets**, or **spreadsheets**, which represent another existing base of valuable information for potential expert systems. VP-Expert allows you to use existing spreadsheet data as the

```
!The file wp14.kbs will let you add a new
!         word processing record to the database wpdbl.
!         The field names in the database file are:
!         Word_proce, Networked, Support, Cost.
!         The user will loop until he/she enters a ?.

ACTIONS
          DISPLAY "This expert system WPHELP will let you add a
new word processing record to the database wpdbl.dbf.

Press the Spacebar to start the consultation.~"
          CLS
          WHILEKNOWN More
     (C3)----> RESET ALL
               DISPLAY "Please enter the new record."
               FIND Word_proce ]
               FIND Cost       ]  <---(C1)
               FIND Support    ]
               FIND Networked  ]
               DISPLAY "The new record is:
          Word_processor : {Word_proce}
                    Cost : {Cost}
                 Support : {Support}
               Networked : {Networked}
               "
               FIND Update
               FIND More
          END
          DISPLAY "Press the Spacebar to exit WPHELP.~"
          CLS;
RULE 1
          IF Reply = Yes
               THEN Update = On
                    APPEND wpdbl1 <-------------------(C2)
                    DISPLAY "The record has been added."
                    DISPLAY " ";
RULE 2
          IF Go = Yes
               THEN More = Yes;

ASK Word_proce: "Name of the word processor?";

ASK Cost: "Cost?";

ASK Support: "Support? Enter good or fair.";

ASK Networked: "Networked? Enter yes or no.";

ASK Reply: "Is this data correct?";
CHOICES Reply: Yes, No;

ASK Go: "Do you want to continue?";
CHOICES Go: Yes, No;
```

FIGURE M5.28 PROGRAM CODE FOR ADDING A DATABASE RECORD (wp14.kbs)

front end to an expert system application. Files created with Lotus 1-2-3 or Symphony can be used in VP-Expert applications. You can access the data, manipulate it in the knowledge base, and write it back to the spreadsheet. Files created with Lotus 1-2-3 Versions 2.0 and 2.01 can be accessed but not written to at this time (except with CCall).

Overview

Consider some background issues before using a spreadsheet in a VP-Expert application. First, the spreadsheets cannot have blank cells. VP-Expert will skip over blanks, causing the data to be aligned incorrectly. When data is being written to a spreadsheet, data types must match. If a numeric field is written to a text cell, the results will be treated as labels; hence, no calculations can be made. For the reverse situation, writing text to a numeric field will cause the result to be treated as a zero. Structural changes, moreover, cannot be made to the spreadsheet. Only cells that contain data can receive data. In other words, you can not write to new cell locations. If the data to be written exceeds the existing cell length, the data will be truncated. Despite these limitations, the ability to use certain spreadsheet files as a front end to an expert system is useful.

Reading from a Spreadsheet

We will use our familiar data on word processing and access it from a spreadsheet. The data shown in Figure M5.29 is stored in a Lotus 1-2-3 Version 1A file. The cell locations are marked by the usual column/row notation. For example, word processor A is located in cell A2, and the cost of product A is in cell D2. The column headings, WORD_P, NETWORK, SUPPORT, and COST, are in cells A1, B1, C1, and D1, respectively. Spreadsheet labels are case-sensitive, so COST and Cost are not the

```
A7:                                                                    READY

        A          B          C          D          E          F          G          H
1   WORD_P     NETWORK    SUPPORT    COST
2   A          NO         FAIR            200
3   B          NO         GOOD            200
4   C          YES        GOOD            200
5   D          YES        FAIR            200
6   E          YES        GOOD            500
7
8
9
10
11
12
13
14
15
16
17
18
19
20
13-Sep-94   08:14 PM                                           NUM CAPS
```

FIGURE M5.29 WORD PROCESSING INFORMATION IN A WORKSHEET (wpws.wks)

same headings. We have chosen to capitalize the headings and the product names. All other cells are lowercase. The COST cells are, of course, numeric. The spreadsheet itself is called wpws.wks.

As usual, we will use an example from our Sample File directory to show how to access worksheets or spreadsheets from within VP-Expert. We will use file wp15.kbs to read the worksheet wpws.wks, which are both in the Sample File directory. If you have access to a printer, get to the Main Menu of VP-Expert and follow the familiar steps to run a consultation. (If you do not have a printer, delete the PRINTON and PRINTOFF commands in wp15.kbs.)

7Path (C:\VPX_samp:)

6FileName (wp15)

4Consult

  2Go

  (Remain in the Consult Menu)

The consultation will pull the data from the spreadsheet wpws.wks, store it in memory, and display it on the screen. Figure M5.30 shows the results of your consultation with wp15.kbs.

```
Please wait. The worksheet is being loaded.

1 A no fair 200
2 B no good 200
3 C yes good 200
4 D yes fair 200
5 E yes good 500

    Press the SPACEBAR to exit WPHELP

[ RULES ]
More = yes CNF 100
Finding value for More
Testing 1
RULE 1 IF
Cword_p[X] <> UNKNOWN
THEN
More = yes CNF 100
Finding value for Cword_p[X]

[ FACTS ]
More    yes CNF 100
X    3 CNF 100
More    yes CNF 100
X    4 CNF 100
More    yes CNF 100
X    5 CNF 100
More    yes CNF 100
X    6 CNF 100
```

FIGURE M5.30 LISTING OF WORKSHEET FILE (wpws.wks) BY PROGRAM wp15.kbs

The four columns of information from the spreadsheet are displayed on the screen along with a numbering of rows (1, 2, 3, 4, and 5). How does VP-Expert accomplish this? The answer is the WKS, or read worksheet clause. This instruction has three parts (parameters) after the key word WKS. The first parameter identifies the VP-Expert variable that will receive the spreadsheet information. The second part tells VP-Expert the location of the desired data in the spreadsheet. The third part identifies the name of the spreadsheet to be accessed. Thus, to read the four columns of data, the following four clauses are used:

```
WKS Cword_p, A2..A6, wpws
WKS Cnetwork, B2..B6, wpws
WKS Csupport, C2..C6, wpws
WKS Ccost, D2..D6, wpws
```

The first part of each command is WKS, the second is the VP-Expert variable, the third is the spreadsheet range, and the last is the spreadsheet name without the wks extension. Note that we have named our VP-Expert variables by inserting an upper case C in front of the corresponding spreadsheet column name. For readability, we chose lowercase for the rest of the variable names. The four variables, Cword_p, Cnetwork, Csupport, and Ccost, must each contain five pieces of data. To identify each part, we place a number (1, 2, 3, 4, or 5) in square brackets after the variable name. For example, after the WKS clauses are executed,

```
Cword_p[1] = A
Cword_p[2] = B
Cword_p[3] = C
Cword_p[4] = D
Cword_p[5] = E
```

We have similar notations for the other variables. The information in row 2 is:

```
Cword[2]     = B
Cnetwork[2] = no
Csupport[2]  = Good
Ccost[2]      = 200
```

Variables with a name followed by a number in square brackets are known as arrays, dimensioned variables, or subscripted variables. When a variable name appears after the WKS keyword, the system assumes that these variables are arrays. We can examine these arrays because they are still in memory after a consultation is run.

You should still be in the Consult Menu, where your screen appears similar to that in Figure M5.31. (If you returned to the Main Menu, run the consultation again.) At the end of the consultation, we can examine the value of variables by using the 4Variable option from the Consult Menu. Select the 4Variable option. Your screen should appear as Figure M5.31.

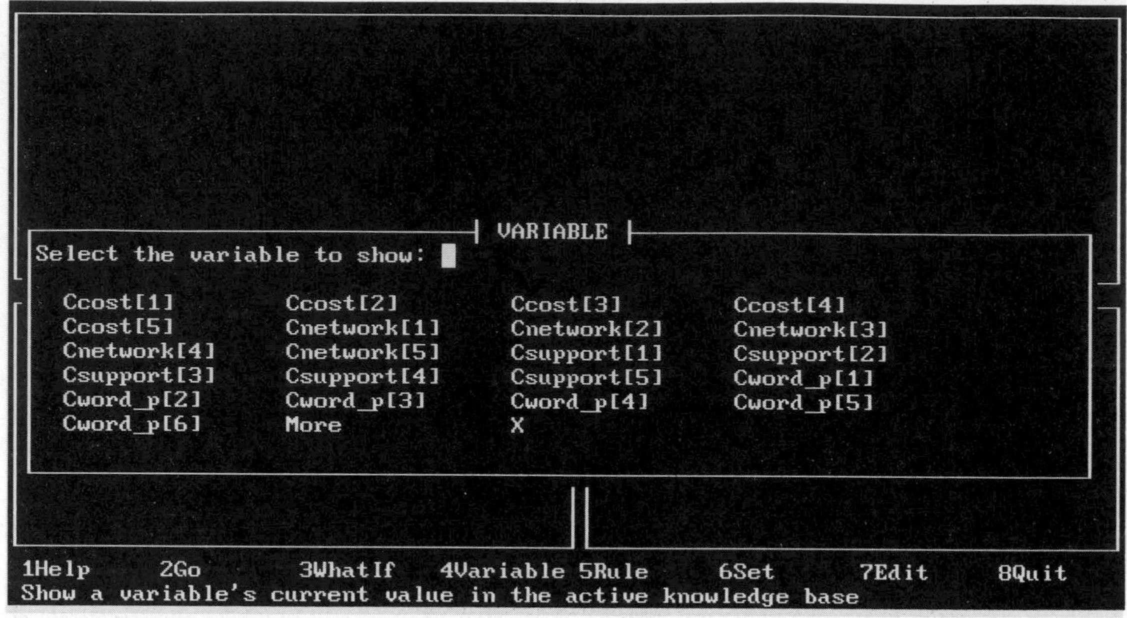

FIGURE M5.31 CHECKING THE VALUE OF VARIABLES AFTER A CONSULTATION WITH wp15.kbs

Select the variable, Cword_p[3], by pressing the arrow keys until the light bar is over this variable. Press the Enter key to verify that the value of

Cword_p[3] = C.

Check the following list of variables by selecting 4Variable, paging down, and selecting the variable:

Cword_p[5]	= E
Csupport[2]	= Good
x	= 6
More	= Undefined

The VP-Expert system will display these values through the 4Variable option of the Consult Menu.

Now let's read the code for wp15.kbs. Select the 7Edit option of the Consult Menu. Use the F10 and End keys to obtain a printout of wp15.kbs, as shown in Figure M5.32.

We have four specific comments to make about this program. From Comment 1 (C1) of Figure M5.32, we see the four WKS clauses that read the column values into arrays or subscripted variables. The ranges tell the computer where to find the data in the spreadsheet. Note in Comment 2 that the variable X is set to 1 so that we can display

```
!The file wp15.kbs will access information
!         from a worksheet file called wpws.wks.
!         The column headings in the worksheet are:
!         WORD_P, NETWORK, SUPPORT, COST
!         The range is a1 to d6.

ACTIONS
          DISPLAY "This expert system WPHELP will read all the
data from a worksheet file (wpws.wks), display it on the screen
and send it to the printer.  Turn on the printer.

Press the Spacebar to start the consultation.~"
          CLS
          DISPLAY " "
          DISPLAY "Please wait.  The worksheet is being loaded.
"
          PRINTON
      (C1)  WKS Cword_p, a2..a6, wpws
          WKS Cnetwork, b2..b6, wpws
          WKS Csupport, c2..c6, wpws
          WKS Ccost, d2..d6, wpws
      (C2)  X = 1
          WHILEKNOWN More
               DISPLAY "{X} {Cword_p[X]} {Cnetwork[X]} {Csupport[X]} {Ccost[X]}
      (C3)     X = (X + 1)
               RESET More
               FIND More
          END
          PRINTOFF
          DISPLAY " "
          DISPLAY "       Press the Spacebar to exit WPHELP.~"
          CLS;

RULE 1
      (C4)  IF Cword_p[X] <> UNKNOWN
               THEN More = yes;
```

FIGURE M5.32 PROGRAM CODE FOR READING A WORKSHEET FILE WITH wp15.kbs

the first elements in each of the worksheet rows. After this is displayed, we increase X by 1 (Comment 3). When the More variable is RESET followed by a FIND, then Rule 1 is fired (Comment 4). Because Cword_p[2] = B, the rule is fired and the variable More is assigned the value Yes. Because the END command is next, the system will test the WHILEKNOWN More condition. This is found to be true, so the loop is executed again. This continues as long as the next word processing record in the spreadsheet exists. When the fifth word processor has been read, the X variable is increased to 6. The value of Cword_p[6], however, is UNKNOWN so that the variable More is not changed. Hence, the WHILEKNOWN test fails. This is how VP-Expert can display the worksheet data from memory on the screen.

Let's move to another program example in which the data in the spreadsheet is converted to information on the display screen. In addition, this example will allow the user to select an additional, but optional, printer output of the spreadsheet values. The program wp16.kbs will use wpws.wks to format the data into information on the screen. Run a consultation with wp16.kbs from your Sample File directory to see this display. Be sure that the printer is on before running the consultation with wp16.kbs.

Select Yes to the question concerning an additional printout. At the end of the consultation, your screen and printer output should resemble Figure M5.33.

Note that this printout has a title of

WORD PROCESSING INFORMATION
from Worksheet wpws.wks

The reformatting has inserted phrases such as "has good support" and "costs $200." This optional printout is a plus to any application. Converting data to information is still another asset (even though the phrase "networked is no" is awkward).

Let's analyze the program code for wp16.kbs. Use the 7Edit option of the Consult Menu. Once in the editor, use the F10 and End keys to initiate a printout. Exit the editor with ALT-F8. The printout of wp16.kbs should be identical to Figure M5.34.

We have five comments to make about this code. First, four WKS clauses (C1) read the data from designated ranges into subscripted variables. We see from Comment 2 that a FIND Print clause initiates a search for a value of Reply. This fires the question to the user concerning a printout. If the Reply is Yes, then the PRINTON command is executed and DISPLAY statements in Rule 2 print the title. Within the WHILEKNOWN More loop (C4), DISPLAY statements reformat the variable data into information. The variable X is used to indicate which value of the subscripted variable is to be displayed. When the value of Cw[X] becomes UNKNOWN (X = 6), then the loop test fails and the consultation (C5) ends.

We have seen how the use of WKS clauses with subscripted variables allows access to spreadsheet data. The use of DISPLAY statements with proper formatting allows the transition of spreadsheet data into readable information. In the next section, we will discuss writing data back to the spreadsheet file.

Writing to a Spreadsheet

Now we will explore how VP-Expert writes new values back to the spreadsheet. We will use the PWKS, or put worksheet, clause to accomplish this task.

Once we have the spreadsheet information in computer memory, we can manipulate it according to any rules we put in the knowledge base. Many times, we will need to write changed values back to the spreadsheet. We use the PWKS clause to do this. Run wp17.kbs from your Sample File directory. This program allows you to change the values for the Cost column in the wpws1.wks file (a copy of wpws.wks). Run a consultation and change the cost of word processor B from $200 to $300. Enter the same values for the other products.

After running a consultation with wp17.kbs, obtain a printout of the program code. Your program listing should appear as Figure M5.35.

```
                    costs $200
                    networked is yes

            Row 5 : Word processor E has
                    good support
                    costs $500
                    networked is yes


        Press the SPACEBAR to exit WPHELP.
```

```
[ RULES ]
More = yes CNF 100
Finding value for More
Testing 1
RULE 1 IF
Cw[X] <> UNKNOWN
THEN
More = yes CNF 100
Finding value for Cw[X]
```

```
[ FACTS ]
More = yes CNF 100
X = 3 CNF 100
More = yes CNF 100
X = 4 CNF 100
More = yes CNF 100
X = 5 CNF 100
More = yes CNF 100
X = 6 CNF 100
```

WORD PROCESSING INFORMATION
from Worksheet wpws.wks

Row 1 : Word processor A has
fair support
costs $200
networked is no

Row 2 : Word processor B has
good support
costs $200
networked is no

Row 3 : Word processor C has
good support
costs $200
networked is yes

Row 4 : Word processor D has
fair support
costs $200
networked is yes

Row 5 : Word processor E has
good support
costs $500
networked is yes

FIGURE M5.33 OPTIONAL PRINTOUT FROM WORKSHEET (wp16.kbs)

```
!The file wp16.kbs will access data from a worksheet (wpws.wks).
!          Output of information is to the screen
!          with optional printer output.

ACTIONS
          DISPLAY "This expert system WPHELP will read all the
data from a worksheet file (wpws.wks) and display the rows as
INFORMATION on the screen and allow an optional printout.

Press the Spacebar to start the consultation.~"
          CLS
          DISPLAY " "
          DISPLAY "The data will take time to load.
          "
   (C2) → FIND Print
          WKS Cw, a2..a6, wpws
   (C1) → WKS Cn, b2..b6, wpws
          WKS Cs, c2..c6, wpws
          WKS Cc, d2..d6, wpws
          X = 1
          WHILEKNOWN More
               DISPLAY "        ROW {X} : Word processor {Cw[X]} has
                    {Cs[X]} support
                    costs ${Cc[X]}
                    networked is {Cn[X]}                    (C4)
                    "
                    X = (X + 1)
                    RESET More
                    FIND More
               END
               PRINTOFF
               DISPLAY " "
               DISPLAY "        Press the Spacebar to exit WPHELP.~"
               CLS;

RULE 1
   (C5) →      IF Cw[X] <> UNKNOWN
                    THEN More = Yes;

RULE 2                                                   (C3)
               IF Reply = Yes
                    THEN Print = Yes
                    PRINTON
                    DISPLAY " "
                    DISPLAY "     WORD PROCESSING INFORMATION"
                    DISPLAY "       from Worksheet wpws.wks"
                    DISPLAY " "
               ELSE Print = No;

ASK Reply: "In addition to the screen display,
do you want a printout of the worksheet?";

CHOICES Reply: Yes, No;
```

FIGURE M5.34 PROGRAM CODE FOR wp16.kbs

Let's make three general points about this code. First of all, although unnoticeable to the user, the WKS clauses had been changed (C1). Instead of ranges, we used the column headings that appeared in row 1 of the spreadsheet file. The syntax is

COLUMN = column_name.

```
!The file wp17.kbs will access data from a worksheet (wpws1.wks).
!         Changes in cost will be written back to
!         the worksheet file.
!

ACTIONS
          DISPLAY "This expert system WPHELP will read all the
data from a worksheet file (wpws1.wks) and write back changes for
the cost of a word processor.

Press the Spacebar to start the consultation.~"
          CLS
          DISPLAY " "
          DISPLAY "The data will take time to load.
"
        ┌ WKS Cw, COLUMN = WORD_P,   wpws1
        │ WKS Cn, COLUMN - NETWORK, wpws1
 (C1)──►│ WKS Cs, COLUMN = SUPPORT, wpws1
        └ WKS Cc, COLUMN = COST,     wpws1
          X = (1)
          WHILEKNOWN More
               DISPLAY "Word processor {Cw[X]} costs {Cc [X]}.
"
               RESET Current_cost
               FIND Current_cost
 (C2)────────► Cc[X] = (Current_cost)
               X = (X + 1)
               RESET More
               FIND More
          END
 (C3)───► PWKS Cc, COLUMN = COST, wpws1
          DISPLAY " "
          DISPLAY "Press the Spacebar to exit WPHELP.~"
          CLS;

Rule 1
          IF Cw[X] <> UNKNOWN
               THEN More = Yes;

ASK Current_cost: "Enter the displayed cost or a new one.";
```

FIGURE M5.35 PROGRAM CODE FOR wp17.kbs

If we have row names or range labels, we could use these with ROW = name and NAMED = range name. In Comment 2, we observe that the user's response to cost is stored in the VP-Expert variable Current_cost. Current_cost is then assigned to Cc[X] in memory. Parentheses are needed so that the contents of Current_cost is transferred instead of the words "Current_cost." The new information now contained in Cc[X] in memory is transferred to the spreadsheet with the PWKS clause (C3). This clause writes the information in the Cc variable to the Cost column in the spreadsheet wpws1.wks. You can think of PWKS as put to the worksheet.

There is an interesting problem here. If you observed the Facts window of the consultation, you observed new values for the X in Cc. What VP-Expert really does is append or add the new values to the column of old values. To place the new values where they belong, you have to open the spreadsheet and use spreadsheet commands. This is awkward, but it is the only way available now for writing to a spreadsheet. Hopefully, by the time you read this, this will have changed. Again, we make the point that once spreadsheet data is available to VP-Expert, it can be utilized in the knowledge base. The resulting information can be written back to the spreadsheet (albeit awkwardly). All the subscripted information can be used in rules. For example,

```
If Cs[x] = Good AND Cn[x] = yes
   THEN Budget = Cc[x]
      Maintenance = 0.3 * (Cc[x]);
```

Networked environments carry larger maintenance overheads. The spreadsheet information can be part of the rule base. The new results, Budget or Maintenance, can be written back to the spreadsheet (if they existed there to start with). It is possible, moreover, to change the format of a spreadsheet cell. The command for achieving this is

```
WFORMAT clause.
```

Three numeric data types can be specified with this clause, fixed decimal (FIXED), scientific format (SCI), and currency (CURR). Thus, the clause

```
WFORMAT CURR, 4
```

would change the Cc (cost) column to four decimals of the currency format. This feature can be quite useful when formatting reports from the spreadsheet.

We have seen how the WKS clause reads data from a spreadsheet using arrays or subscripted variables. The PWKS puts changed information back into the spreadsheet. The numeric format of cells can be changed with the WFORMAT clause. We have seen how VP-Expert can be the front end to both database and spreadsheet files. In the next module, we will take a closer look at confidence factors for expert systems in VP-Expert.

Key Points

- VP-Expert can access information from selected microcomputer database and spreadsheet files. The most recent products and version numbers are contained in readme.vpx in C:\vpx3ed.
- Data from database and spreadsheet files can be incorporated into rules in the knowledge base. Results of a consultation can be written back to existing fields or cells in the database or spreadsheet, respectively.
- The large amounts of data in existing database and spreadsheet files foster the development of expert system applications.
- The commands for database usage are MENU, GET, PUT, and RESET.
- The MENU clause generates a list of choices from a column in the database file to accompany a question in an ASK statement. The user's reply from the menu is stored in a variable.

- The GET command is used to transfer data from a database into variables in computer memory.
- The PUT clause is used to transfer data from variables in computer memory to database records.
- The MRESET clause is used to unallocate or release the memory required for the MENU clause.
- Database field names and VP-Expert variable names should be different.
- The WHILEKNOWN-END loop structure allows the instructions in the loop to be executed once. At the end of the loop, a test is made on the WHILEKNOWN variable. If the variable has a value, the loop is executed again. If the variable has no value (UNKNOWN), then the loop instructions are not repeated.
- The GET clause is placed inside a WHILEKNOWN-END loop to read or transfer data from the database record into memory.
- The RESET clause assigns a variable the value of UNKNOWN. When the FIND command comes from a variable that is UNKNOWN, the inference engine initiates a search for a known value for the variable.
- For spreadsheets, the commands are WKS, PWKS, and WFORMAT.
- The WKS command sends data from spreadsheet locations to arrays or subscripted variables in computer memory. Data may be transferred by columns, rows, or range names.
- Arrays or subscripted variables will have a numeric value or variable inside square brackets. For example, Cost[3] is a subscripted variable.
- All designated spreadsheet information is read into computer memory at one time. WHILEKNOWN-END loop is not needed as it is with database access.
- A WHILEKNOWN-END loop may be used with a subscripted variable to display or print information.
- Spreadsheet data is transferred from computer memory to spreadsheet locations by the PWKS clause.
- The WFORMAT clause may be used to change the format of an existing spreadsheet cell.
- The PRINTON clause sends the contents of any DISPLAY statements to the printer. The PRINTOFF clause cancels any additional printing of DISPLAY statements to the printer.

Exercises

1. Which of the following commands moves data from a database record to computer memory?
   a. PUT
   b. GET
   c. GWKS
   d. PWKS
   e. READ

2. Which of the following commands transfers data from computer memory to a database record?
   a. PUT
   b. GET
   c. GWKS
   d. PWKS
   e. WRITE
3. Using Figure M5.1, write the command to generate the following menus: (Use Reply as the VP-Expert variable.)
   a. A B C D E
   b. No Yes
   c. Fair Good
   d. 200 500
4. From Figure M5.1, what data is retrieved by the following? (Assume a WHILEKNOWN-END loop.)
   Reply = no
   a. Get Reply = Networked, wpdb, ALL
   b. Get Reply = Networked, wpdb, Cost
   c. Get Reply = Networked, wpdb, Networked
   d. Get Reply = Networked, wpdb, Word_proce
5. What data is retrieved by each of the following commands? (Assume a WHILEKNOWN-END loop and that All means the next record.)
   a. Get All, wpdb, ALL
   b. Get All, wpdb, Word_proce
   c. Get All, wpdb, Support
   d. Get All, wpdb, Cost
   e. Get All, wpdb, Networked
6. What data is retrieved by each of the following commands? (Assume a WHILEKNOWN-END loop.)
   Reply = Yes
   Answer = 200
   a. Get Reply = Networked AND Answer=Cost, wpdb, Word_Proce
   b. Get Reply = Networked OR Answer=Cost, wpdb, Word_proce
   c. Get Reply = Networked AND Answer=Cost, wpdb, Support
   d. Get Reply = Networked OR Answer=Cost, wpdb, Support
7. From Figure M5.1, what data is retrieved by the following commands? (Assume a WHILEKNOWN-END loop.)
   a. Reply = 200
      Get Reply = Cost, wpdb, Word_proce
   b. Reply = yes
      Get Reply = Networked, wpdb, Word_proce
   c. Reply = Fair
      Get Reply = Support, wpdb, Word_proce
   d. Reply = E
   e. Get Reply = Word_proce, wpdb, Word_proce

8. Identify each of the following as true (T) or false (F):
   a. The RESET clause removes any value(s) assigned to a named variable.
   b. The RESET command assigns the value UNKNOWN to the named variable.
   c. A RESET ALL is illegal in VP-Expert.
   d. The RESET clause cannot be used with spreadsheet or worksheet variables.
9. Identify each of the following as true (T) or false (F):
   a. A database variable and a VP-Expert variable can have the same name without affecting the consultation.
   b. If a GET clause does not find database records/fields, then all database record names are assigned the value UNKNOWN.
   c. The first parameter of the GET clause is either the keyword ALL or a database rule.
   d. A GET clause can only retrieve all fields or just one field in a database.
10. Identify each of the following as true (T) or false (F):
    a. The format of the WHILEKNOWN clause is WHILEKNOWN database-name.
    b. A WHILEKNOWN loop always executes once.
    c. The WHILEKNOWN loop ends with the STOP command.
    d. A RESET command does not affect the WHILEKNOWN clause.
11. Identify each of the following as true (T) or false (F):
    a. THE WKS clause reads spreadsheet data into VP-Expert arrays or dimensioned variables.
    b. The command WKS A, B8, C loads data from cell B8 of the A worksheet into the VP-Expert variable C.
    c. A range of cells cannot be included in the WKS clause.
    d. Worksheet data can only be accessed by rows in VP-Expert.
12. Identify each of the following as true (T) or false (F):
    a. The PWKS clause transfers data from VP-Expert variables to spreadsheets.
    b. The clause PWKS A,B,C will transfer the value of A to location B in spreadsheet C.
    c. Without the WFORMAT clause, data being transferred into a worksheet must be the same data type as data stored in corresponding worksheet cells.
    d. PWKS is the opposite of the WKS command.

Assignments

1. We need to clean up the Sample File directory after running Text tree traces. At the C:\VPX_samp prompt, use the following DOS commands:
   a. To delete the trace files, enter ERASE *.TRC.
   b. To verify that these trace files are gone, enter DIR *.TRC.
2. Using a printout of wp8.kbs:
   a. Circle the expert system variables. Draw a rectangle around the database variables.
   b. Repeat (a) for wp10.kbs.
   c. Repeat (a) for wp17.kbs.

3. Make a copy of wp13.kbs by using the following DOS command at the C:\VPX_samp subdirectory:
   copy wp13.kbs my13.kbs
   The new name of wp13.kbs is my13.kbs. In the editor, make these three changes:
   a. On the top line, delete the letters wp13.kbs and replace with my13.kbs.
   b. Delete the instruction CLOSE wpdbl with CTRL and Y.
   c. Run a consultation with my13.kbs. What are the results?
4. Make a copy (named my15.kbs) of wp15.kbs with the DOS copy command. Use the editor to change X = (X + 1) to X = X + 1.
   a. Will the program still run correctly?
   b. Run a consultation with my15.kbs and record the results.
5. Make a copy of wp11.kbs under the name my11.kbs. Using the editor, change the condition of Rule 1 to
   IF Support <> UNKNOWN
   a. Will the program still run correctly?
   b. Run a consultation with my11.kbs and record the results.
6. Using my11.kbs and the VP-Expert editor, change the condition of Rule 1 to
   IF Kbsupport <> UNKNOWN
   a. Will the program run correctly?
   b. Run a consultation and record the results.
7. Make a copy of wp10.kbs and call it my10.kbs. Add rules to the knowledge base that predict the annual maintenance cost to be 30% of the cost of networked word processors. Additional coding will be necessary.
8. Make a copy of wp16.kbs under the file name my16.kbs. Add rule(s) to the knowledge base that predict the projected annual savings to be the annual cost times
   a. 25% for "good" networked word processors
   b. 15% for "fair" networked word processors
   c. 8% for all others
   d. Additional coding will be necessary.
9. Use the VP-Expert Help Facility to obtain a printout on each database-related instruction:
   a. PUT
   b. MENU
   c. GET
   d. MRESET
10. Use the VP-Expert Help Facility to obtain a printout on each worksheet-related instruction:
    a. Subscripted variables
    b. WKS
    c. PWKS
    d. WFORMAT
11. Use the VP-Expert Help Facility to obtain a printout on each of the loop commands:
    a. FOR
    b. WHILEKNOWN
    c. WHILETRUE
12. In a database, to reorder a list of files, an "index" is made for a certain field. The

VP-Expert clause INDEX makes use of existing dBASE index files to fetch records in less time.

a. Use the Help Facility to obtain a printout of INDEX.
b. Explain the syntax of INDEX and how it is used in a .kbs program.

13. Copy wp9.kbs as loop1.kbs and again as loop2.kbs.
    a. Rewrite loop1.kbs using a FOR loop. (Assume that the number of records is 4.)
    b. Rewrite loop2.kbs using a WHILETRUE loop.

Module 6

Using Confidence Factors

One of the features of an expert system is its ability to deal with uncertainty. From our discussion of the MYCIN expert system in Chapter 6, we saw that one of the useful techniques for dealing with uncertainty is the use of certainty factors. Certainty factors are a measure of belief (or disbelief) or confidence in an assertion.

Like many other expert system shells, VP-Expert uses certainty factors. VP-Expert uses the term confidence factor (CNF) *to refer to certainty factors.*

In this module, we will discuss how confidence factors are assigned to both the conditions and the conclusions of a rule. We will also discuss how threshold values are used to determine if a rule is to be fixed and how the results of a consultation can be sorted using confidence factors.

Confidence Factors for Rules with AND Conditions

The user enters confidence factors for rule conditions during a consultation with a knowledge base. The CNF is assigned at the time the user chooses a value for a variable; the knowledge base requests the confidence factor in an ASK statement. The selected value of the variable and its assigned confidence factor are then displayed in the Facts window during the consultation.

To review an activity from Module 4, the user can enter a CNF when making a choice presented by an ASK statement by pressing the HOME key before making a selection. After the HOME key is pressed, a highlighted area will appear on the screen; then the user can enter an integer from 0 to 100.

The number entered is a measure of the level (percentage) of confidence the user has in that selection. A value of 0 indicates absolutely no confidence; a value of 100 indicates complete confidence. If a value less than 40 is entered, VP-Expert assumes a lack of confidence, and indicates this fact.

The default value for a CNF is 100. If a selection is made with no CNF entered, then the selection is assigned a CNF of 100.

After the CNF has been entered, pressing the ENTER key binds the indicated CNF to the chosen selection. If the user makes an error in entering a CNF, he or she can repeat the process and the new CNF value will replace the incorrect one. For example, suppose that we have selected Yes with a CNF of 80 in response to a query from the consultation. We then realize that we should have selected No. We can change the selection and, in doing so, the CNF of 80 will be deleted from the Yes selection.

In expert systems development, a rule will often have more than one condition and the conditions will be connected with the conjunction AND. (We have previously discussed this situation in Chapter 6.) When VP-Expert encounters multiple conditions connected by AND, it first checks to see that all of the conditions are satisfied. If so, VP-Expert assigns a CNF to the set conditions by using the minimum of the confidence factors assigned to the conditions. For example, if we have the following rule:

IF condition 1 CNF1
AND condition 2 CNF2
AND condition 3 CNF3
THEN conclusion;

the CNF for the set of conditions would be

CNF = minimum (CNF1, CNF2, CNF3)

This technique seems intuitively reasonable because we expect the confidence level to be no higher than the smallest CNF.

After VP-Expert has determined the CNF for the condition, it is then concerned with the CNF of the rule as a whole. To do this, VP-Expert must inspect the CNF of the conclusion. The expert system developer enters the confidence factor of the conclusion when the knowledge base is compiled. Determining the CNF for the conclusion is the task of either the system developer or the human expert. As with the conditions' confidence factors, nonspecification of a CNF causes VP-Expert to assign a CNF of 100 to the conclusion.

To calculate the CNF for the entire rule, VP-Expert uses the product of the CNF for the condition and the CNF for the conclusion. That is,

CNF(rule) = CNF(condition)CNF(conclusion)

From this example, suppose that CNF1 = 70, CNF2 = 90, CNF3 = 85, and that the CNF of the conclusion is 95. First, VP-Expert finds the CNF of the AND condition

CNF(condition) = minimum (70, 90, 85) = 70

Then,

$$\begin{aligned}\text{CNF(rule)} &= (70)(95)\\ &= 6650/100\\ &= 67\end{aligned}$$

VP-Expert divides this product by 100 (to obtain a value between 0 and 100) and rounds that number off, so that a CNF of 67 is assigned to the rule. Again, this seems reasonable. Because our confidence level for the condition is less than 100%, we would expect to have a lower confidence in the rule. If the CNF of the condition is 100, then the CNF of the rule equals the CNF of the conclusion. Likewise, if the CNF of the conclusion is 100, then the CNF of the rule is the same as the CNF of the condition.

Let's illustrate this discussion by running a consultation with wp18.kbs. As before, boot up your system and enter VP-Expert. Select wp18.kbs from the subdirectory.

During the consultation, select Yes for the variable Afford and assign a CNF of 85. Select Yes for the variable Help and assign a CNF of 75. Finally, select Yes for both the variable Time and the variable Need, but do not assign a CNF. After you have made your final selection and run the consultation, your terminal screen should look like Figure M6.1.

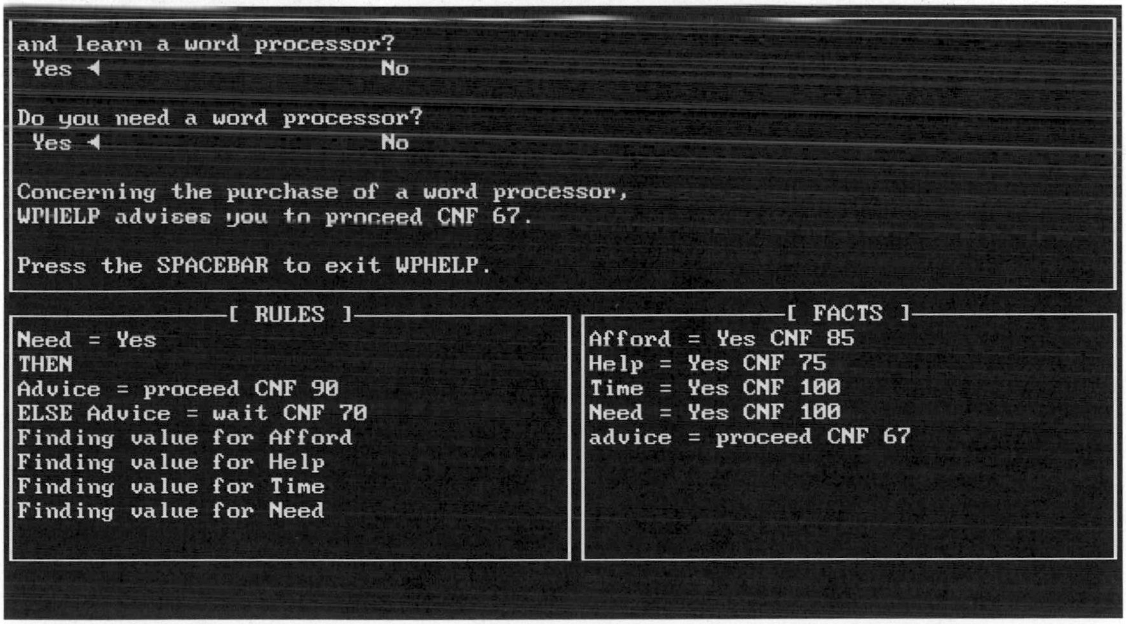

```
and learn a word processor?
 Yes ◄                      No

Do you need a word processor?
 Yes ◄                      No

Concerning the purchase of a word processor,
WPHELP advises you to proceed CNF 67.

Press the SPACEBAR to exit WPHELP.

-------------[ RULES ]-------------      -------------[ FACTS ]-------------
Need = Yes                               Afford = Yes CNF 85
THEN                                     Help = Yes CNF 75
Advice = proceed CNF 90                  Time = Yes CNF 100
ELSE Advice = wait CNF 70                Need = Yes CNF 100
Finding value for Afford                 advice = proceed CNF 67
Finding value for Help
Finding value for Time
Finding value for Need
```

FIGURE M6.1 FINAL CONSULTATION SCREEN FOR wp18.kbs

You should note several things about what you see on the screen. First, in the Consultation window, VP-Expert has advised us to proceed with a CNF of 67. This calculation was made using the CNFs that you entered.

Second, note that the CNF values that you entered for the Afford and Help variables are shown in the Facts window. You did not enter a CNF for either the Time or Need variable, so VP-Expert assigned a CNF of 100 to both of them. The calculated CNF for the rule is 67; you should verify this result by doing the calculations we just described.

Now, run another consultation of wp18.kbs. This time, make the same selections and assign the same confidence factors for the Afford (Yes, 85) and Help (Yes, 75) variables, as in the previous consultation. This time, select No for the Time variable but do not assign a CNF. After making the selection for the Time variable, your screen will appear, after the consultation, as shown in Figure M6.2.

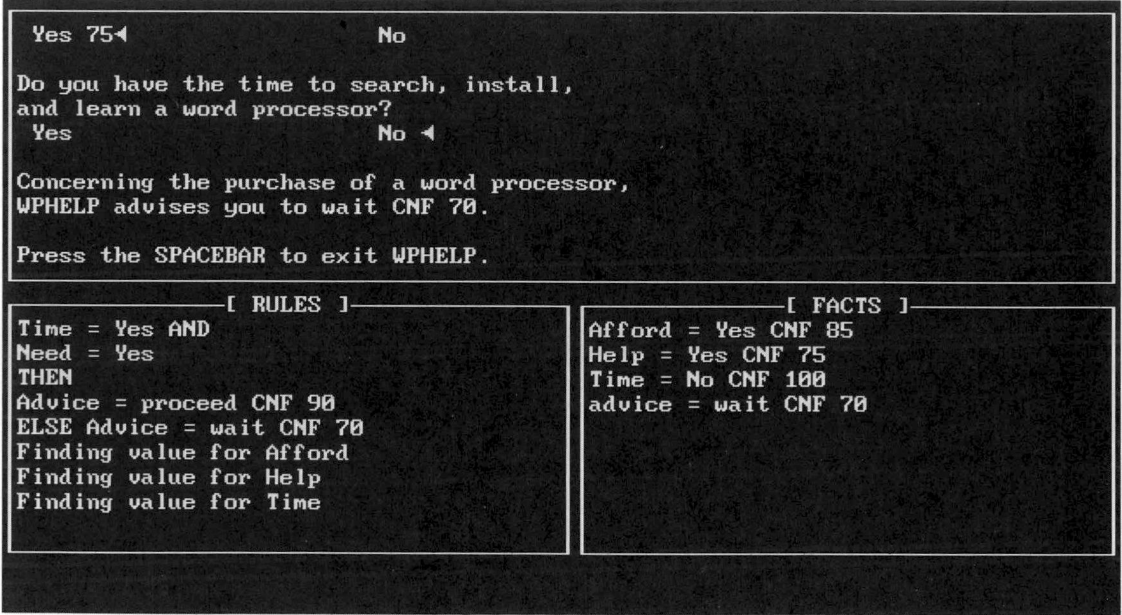

FIGURE M6.2 FINAL CONSULTATION SCREEN FOR wp18.kbs WITH TIME = NO

Several things are different about this screen and the screen in Figure M6.1. First, VP-Expert did not ask you about the need for a word processor. The consultation ended as soon as you selected No for the Time variable because VP-Expert was working with a rule that had an AND condition. When you entered No, VP-Expert interpreted the condition of the rule to be false, with the resulting advice being to wait on the selection of the word processor. Inspecting the Rules window, you will note that the rule has an ELSE segment attached to it. Whatever is contained in the ELSE segment is concluded if the test of the condition fails, as it did in this consultation. Note that the

ELSE segment had a CNF of 70. The CNF of the result did not change because the CNF of the Time variable defaulted to 100.

We wish to make three additional comments about wp18.kbs. To help you understand these comments, let's look at a program listing for wp18.kbs, shown in Figure M6.3.

```
!The file wp18.kbs has one compound AND
!         IF-THEN-ELSE rule
!         with confidence factors.

!ACTIONS BLOCK

ACTIONS
      DISPLAY "This expert system WPHELP will advise you whether you need to buy
a word processor soon or wait.

Press the Spacebar to start the consultation.~"
          CLS
          FIND Advice
          DISPLAY "Concerning the purchase of a word processor, WPHELP advises
          you to {#Advice}.                                    <-- (C3)

Press the Spacebar to exit WPHELP.~"
          CLS;

!RULES BLOCK
RULE 1
IF        Afford = Yes
          AND Help = Yes
          AND Time = Yes
          AND Need = Yes
THEN Advice = proceed CNF 90 }
ELSE Advice = wait CNF 70;   }  <-- (C2)

!STATEMENTS BLOCK
(C1) --> ASK Afford: "Can you afford the money to buy a word processor at this time?";

CHOICES Afford: Yes,No;
(C1) --> ASK Help:  "Do you have some word processing help—people, tutorials,
experience, etc.—available to help you?";

CHOICES Help:  Yes, No;

(C1) --> ASK Time: "Do you have the time to search for, install, and learn a word processor?";

CHOICES Time: Yes, No;

(C1) --> ASK Need: "Do you need a word processor?";

CHOICES Need: Yes, No;
```

FIGURE M6.3 PROGRAM CODE FOR wp18.kbs

First, there are four CHOICES statements, one for each variable (C1). The range of the choice is the same for each variable (Yes, No). Each CHOICES statement is preceded by an ASK statement, which prompts you about making your selection.

Second, the rule in the program has a CNF assigned to both the THEN and ELSE parts of the rule (C2). Notice that the sum of the two CNFs is not 100 but is greater than 100. We discussed the rationale for this occurrence when we described the MYCIN certainty factor scheme in Chapter 6. Both of these CNFs are entered by the expert system developer, the human expert, or both.

The third comment (C3) relates to the use of the # symbol in the DISPLAY statement. The use of this symbol tells VP-Expert to display the final value of the CNF when displaying its recommendation on the screen.

Confidence Factors for Rules with OR Conditions

Multiple conditions in a rule may also be connected by OR. When this occurs, VP-Expert uses a different technique to calculate the CNF of the condition. VP-Expert checks the conditions in order; if it finds a condition that is true, it continues to check the other condition(s).

For example, suppose we have the following rule:

IF Condition 1 CNF1
OR Condition 2 CNF2
THEN Conclusion CNF5

If both conditions in the rule are found to be true, then the CNF of the condition is given by

CNF1 + CNF2 - (CNF 1 * CNF2).

In the rule, suppose that CNF1 = 80, CNF2 = 60, and CNF5 = 90. For the condition, we have

CNF = 80 + 60 - (80 * 60)/100
= 140 - 48 = 92
CNF (rule) = 92 * 90 = 83

This is the result we would expect, based on our discussion about the final CNF of the rule.

A rule may have more than two true conditions. In this case, VP-Expert uses the results of applying the OR formula to the first two conditions as a CNF to be used with the CNF of the third condition. For example, if our rule looked like the following:

IF Condition 1 CNF1
OR Condition 2 CNF2
OR Condition 3 CNF3
THEN Conclusion CNF6;

then VP-Expert would proceed as follows:

1. Calculate the CNF for Condition 1 OR Condition 2.
2. Use this result with the CNF for Condition 3 in the OR formula.

The result of the second step would be the CNF for the whole condition. If more conditions connected by OR existed in the rule, VP-Expert would continue to apply the formula in a similar way.

Let's illustrate this calculation with a sample knowledge base. Your system may be ready to run a consultation. If so, choose wp19.kbs as the file you wish to consult. If not, boot your system, load VP-Expert, and proceed as before.

You will be prompted to choose none, one, or more than one Activity. As before, assign a CNF to each selection (using the HOME and ENTER keys). For this consultation, select Letters with CNF 80, select Resumes with CNF 70, and Notices with CNF 85. Press the End key. Your screen should look like Figure M6.4.

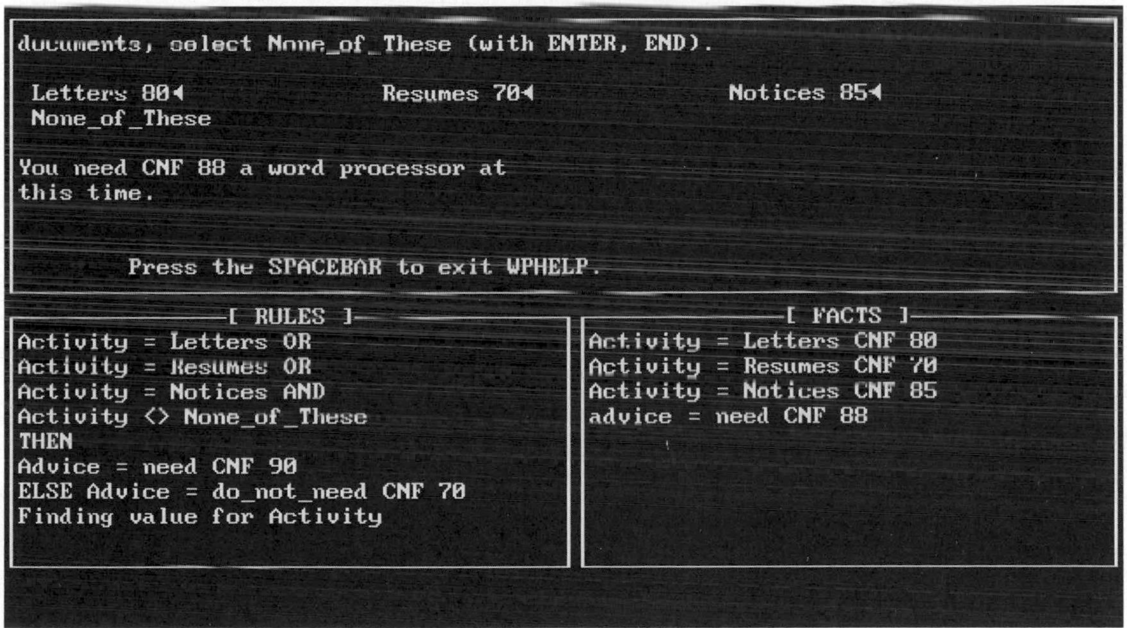

FIGURE M6.4 FINAL CONSULTATION SCREEN FOR wp19.kbs

You should note several things about this screen. As before, the CNFs that you entered for the different activities are shown in the Facts window. Second, the CNF has been calculated at 89 for the rule. How did VP-Expert arrive at this figure?

First, it used the two conditions, Letter (CNF 80) and Resume (CNF 70) in the OR formula.

$$\begin{aligned} CNF &= 80 + 70 - (80 * 70)/100 \\ &= 150 - 56 \\ &= 94 \end{aligned}$$

Next, it used this result (94) with the third condition. Notice (CNF 85) in the OR formula.

$$CNF = 94 + 85 - (94 * 85)/100 = 99$$

$$\begin{aligned} CNF\ (rule) &= 99 * 90 \\ &= 89 \end{aligned}$$

To complete our work with wp19.kbs, let's use the program code in Figure M6.5. Some comments need to be made about the code.

```
!The file wp19.kbs has a PLURAL variable,
!         Activity, in a compound OR IF-THEN-ELSE-BECAUSE
!         rule with confidence factors.

!ACTION BLOCK

ACTIONS
          DISPLAY "WPHELP will advise you whether you need a word processor or not.

Press the Spacebar to start the consultation.~"
          CLS
          FIND Advice
          DISPLAY "You {#Advice} a word processor at this time.

   (C3)   Press the Spacebar to exit WPHELP.~";

!RULES BLOCK

RULE 1
IF Activity = Letters
OR Activity = Resumes
OR Activity = Notices                    (C2)
THEN Advice = need CNF 90
ELSE Advice = do_not_need CNF 70
BECAUSE"If you generate frequent letters, resumes, notices, or papers, then you need a
          word processor.";

!STATEMENTS BLOCK

ASK Activity: "Which of the following documents do you need to write for your business?

Select your answers by pressing the Enter key. Ater you have made all your selections,
press the End key. If you don't need to write any of these documents, select NONE-OF-
THESE (with Enter, End).";

CHOICES Activity: Letters, Resumes, Notices, None_of_These;

Plural: Activity;  <-(C1)
```

FIGURE M6.5 PROGRAM CODE FOR wp19.kbs

You also may have recognized that Activity is a PLURAL variable (C1); PLURAL variables were discussed in Module 3. You should not assume, however, that all conditions using OR represent PLURAL variables. Indeed, it is valid to have conditions in which different variables are connected with OR. In this case, VP-Expert calculates the CNF of the condition using the OR formula.

Our second comment concerns the THEN-ELSE part of the rule. The CNFs will come into the picture when the final CNF of the rule is calculated. In particular, we have already seen the effect of the CNF of the THEN part of the rule in determining the final CNF for the rule. The CNF of the ELSE will be used only if the rule condition is found to be false (none of the conditions are satisfied).

Our final comment on this program pertains to the display of the final results. The # symbol tells VP-Expert that the CNF of the final result is to be displayed.

Confidence Factors for Rules with AND/OR Conditions

The condition part of a rule could include the use of both the AND and OR connectives. For example, we might have a rule in the form of Rule 1:

```
RULE 1
IF Activity  =  Letter OR
   Activity  =  Resume AND
       Cost  =  Low
THEN Word_processor = D CNF 90;
```

This rule is different from the previous examples in that the condition has both AND and OR connectives. How does VP-Expert deal with the situation?

To calculate the CNF for the condition, VP-Expert takes the rule and splits it into two rules as follows:

```
RULE 1A
IF Activity = Letter AND
Cost = Low
THEN Word_processor = D CNF 90;

OR

RULE 1B
IF Activity = Resume AND
Cost = Low
THEN word_processor = D CNF 90;
```

The AND and OR formulas are then used to calculate the CNF of the condition.

As an example, let's suppose that the user has entered confidence factors so that the rules (after separation) appear as follows:

Rule 1A
IF Activity = Letter CNF 90 AND
Cost = Low CNF 80
THEN Word_processor = D CNF 90;

OR

Rule 1B
IF Activity = Resume CNF 70 AND
Cost = Low CNF 80
THEN word_processor = D CNF 90;

VP-Expert applies the AND formula to the condition that yields a CNF of 80. The CNF for Rule 1A is then calculated as

$$\begin{aligned} CNF &= 80 * 90 \\ &= 7200/100 \\ &= 72 \end{aligned}$$

For Rule 1B, the CNF of the condition is the minimum of 70 and 80, which is 70. The CNF for Rule 1A is then calculated as

$$\begin{aligned} CNF &= 70 * 90 \\ &= 6300/100 \\ &= 63 \end{aligned}$$

The CNFs for the two rules are then put into the OR formula to yield

$$\begin{aligned} CNF &= 72 + 63 - (72 * 63)/100 \\ &= 135 - 4536/100 \\ &= 135 - 45 \\ &= 90 \end{aligned}$$

Hence, the CNF of the original rule for our example is 90.

Threshold Values

VP-Expert allows the expert system developer to set a minimum, or ***threshold,*** value to determine when a rule condition is to be considered true. The assigned threshold value is compared to the rule condition CNF. If the condition CNF is less than the threshold value, the condition is considered false; that is, the rule condition fails.

The default value for the threshold in VP-Expert is CNF 50. The threshold value can be assigned with the use of a TRUTHTHRESH clause in the ACTIONS block with the following format:

TRUTHTHRESH = <value>

where value is a number between 0 and 100. A threshold value may be assigned to a particular rule by placing a TRUTHTHRESH clause after the THEN part of a value. A potential danger in assigning a threshold value to a particular rule is that all rules following that rule will also use the same threshold value, which may not be desirable.

While a threshold value can cause a rule *condition* to fail, the threshold value does not cause a *rule* to fail. This is true even though the final CNF of the rule is less than the threshold value.

Let's illustrate the use of the threshold value with an example. If you are at the 2Go option after running wp19.kbs, select 8Quit to return to the Main Menu, then select 6FileName and choose wp20.kbs. If you have exited from VP-Expert, return to it. Now, run a consultation with wp20.kbs. When prompted, enter the following selections and associated CNF for each selection. For this example, select Letters with CNF 80, Resumes with CNF 70, and Notices with CNF 85, for the Activity variable. After making the selections, press End to end the consultation. Your screen should appear like the one in Figure M6.6.

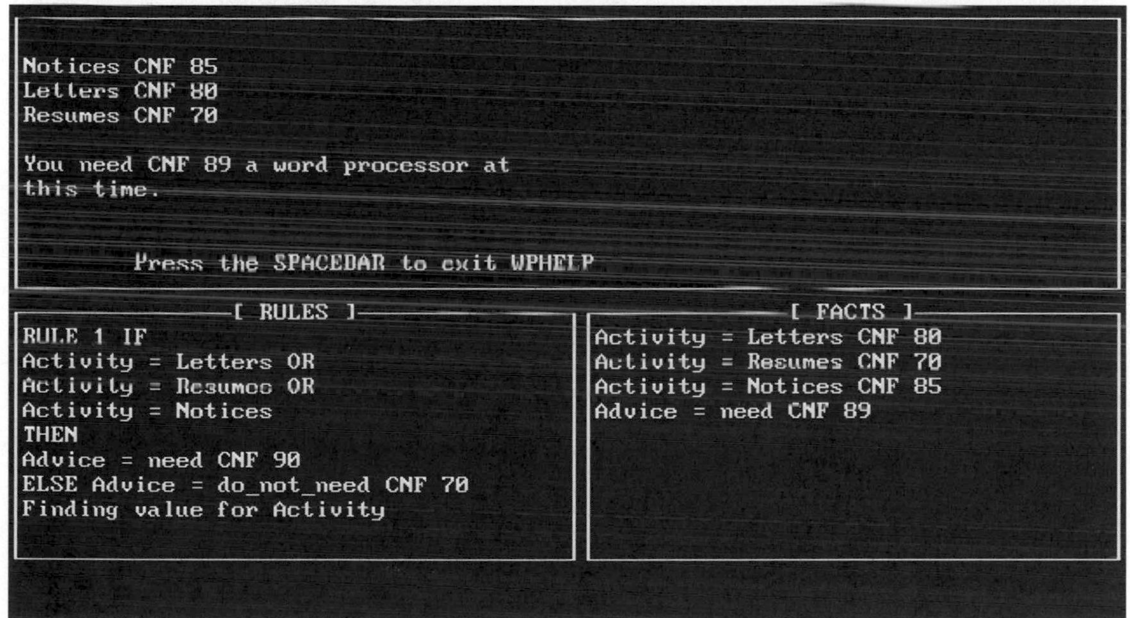

FIGURE M6.6 FINAL CONSULTATION SCREEN FOR wp20.kbs RUN 1

The result of the consultation is that you need a word processor with CNF 89. (Applying the OR formula to the rule condition yields a final CNF of 99, and the product of 99 and 90, the CNF of the conclusion, yields a final rule CNF of 89.) Everything appears in order.

Now, run a second consultation on wp20.kbs. Enter only the selection Letters with CNF 80. The resulting advice, shown in Figure M6.7, is that you do not need a word processor. This occurs because the rule has a truth-threshold higher than the resulting CNF because the rule CNF is 70.

```
The following documents are sorted in descending
order by their confidence factors.

Letters CNF 80

You do not need CNF 70 a word processor at
this time.


        Press the SPACEBAR to exit WPHELP.

[ RULES ]                                  [ FACTS ]
RULE 1 IF                                  Activity = Letters CNF 80
Activity = Letters OR                      Advice = do_not_need CNF 70
Activity = Resumes OR
Activity = Notices
THEN
Advice = need CNF 90
ELSE Advice = do_not_need CNF 70
Finding value for Activity
```

FIGURE M6.7 FINAL CONSULTATION SCREEN FOR wp20.kbs RUN 2

Next, run a third consultation on wp20.kbs. For this consultation, enter only the choice, Letters, with CNF 95. The results are shown in Figure M6.8. This time, the advice is that you need a word processor with CNF 85.

Was a threshold value used? Look at Figure M6.9, the program code for wp20.kbs. Note that there is a TRUTHTHRESH clause with a value of 85 (C1). Thus, if the condition CNF was less than 85, the rule condition would fail.

Results Sorted by Confidence Factors

When PLURAL variables are used in rules, it is often useful to be able to see a **ranking** of the variable values by their CNFs. It is then easy to see the relative weights the user placed on the variable values.

Our consultation with wp20.kbs illustrates this feature. Refer back to Figure M6.6 and note that the rules of the variable activity are shown in the Results window in the upper-left corner. The values are ranked (sorted) by CNF from top to bottom.

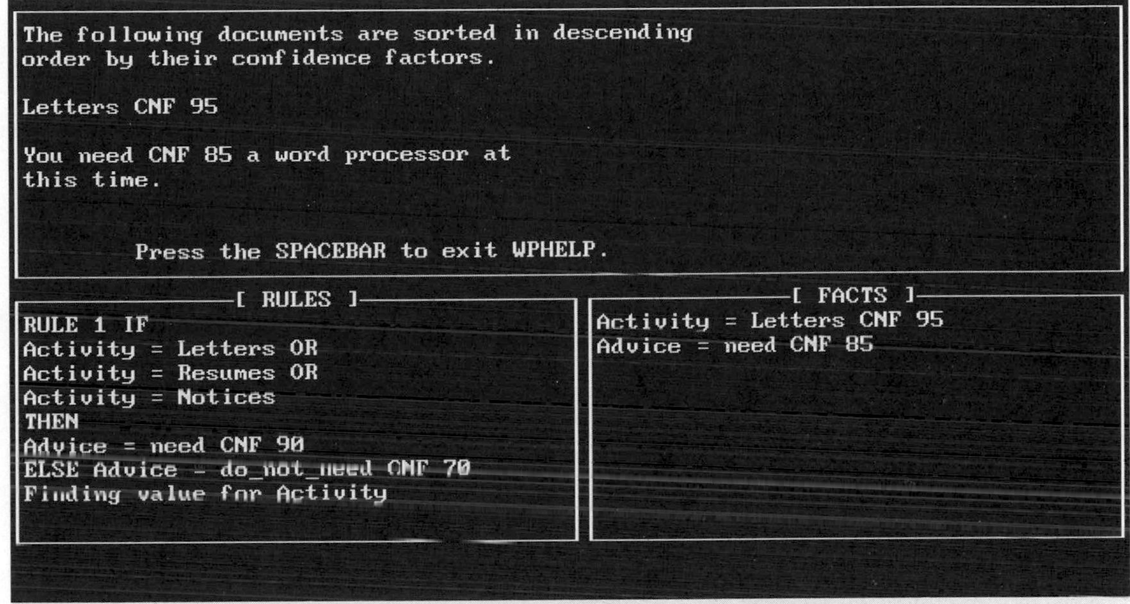

FIGURE M6.8 FINAL CONSULTATION SCREEN FOR wp20.kbs RUN 3

This feature is invoked by the use of the SORT command. The format for the command is

SORT <plural variable>.

It is placed in the ACTIONS block after the FIND command (C2), as shown in Figure M6.9. The results of the sort are seen with the use of a DISPLAY statement, which is shown in the declaration immediately below the SORT clause. See C2 on Figure M6.9.

```
!The file wp20.kbs has a PLURAL variable,
!         Activity, in a compound OR IF-THEN-ELSE-BECAUSE
!         rule with confidence factors.  TRUTHTHRESH and SORT
!         are also used.

!ACTIONS BLOCK

ACTIONS
          DISPLAY "WPHELP will advise you whether you need a word processor or not.

Press the Spacebar to start the consultation.~"
          CLS
          TRUTHTHRESH = 85 <--(C1)
              FIND Advice
             { SORT Activity
      (C2)-->{ DISPLAY "The following documents are sorted in descending order by
              their confidence factors.

              {#activity}"
          DISPLAY " "
          DISPLAY "You {Advice} a word processor at this time.
          Press the Spacebar to exit WPHELP.~"
          CLS;
!RULES BLOCK
RULE 1
IF        Activity = Letters
          OR Activity = Resumes
          OR Activity = Notices
THEN      Advice = need CNF 90
ELSE      Advice = do_not_need CNF 70
BECAUSE   "If you generate frequent letters,resumes, or notices, then you need
          a word processor.";

!STATEMENTS BLOCK

ASK Activity :  "Which of the following documents do you need to produce?  Select your
answer(s) by pressing the Enter key.  After you have made your selections, press the
Enter key.
";

CHOICES Activity: Letters, Resumes, Notices, None_of_these;

Plural: Activity
```

FIGURE M6.9 PROGRAM CODE FOR wp20.kbs

Key Points

- VP-Expert makes use of the certainty factors concept through the use of confidence factors (CNFs).
- Confidence factors are integers between 0 and 100 that represent a level (percentage) of confidence. CNF 100 represents absolute confidence; and CNF 0 represents absolutely no confidence. Most CNF values occur between the extreme values.
- The user can assign confidence factors to rule conditions by using the Home key and entering a CNF value when a value is selected for a variable during a consultation.
- If the user does not assign a CNF value, VP-Expert assigns a default value of 100.
- If there are multiple conditions connected by AND, VP-Expert assigns a final CNF for the condition by using the minimum value of the set of CNFs.
- If there are multiple conditions connected by OR, VP-Expert assigns the final CNF

for the condition by using the formula CNF1 + CNF2 - (CNF1 * CNF2), where CNF1 and CNF2 are the confidence factors associated with condition 1 and condition 2. If there are more than two conditions connected by OR, VP-Expert applies the formula to the first two conditions, then uses that result with the third condition, and so on.

- The expert system developer and/or the human expert enters the confidence factor for the conclusion of a rule.
- The final CNF for a rule is the product of the CNF for the condition and the CNF for the conclusion.
- For conditions that have both AND and OR connectives, VP-Expert splits the rule into two or more rules and applies the AND and OR formulas as appropriate.
- The expert system developer can set threshold values for the minimum acceptable CNF for a rule condition to be true. The default threshold value is 40.
- The threshold value is entered with a TRUTHTHRESH clause in the ACTIONS block or after the THEN part of an individual rule. In the latter case, all rules following the TRUTHTHRESH entry are affected.
- A rule condition can fail because of a threshold value, but the threshold value cannot cause a rule to fail.
- The SORT clause is used to sort the values of a plural variable by the CNFs of the values in descending order.
- The SORT clause is entered in the ACTIONS block after the FIND command.

Exercises

1. What are confidence factors (CNFs)?
2. What is the default value for a CNF?
3. How does VP-Expert calculate the CNF for multiple conditions connected by AND?
4. How does VP-Expert calculate the CNF for multiple conditions connected by OR?
5. What are threshold values? How are they used?
6. Why is the SORT command useful? Under what conditions is it used?
7. How is the final CNF for a rule calculated?
8. How are confidence factors entered for conditions?
9. How are confidence factors entered for conclusions?
10. Given the following rule, calculate the final CNF.

```
IF A CNF 80 AND
   B CNF 70 OR
   C CNF 90
THEN D CNF 85;
```

11. Calculate the condition CNF for the following rule:
    IF J CNF 60 OR
       K CNF 70 OR
       L CNF 90
    THEN M CNF 40;
12. Given the following rule, calculate the final CNF for the rule:
    IF condition 1 CNF 90 OR
       condition 2 CNF 100 AND
       condition 3 CNF 70
    THEN conclusion CNF 80

Assignments

1. Copy wp18.kbs to your Data Disk as my18.kbs. Modify wp18.kbs so that it has a threshold of 80. Run a consultation after the modification. What are the results?
2. Copy wp19.kbs to your Data Disk as my19.kbs. Modify my19.kbs so it can sort the PLURAL variable values and has a threshold value of 90. Run a consultation and enter the following CNFs:
   value CNF
   Letters 65
   Resumes 70
   Notices 75

   Describe the results of the consultation.
3. Talk to a person who is knowledgeable (an expert) about word processing packages. Using wp6.kbs as a basis, expand the knowledge base to include other appropriate criteria. Add rules and confidence factors for both conditions and conclusions. Test your expert system on several people who either are ready to purchase a word processing package or are interested in doing so. Use their comments to improve your system.

Module 7

Implementing The User Interface

An expert system is composed of three parts: the knowledge base, the inference engine, and the user interface. In this module, we will concentrate on the features of VP-Expert that help us implement and improve the user interface. Even if the knowledge base is accurate and the inference engine is flawless, an expert system without good communications between the intended user and the computer system minimizes its potential.

In this module, we will examine how VP-Expert formats the screen. We will see how text is displayed, how windows are created, how color can be used, and the potential for graphics output. Commands for effectively controlling the printer, the run-time environment, and a mouse are included. Finally in this module, we will run five new versions of old programs to describe many of the major points. The program examples appear at the end of this module because certain commands were used before and many key points can be illustrated in one program.

Formatting the Screen Display

The screen is the major method of communications in most computer-based systems. How text is displayed on the screen, how the display is partitioned, and how color is used can add to the communications. Let's consider the display of text first.

Text Display

We have already had many examples of the use of the DISPLAY clause to send any text enclosed within quotes to the screen. Both uppercase and lowercase can be sent to the screen simply by inserting the desired text between the quotes. An appropriate mixture is necessary because all-uppercase text is difficult to read. More characters can be displayed than appear on the keyboard. These characters, including numerals, lowercase, uppercase, and special symbols, on IBM-compatible microcomputers use the *American Standard Code for Information Interchange (ASCII)* format. This is a one-to-one correspondence between the numbers 0 to 255 and a set of corresponding symbols for IBM-compatible microcomputers. (Some ASCII codes are for control rather than display.) See the table of ASCII characters in Appendix B.

To display a character or symbol not found on the keyboard, the CHR command is used. The CHR command is

```
CHR number, variable
```

After the command is executed, the symbol corresponding to the number is placed in the variable name. Then the DISPLAY statement can be used to print the symbol to the screen. For example,

```
CHR 65, Symbol
DISPLAY "{Symbol}"
```

would display A on the screen because the ASCII value of A = 65. If the number in the example were 240, the equivalence symbol = would appear.

To reverse this process, use the ASC clause. The syntax for this command is

```
ASC character-value, variable
```

The ASC clause places the numeric value of the ASCII character (character-value) in the specified variable. To display the value associated with A, we write

```
ASC A, number_value
DISPLAY "{number_value}"
```

to display the ASCII value of A or 65 on the screen. ASC and CHR are almost perfect inverses of each other with the exception that CHR will accept a non-integer input (37.89) and truncate to an integer (37).

Another name for text or groups of characters is a string. A **string** is composed of ASCII characters and has a length. For example, the string, expert, has a length of six. If we count the letters starting at zero, then the letter p is in position 2 of the string expert. VP-Expert uses the following clauses to manipulate or handle strings:

```
UPPER     variable
LOWER     variable
LENGTH    variable1, variable2
CONCAT    string, variable
POS       string1, string2, variable
SUBSTR    string, variable, integer1, integer2
```

The UPPER clause converts a text string to all uppercase letters, whereas the LOWER command converts all text to lowercase. The LENGTH command places the length of the string variable1 in variable2. For example,

```
last_name = Smith
LENGTH last_name, this_many
DISPLAY "The string {last_name} stored in the
         variable  last_name has {this_many}
         characters."
```

would tell us that Smith has five characters or that the length of the variable last_name is 5.

The CONCAT clause helps us to "add strings." This clause attaches the first string to the end of the given variable. For example, the commands

```
last_name1 = Smith
last_name2 = field
CONCAT last_name1, last_name2
DISPLAY "{last_name2}"
```

would combine the two strings and print Smith field to the screen. If the CONCAT clause was changed to CONCAT last_name2, last_name1, then field Smith would be printed to the screen by the clause

```
DISPLAY "{last_name1}".
```

The POS command is used to search a given string for a matching substring within. If the first string is found within the second string, then its start position or offset is placed in the third parameter, a variable. If the substring does not match, then the variable is not set. If the substring occurs more than once within a given string, then the offset of the first match is identified. For example, the substring VP is found in the string VP-Expert starting at position 0, whereas the substring Ex is found at position 3. The substring vp is not found in the string, however, because of the difference in case. These three examples are written, respectively, as follows:

```
POS "VP", "VP-Expert", position1
POS "Ex", "VP-Expert", position2
POS "vp", "VP-Expert", position3
```

The values of position1, position2, and position3 are 0, 3, and unassigned.

If you use variable names with assigned values, use the indirect addressing symbol @. If we assign the string VP-Expert to a variable called Product_name and the substring Ex to a variable called Part_of, then the second example above becomes:

```
Product_name = VP-Expert
Part_of = Ex
```

```
POS @Part_of, @Product_name, position
DISPLAY "The substring {Part_of}
      starts at position {position}
      of the string {Product_name}.";
```

The DISPLAY command would output "The substring Ex starts at position 3 in the string VP-Expert." The POS clause is best used in conjunction with the CONCAT and SUBSTR.

The SUBSTR clause will copy a portion of a string, starting in a given position (offset) of a specified (substring) length into a variable. The syntax of the command is SUBSTR string, variable, start_position, length. These four parameters have the following meanings:

string	The string from which a substring is copied
variable	The name of the variable to contain the copied substring
start_position	The offset (in the string) of the substring to be copied
length	The length (not zero-based) or the number of letters to be copied

To copy the substring Ex (which starts at offset 3 of length 2) from the string VP-Expert and store the result in target, use the following syntax:

```
SUBSTR "VP-Expert", target, 3, 2
```

To assign the string VP-Expert to a variable Product_name, use the following:

```
Product_name = VP-Expert
SUBSTR @Product_name, target, 3, 2
DISPLAY "The letters {target} make up a
   substring of the string {Product_name}.
   The letters start at position 3 for a
   length of 2.";
```

These string manipulation clauses—UPPER, LOWER, CONCAT, POS, and SUBSTR—are particularly useful in extracting initials from name fields, manipulating social security numbers (strings), handling part numbers, and related applications.

To continue our discussion of text-related issues, the DISPLAY statement has several interesting features. Values of variables can be sent to the screen by enclosing the variable within curly brackets. If a pound sign (#) is placed within the curly brackets, then the confidence factor associated with that variable is displayed on the screen. If an integer precedes the variable inside the curly brackets, then the screen will right-align that value of the variable in the designated number of spaces. The following examples illustrate these points:

```
Cost = 199.99
DISPLAY "Cost is {Cost}"     Cost is 199.99
DISPLAY "Cost is ${Cost}"    Cost is $199.99
DISPLAY "Cost is ${#Cost}"   Cost is $199.99
                                     CNF 95
DISPLAY "Cost is ${8Cost}"   Cost is $  199.99
```

The FORMAT command can be used to change the formatting of the display of a numeric value of a variable. If

```
FORMAT Cost, 7.3
```

were issued, then the DISPLAY command would yield

```
199.990
```

The first number after Cost identifies the number of places (including the decimal point) and the last number tells how many of the places are to the right of the decimal. In our example, of the seven places, three are to the right of the decimal.

The DISPLAY clause has another interesting option. If a tilde (~) appears at the end of the string before the last quote, then the system will pause until the user presses a key. This is useful when scrolling long screens of text at the display terminal. We have used this option in the DISPLAY clause many times.

The placement of the text from a DISPLAY clause is important. In general, the display format of a screen is 25 rows by 80 columns. The choice of 40 columns is also available on the IBM-PC. The video screen is laid out with 0,0 as the upper-left corner and 24,79 as the lower-right corner. If position x,y is on the screen, then x stands for the number of rows down, and y indicates the number of columns across, as shown in Figure M7.1.

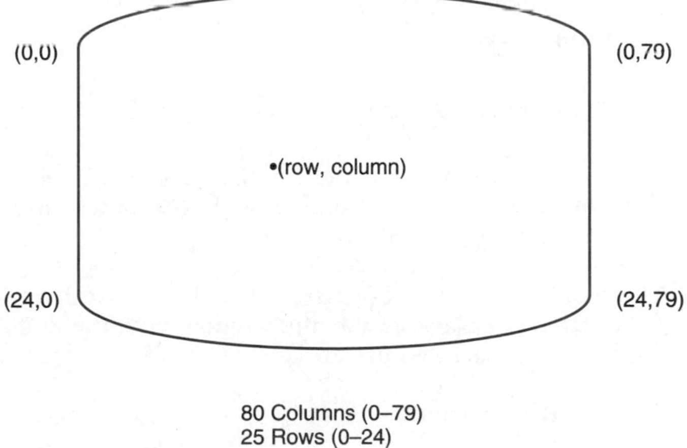

FIGURE M7.1 SCREEN LAYOUT FOR TEXT MODE

The other corners of the screen are 24,0 (lower-left) and 0,79 (upper-right). The WOPEN and LOCATE commands will help us place text where we want it on the video screen.

Windows

Windows, or rectangular boxes, are one effective way of organizing the layout of a screen. You have already seen the effective use of three windows, one each for the Consultation, Rules, and Facts windows, from the Main Menu. These boxes or windows help to organize information on the screen. VP-Expert has several commands, WOPEN, ACTIVE, and LOCATE, to facilitate the windowing effects for the user interface. Windows allow the screen to be partitioned in almost any arrangement. Consider the different window configurations in Figure M7.2.

In screen (a), we have a single window centered on the screen. Overlapping windows are shown in screen (b). The window arrangement of screen (c) is that of the VP-Expert consultation environment, with one large window across the top and two smaller ones underneath. Small horizontal windows are shown in screen (d), whereas vertical windows appear in screen (e). Finally, the pull-down, or pop-up, menus are shown in screen (f). These windows pull down when the user wants more information.

VP-Expert allows up to nine windows on the screen at one time. To create a window, use the WOPEN clause with the following parameters:

```
WOPEN window#,
   row_start,
   column_start,
   number of rows,
   number of columns,
   background_color#
```

For example,

```
WOPEN 1,4,6,8,30,background_color#
```

would draw a window number 1 with its upper-left corner at 4,6. The window would be eight rows deep and 30 columns wide. (Color will be discussed later.) The window is illustrated in Figure M7.3.

The window numbers can range from 1 to 9. Nothing other than the background color can be displayed inside the window until the ACTIVE clause is executed. The format of the clause is

```
ACTIVE window#.
```

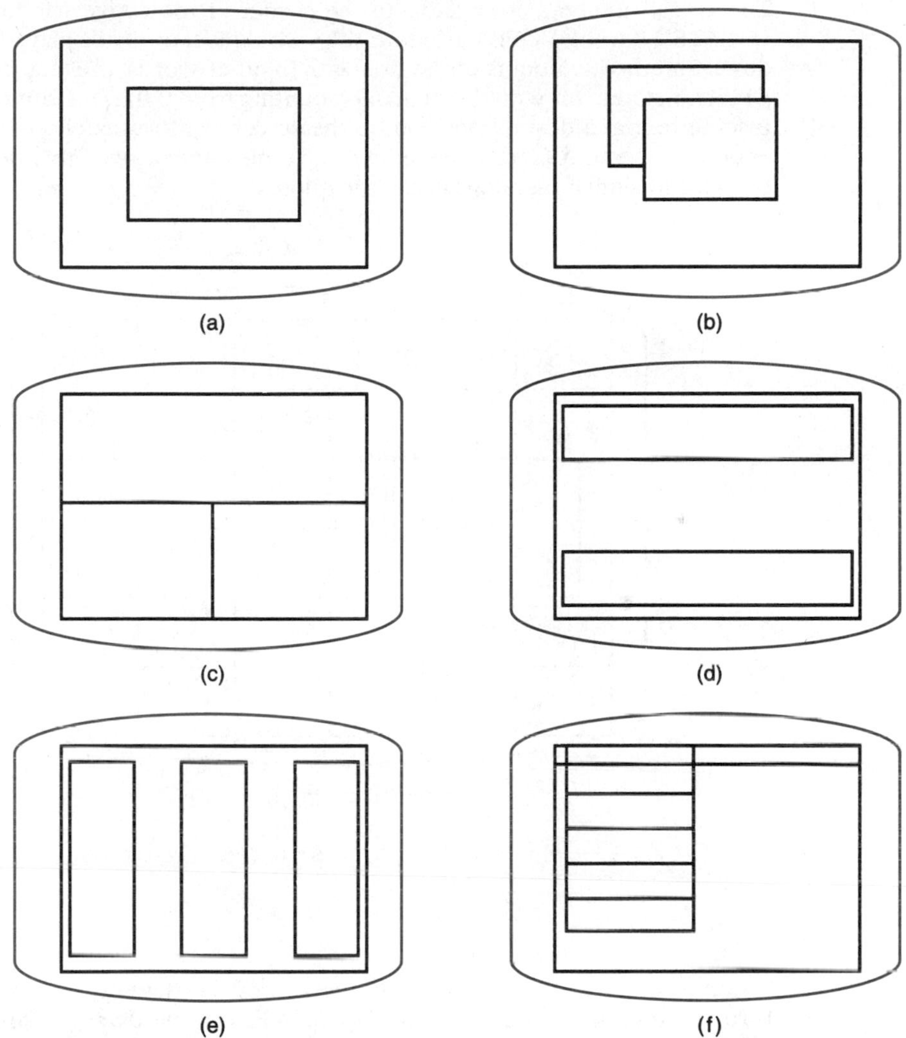

FIGURE M7.2 WINDOWING ARRANGEMENT FOR DISPLAY SCREENS

Only one window can be active at a time even though several may be on the display screen. A WOPEN clause for window 1 must precede a corresponding ACTIVE clause for window 1.

Once WOPEN and ACTIVE clauses (with the same window number) have been executed, the LOCATE command can then be used to position the display of text within the active window. The format for this command is

LOCATE row_number, column_number.

These two numbers are relative to the corner of the active window. For example, LOCATE 2,3 from WOPEN in Figure M7.3 would locate any displayed text at two rows down and three columns across from 4,6. In other words, the text from a subsequent DISPLAY statement would start at 6,9 (counting from 0,0). This command is helpful in placing text at a desired position on the screen. Unfortunately, when the inference engine uses the ASK statements, we have no control over this display except for blanks and similar markings inside the quotes.

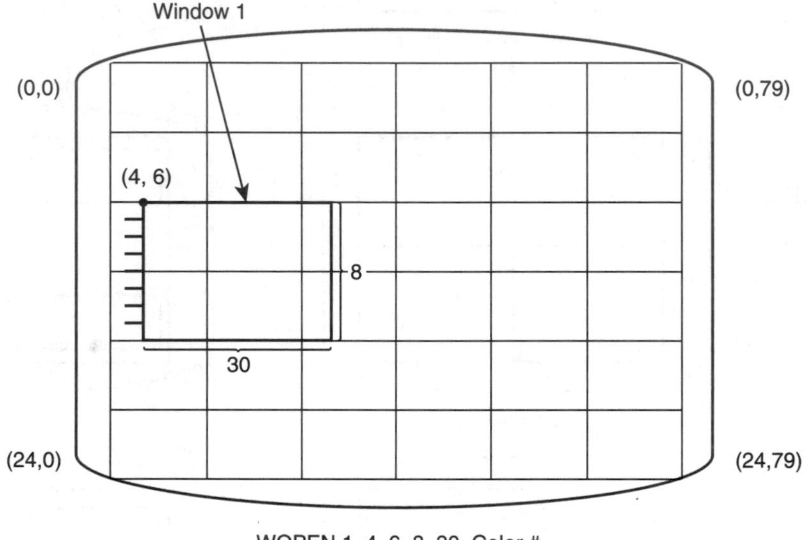

FIGURE M7.3 THE WINDOW CREATED BY WOPEN 1,4,6,8,30,BACKGROUND_COLOR#

Color

Whether color can be added to the user interface depends on the availability of a board (hardware) inside the microcomputer. This hardware is sometimes called a video adapter, graphics board, or graphics adapter. If your consultations have been in color, then your computer has a graphics adapter. If you do not have color display, then your screen display shows white letters on a black background, or amber or green characters on a black background. Text that involves color will be displayed in black and white on a monochrome system.

Several de facto standards in graphics adapters have evolved over the years. They are listed below:

Color Graphics Adapter (CGA)

Enhanced Graphics Adapter (EGA)

Video Graphics Adapter (VGA)

Super Video Graphics Adapter (SVGA)

Enhanced Super Video Graphics Adapter

Hercules Graphics Card (HGA)

Even though there are many more types of hardware graphics boards, these are the most common. Unfortunately, each board works differently. The first three in the list are from International Business Machines (IBM) and are listed in ascending order of capability. Because the CGA has been around the longest, we will comment on this adapter relative to VP-Expert.

The CGA allows color and text to be displayed in both 25 by 80 and 25 by 40 modes. You have a choice of 8 background colors and 16 foreground colors. A background color can be thought of as the color of the writing surface (display screen); the foreground color is the color of the pen or ink used to write letters on the screen. If one selects the foreground and background color to be the same, then the text on the screen is invisible (a tactic useful for entering passwords). The color names and numbers are given in the following table:

COLOR NUMBER	FOREGROUND COLOR	BACKGROUND COLOR
0	Black	Black
1	Blue	Blue
2	Green	Green
3	Cyan	Cyan
4	Red	Red
5	Magenta	Magenta
6	Brown	Brown
7	White	White
8	Gray	
9	Light Blue	
10	Light Green	
11	Light Cyan	
12	Light Red	
13	Light Magenta	
14	Yellow	
15	Light White	

Because the monitors, or video screens, vary in quality and contrast, red on one screen may not appear as the same red on another screen. Two commands set the background and foreground colors in Text mode:

```
BKCOLOR = background_color# (0-7)
COLOR = foreground_color# (0-15)
```

The two clauses BKCOLOR = 0 and COLOR = 14 would write yellow letters on a black background. Be sure to check the physical appearance of the colors on the screen. Certain combinations are difficult to read. Yellow or white on blue or black usually gives a good contrast. Of course, no colored text is seen on the foreground until a

DISPLAY statement is executed. Windows on the screen can have different background colors, as you have seen in the consultation runs with VP-Expert.

Graphics

Graphics is another potential enhancement to the user interface. By *graphics,* we mean the ability to draw circles, lines, polygons, and so on, on the screen. Instead of controlling characters on the screen, we need to control finer portions of the screen called picture elements or **pixels**. The more pixels, the better the quality of the displayed graphic. Text can also be displayed in Graphics mode. Text and Graphics modes are sometimes confused. Therefore, we will call Text Graphics the mode in which the computer can display symbols and letters, and Graphics mode that in which the computer can draw lines, circles, polygons, and so on (pixel control), in addition to displaying text. Color can be used in any of the modes. Be careful how technical reference manuals refer to **graphics**.

How does the computer display graphics (pixels) on the screen? The answer depends on the graphics board. For example, the CGA uses two common graphics modes: 320 columns by 200 rows (Medium Resolution) and 640 columns by 200 rows (High Resolution). The Medium-Resolution mode allows only certain sets of four colors to be displayed at one time, whereas the High-Resolution mode allows one color and black. The natures of Graphics modes and applications have to be carefully matched. There is a general trade-off between resolution (number of pixels) and number of available colors for screen display (at one time).

The command for invoking a Graphics mode with the proper adapter is given by the clause

GMODE setting#.

The GMODE setting numbers are illustrated in Figure M7.4, with setting descriptions and codes for adapters.

Note that GMODE settings for text (not graphics) are numbers 0 to 3 or 7. Thus, the remaining numbers 4 to 6 or 8 to 19 set the screen to graphics (and text) modes. Once a graphical GMODE setting is executed, a host of commands can be used to draw rectangles, lines, polygons, and other drawing primitives. Some examples are as follows:

PSET	Sets given screen pixel to "on" rather than off
RECTANGLE	Draws a rectangle on the screen
ELLIPSE	Draws an ellipse (or circle) on the screen
FILL	Fills an enclosed figure with a given color
LINETO	Draws a line from the current (pixel) position to a given location

As the graphics commands (or primitives) are executed, the system keeps track of the position where an "imaginary" drawing pen is located. This position is expressed in

Setting	Description	MDA	CGA	MCGA	EGA	VGA	HGC
−1	restore screen to original mode	✔	✔	✔	✔	✔	✔
0	40 × 25 [text], 16 grey	✔	✔	✔	✔	✔	
1	40 × 25 [text, 16/8 color		✔	✔	✔	✔	
2	80 × 25 [text], 16 gray	✔	✔	✔	✔	✔	✔
3	80 × 25 [text], 16/8 color		✔	✔	✔	✔	
4	320 × 200 [graphics], 4 color		✔	✔	✔	✔	
5	320 × 200 [graphics], 4 gray		✔	✔	✔	✔	
6	640 × 200 [graphics], BW		✔	✔	✔	✔	
7	80 × 25 [text], BW	✔	✔	✔	✔	✔	✔
8	720 × 348 [graphics], BW for HGC						✔
13	320 × 200 [graphics], 16 color		✔	✔	✔	✔	
14	640 × 200 [graphics], 16 color			✔	✔	✔	
15	640 × 350 [graphics], BW				✔	✔	
16	640 × 350 [graphics], 4 or 16 color				✔	✔	
17	640 × 480 [graphics], BW				✔	✔	
18	640 × 480 [graphics], 16 color					✔	
19	320 × 200 [graphics], 256 color			✔		✔	

Adapter Code	Description
MDA	Monochome Display Adapter
CGA	Color Graphics Adapter
MCGA	MultiColor Graphics Array
EGA	Enhanced Graphics Adapter
VGA	Video Graphics Array
HGC	Hercules Graphics Card

FIGURE M7.4 THE GMODE SETTING NUMBERS

terms of pixel coordinates of columns and rows. Graphics commands relevant to controlling pen location, the screen, and color are as follows:

MOVETO	Moves the drawing pen to a given x,y location on the screen
GLOCATE	Sets the x,y coordinates in GMODE for a subsequent display of text in Graphics mode with a GDISPLAY
GCOLOR	Sets the current foreground or "drawing pen" color
GBCOLOR	Sets the background color
GCLS	Clears the graphics screen
TMODE	Returns the screen to text mode, the default of VP-Expert screen mode

Some examples will help to unify these commands. The following code segment will set the screen to graphical mode 14 (640 column pixels by 200 row pixels on the MCGA, EGA, or VGA adapters), draw a red (4) rectangle on a blue(1) background, and fill the interior of the rectangle with yellow (14):

```
ACTIONS
   GMODE 14
   GBCOLOR 1
   GCOLOR 4
   RECTANGLE 20,0,380,90
   FILL 21,1,14
   GDLOCATE 2,18
   GDISPLAY "Department"
      :
      :
```

The rectangle, moreover, has its upper-left corner at column 20, row 0 and its lower-right corner at column 380, row 90. Note that the FILL command starts at position 21,1. The last two commands print department in red at position 2,18 on the screen.

Another useful graphics command is the GETCH clause, which pauses execution of the consultation until the user presses a key. When the key is pressed, the ASCII character corresponding to the key pressed is stored in the variable specified by the GETCH command. The GETCH clause used with the GDISPLAY is especially valuable because the ASK statement cannot be used in Graphics mode. Consider this example:

```
GDISPLAY "Type 1 for Integrated word
         processors.
      Type 2 for MacIntosh word
         processors.
      Type 3 for Windows-based word
         processors."
GETCH Reply:
```

The menu-like selections are displayed on the graphics screen with the GDISPLAY instruction. The GETCH clause makes the consultation wait until the user presses a key. The ASCII character the user presses is stored in the variable Reply. Introductory screen comments, moreover, can be programmed as follows:

```
GLOCATE 5,25
GDISPLAY  "Press any key to continue."
GETCH DUMMY;
```

The GETCH command works in both Text (TMODE) and Graphics modes.

Another useful graphics command is the PCX clause, which follows the format

```
PCX filename, x,y
```

where filename is a bitmapped PCX pre-existing file and x,y is a graphics screen location. This instruction will display the graphics contents with the location x,y as its upper-left corner. Multiple PCX files can be displayed on the screen simultaneously, as illustrated by the following code segment:

```
PCX  rose 10,10
PCX  daisy 300,10
PCX  carnation 10,200
PCX  magnolia 300,200
```

Four pictures of flowers would be displayed on the screen. The advantage of using the PCX clause is the time saved in "programming" or drawing the desired picture or graphics. The screen coordinates depend on the GMODE setting.

These graphics can enhance the visual display screen of the user interface. A special graphics library, VP-Graphics, can be added to the VP-Expert system to produce drawings and other presentation graphics. Together, text, windows, color, and graphics can all be used to format the screen display for more effective communications in an expert system application.

Controlling the Printer

The *printer* should not be forgotten in implementing a user interface. The availability of printouts is a vital part of a useful information system. VP-Expert has four commands to control printing:

```
PRINTON
PRINTOFF
PDISPLAY
EJECT
```

We have already used the first two clauses, PRINTON and PRINTOFF. PRINTON will cause all display statements to be sent to the printer until a PRINTOFF clause is encountered. The DISPLAY statements are also still sent to the screen. The text following an ASK statement is not sent to the printer. A message within the program code telling the user to check to be sure the printer is on is required.

PDISPLAY will send any text found in quotes to the printer, not the screen. PDISPLAY does for the printer what DISPLAY does for the screen. The last command, EJECT, will cause the printer to eject a page or roll up the paper until the top of the next page is found. This feature is needed when the printout ends in the middle of a page. Attention to these small details in any application will help the user interface.

Controlling the Run-Time Environment

Another part of the user interface is the *run-time environment* for the expert system. Previously, we have been running our consultations by accessing the Consult Menu from the Main Menu. We can make this easier for the user by hiding some of these consultation features. Six commands (by category) control the run-time environment:

EXECUTE	statement
RUNTIME	statement
SYSTEM	clause
ENDOFF	statement
BLINEON	clause
BLINEOFF	clause

These three statements—EXECUTE, RUNTIME, and ENDOFF—are placed before the ACTIONS block; each is followed by a semicolon. The EXECUTE statement will allow the user to run a selected file from the 4Consult option of the Main Menu without having to enter 2Go of the Consult Menu to run a consultation. When the consultation is over, the EXECUTE command returns the user to the operating prompt, rather than the Main Menu. The RUNTIME statement, which we have already used, will turn off the Results and Facts windows for the consultation. Both of these commands are useful. We have used the windows from the inferences engine because we were interested in learning how an expert system works and how to create one. Many users, however, only want to run a consultation, not create one.

(For those users who have a compiled version of VP-Expert, the VPXRUN command enables the VP-Expert application developer to turn off or suppress the initial VP-Expert copyright screen. The syntax is VPXRUN filename -S where filename is the name of a compiled knowledge base (.kbs) to be executed and -S is a switch to turn the initial copyright screen off.)

The SYSTEM clause is a "hook" feature to VP-Expert programs. This command transfers control to the operating system, or DOS, from within a consultation. The EXIT command typed at the DOS prompt will return control to VP-Expert. After the SYSTEM command is effected, any of the DOS operating system commands (FORMAT, COPY, DIR, etc.) may be used. Neither the SYSTEM or EXIT commands require any parameters.

The next command, ENDOFF, can be used to change the method of selecting an answer to a question through the keyboard. You can use ENDOFF for any question except one involving a PLURAL variable. If only one selection is to be made to answer a question, ENDOFF will turn off the need to press the End key. To make a selection, you may simply select your reply and press the Enter key; the End key is no longer required.

The two VP-Expert clauses, BLINEON and BLINEOFF, act together like a switch. The BLINEOFF command will turn off the display of the bottom menu during a consultation; the BLINEON command turns the display of the bottom menu back on (makes it visible).

All of these commands, EXECUTE, RUNTIME, SYSTEM, ENDOFF, BLINEON, and BLINEOFF, can enhance the user's "runtime environment" for the expert system application.

Using a Mouse

A *mouse* is an alternative input, or control, device for the keyboard. It is a small device attached to your computer system. You can move the cursor around on the screen with the mouse, instead of the arrow keys. To make a selection from a list or menu on the screen, physically move the mouse on a table surface until the cursor is on the desired item, then click, or select, the designated entry. A mouse is faster to use than keyboard arrow keys, but it adds an additional expense and is cumbersome for applications requiring extensive data entry. However, of all possible input devices, the mouse has become the second most important (after the keyboard).

If you have a mouse connected to your computer system, you can use the VP-Expert commands that let you use the mouse for selecting instead of typing from the keyboard. The following commands help you to incorporate a mouse in your user interface:

BUTTON
GETMOUSE
MOUSEON
MOUSEOFF
MOUSEX
MOUSEY

These clauses and statements work only in Graphics mode (settings 4 to 6, 8 to 19). You can find detailed information on each of these commands in the Help facility of VP-Expert.

Exploring Examples

We will now run some sample programs to illustrate the points we have made with VP-Expert commands. Get into VP-Expert and set the Path to the Sample File directory. Our first sample program will display some ASCII characters to the screen. Select wp21.kbs for the 6FileName. After the consultation, obtain a printout of the program code for wp21.kbs. At the end of the consultation, your screen should appear like the one in Figure M7.5.

These symbols are the printable characters from ASCII 50 to ASCII 65. If you examine the program for wp21.kbs (Figure M7.6), you will see some interesting programming structures.

Comment 1 is the CHR command. Notice that we placed the at symbol (@) in front of the variable X, which stands for a number. From C2 you can see the numbers (5 and 3) before the variables in the DISPLAY statement. These numbers tell how many spaces are provided for the display of the variables. Comment 3 identifies a FOR loop that starts at the value 50 and continues until the value is past 65. One is added to the

```
The ASCII value for    50 is   2
The ASCII value for    51 is   3
The ASCII value for    52 is   4
The ASCII value for    53 is   5
The ASCII value for    54 is   6
The ASCII value for    55 is   7
The ASCII value for    56 is   8
The ASCII value for    57 is   9
The ASCII value for    58 is   :
The ASCII value for    59 is   ;
The ASCII value for    60 is   <
The ASCII value for    61 is   =
The ASCII value for    62 is   >
The ASCII value for    63 is   ?
The ASCII value for    64 is   @
The ASCII value for    65 is   A

          Press the SPACEBAR to exit WPHELP.
```

FIGURE M7.5 ASCII CHARACTERS FROM 50 TO 65 FROM wp21.kbs

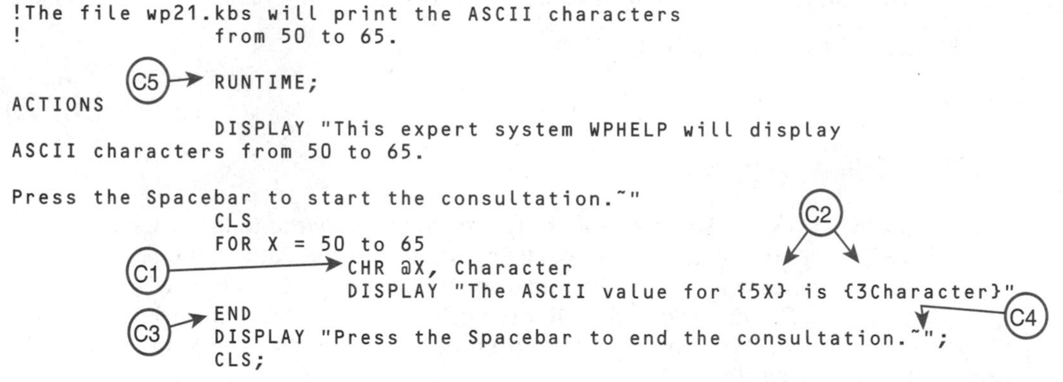

```
!The file wp21.kbs will print the ASCII characters
!               from 50 to 65.

                RUNTIME;
ACTIONS
                DISPLAY "This expert system WPHELP will display
ASCII characters from 50 to 65.

Press the Spacebar to start the consultation.~"
                CLS
                FOR X = 50 to 65
                          CHR @X, Character
                          DISPLAY "The ASCII value for {5X} is {3Character}"
                END
                DISPLAY "Press the Spacebar to end the consultation.~";
                CLS;
```


FIGURE M7.6 PROGRAM CODE FOR DISPLAYING ASCII CHARACTERS (wp21.kbs)

variable X by the system at the END of the loop. In other words, this addition is not coded inside the loop (as the WHILEKNOWN loop does). Comment 4 focuses our attention on the ending message to the user. This message along with the tilde (~) keeps the displayed ASCII characters on the screen until the user presses the Spacebar. Without the last DISPLAY with a tilde, the screen would return to the Main Menu so fast that the user could not read it. The tilde in the DISPLAY statement makes scrolling through paragraphs of text or screen displays easy.

The RUNTIME statement above the ACTIONS block (C5) removes the Rules and Facts window during a consultation. This statement is most useful after an expert system has been debugged and the system is ready for the user.

Let's run a consultation with a window. Exit the Consult Menu and select wp22.kbs from the 6FileName option. Run the consultation and obtain a printout of the code. We have eliminated the EXECUTE statement so the consultation ends at the Consult Menu. The last Consultation screen for wp22.kbs is shown in Figure M7.7.

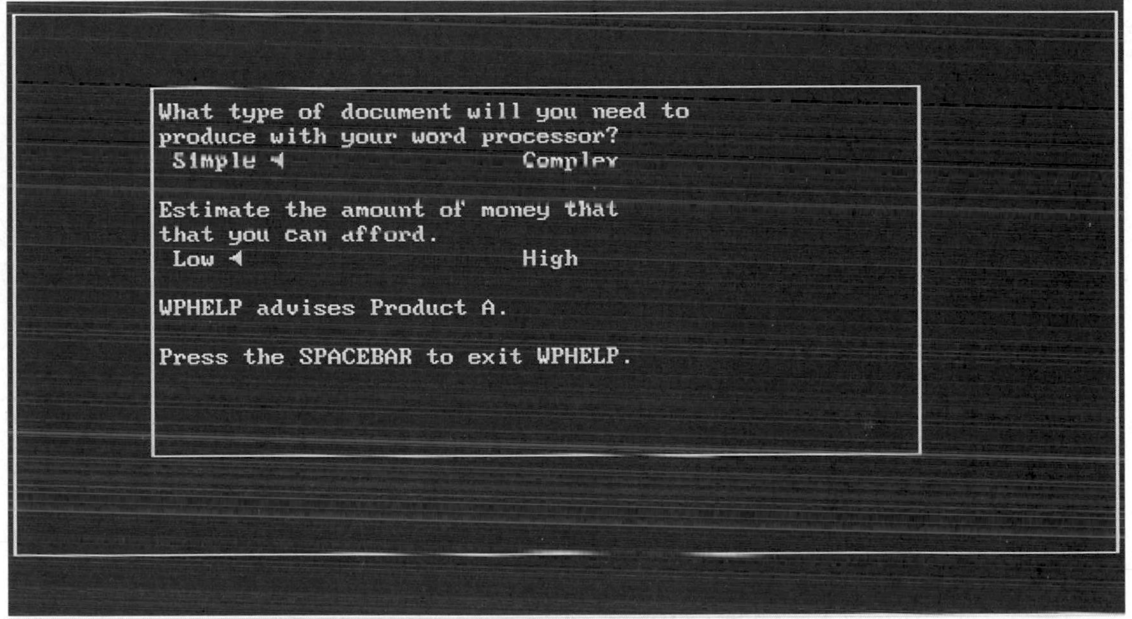

FIGURE M7.7 LAST CONSULTATION SCREEN FOR wp22.kbs

If you have a color monitor, the border is green and the window is blue (background color) with light red (foreground color) letters. The program code for producing this window is detailed in Figure M7.8.

From Comment 1, note the use of the RUNTIME and ENDOFF statements for controlling the run-time environment. The RUNTIME turned off the Results and Facts windows, while the ENDOFF determined the Enter key to be the only key needed to select answers to questions. In Comment 2, we point out that Window 1 was opened, made active, and then closed. In the WOPEN clause, the 1 indicates Window 1. The next two parameters (3 and 10) tell us that the upper-left corner of the window is 3,10. The next number (15) indicates that Window 1 is 15 rows deep; the next parameter (55) says that Window 1 is 55 columns wide. The last parameter (1) tells us that the background color of Window 1 is blue.

```
!The file wp22.kbs is wp1.kbs with a different
!          user interface using color and a
!          window for a run-time environment.

    (C1)→  RUNTIME;
           ENDOFF;

ACTIONS
    (C2)→  WOPEN 1,3,10,15,55,1
           ACTIVE 1
           COLOR = 12      (C3)
           LOCATE 3,5  ←
           DISPLAY "WPHELP will advise you on word processing."
           LOCATE 5,5
           DISPLAY "Press the Spacebar to start the consultation.~"
           CLS
           FIND Word_processor                          (C4)
           DISPLAY "WPHELP Advises {Word_processor}."
           DISPLAY " "                              ↙
           DISPLAY "Press the Spacebar to exit WPHELP.~"
           WCLOSE 1
           CLS;
RULE 1
IF         Functionality = Simple AND
           Cost = Low
THEN       Word_processor = Product_A;

RULE 2
IF         Functionality = Simple AND
           Cost = High
THEN       Word_processor = Product_A;

RULE 3
IF         Functionality = Complex AND
           Cost = Low
THEN       Word_processor = none;

RULE 4
IF         Functionality = Complex AND
            Cost = High
THEN       Word_processor = Product_B;

ASK Functionality: "What type of document will you need to
produce with your word processor?";

CHOICES Functionality: Simple, Complex;

ASK Cost: "Estimate the amount of money that you can afford.";

CHOICES Cost: Low, High;
```

FIGURE M7.8 PROGRAM CODE FOR A WINDOW (wp22.kbs)

This window is where all our text is displayed in the COLOR = 12 clause after the ACTIVE clause sets the text color to light red. In Comment 3, we show the use of the LOCATE command to position the light red text inside the active window. The tilde is necessary in the last DISPLAY statement, Comment 4, so that the screen stays visible for the user.

Our next example will show how we utilize two windows in the user interface. Select wp23 from 6FileName. Run this consultation and then obtain a printout of the program code. The last Consultation screen for wp23.kbs appears in Figure M7.9.

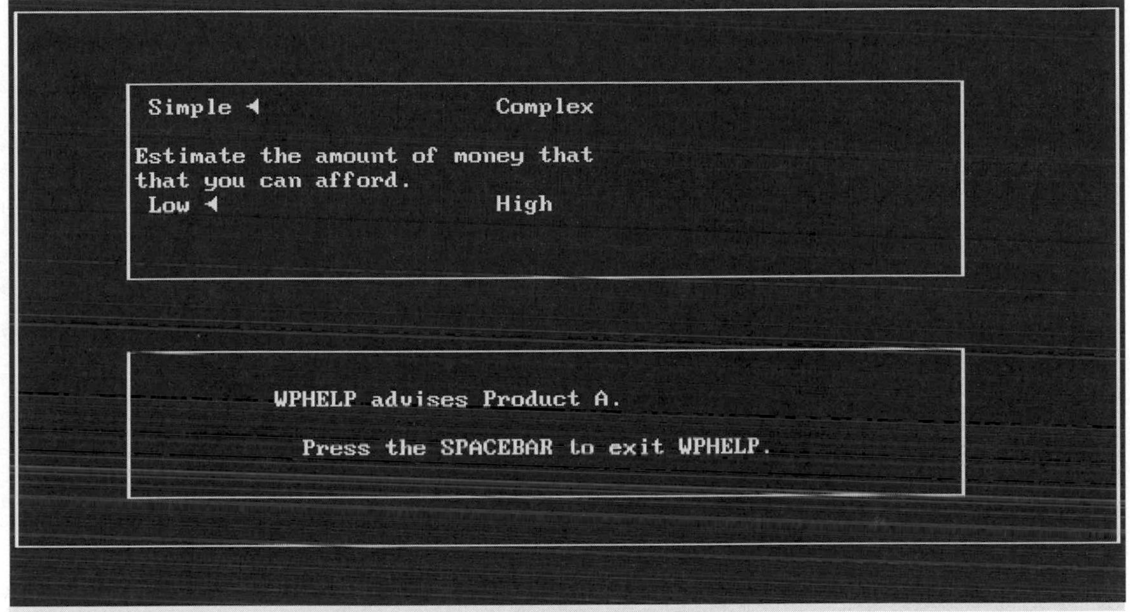

FIGURE M7.9 LAST CONSULTATION SCREEN FOR wp23.kbs

This program advises us on which word processors to select based on type of document and cost. The user interface has two windows, one for messages between the user and expert system, and the other for the advice from the consultation. The two windows partition the display screen horizontally. The program code is detailed in Figure M7.10.

From Comment 1 we see that Window 1 (for questions to the user) is opened, made active, and that text is located inside. The advice window, Window 2, is then made active (C2), and the results of the consultation are displayed. In the WOPEN clause of C1, Window 1 has its upper-left corner at 3,8 and is 8 rows deep and 60 columns wide. The background color (9) is light blue. Note that the clause COLOR = 7 sets the text in Window 1 to white. From the WOPEN of C2, we conclude that Window 2 is located at 14,8 and is 6 rows deep and 60 columns wide. From the 7 in this WOPEN clause and the clause COLOR = 9, we note the background and foreground colors in Window 2 are the reverse of Window 1. Even though two windows appear on the screen, you can write to only one at a time. The last ACTIVE clause identifies where the text from the DISPLAY statement is written.

Our next sample run will show us the use of **overlapping windows** in a consultation. Run wp24.kbs and then obtain a printout of the code. This consultation will retrieve word processing information from the database file wpdb.dbf.

The last Consultation screen is displayed in Figure M7.11. The questions and messages

```
!The file wp23.kbs uses two windows.
          ENDOFF;
          RUNTIME;
ACTIONS
         DISPLAY "This expert system WPHELP will advise you on selecting
a word processor for your IBM compatible microcomputer for your business.

Press the Spacebar to start the consultation.~"
          CLS
          WOPEN 1,3,8,8,60,9
(C1)----> ACTIVE 1
          LOCATE 2,0
          COLOR = 7
          FIND Word_processor
          WOPEN 2,14,8,6,60,7
(C2)----> ACTIVE 2
          COLOR = 9
          LOCATE 1,10
          DISPLAY "WPHELP advises {Word_processor}.
"
          LOCATE 3,7
          DISPLAY "      Press the Spacebar to exit WPHELP.~"
          WCLOSE 1
          WCLOSE 2
          CLS;

RULE 1
IF        Functionality = Simple AND
          Cost = Low
THEN      Word_processor = Product_A;

RULE 2
IF        Functionality = Simple AND
          Cost = High
THEN      Word_processor = Product_A;

RULE 3
IF        Functionality = Complex AND
          Cost = Low
THEN      Word_processor = none;

RULE 4
IF        Functionality = Complex AND
          Cost = High
THEN      Word_processor = Product_B;
ASK Functionality: "What type of document will you need to produce with
your word processor?";

CHOICES Functionality: Simple, Complex;

ASK Cost: "Estimate the amount of money that you can afford.";

CHOICES Cost: Low, High;
```

FIGURE M7.10 PROGRAM CODE FOR TWO WINDOWS (wp23.kbs)

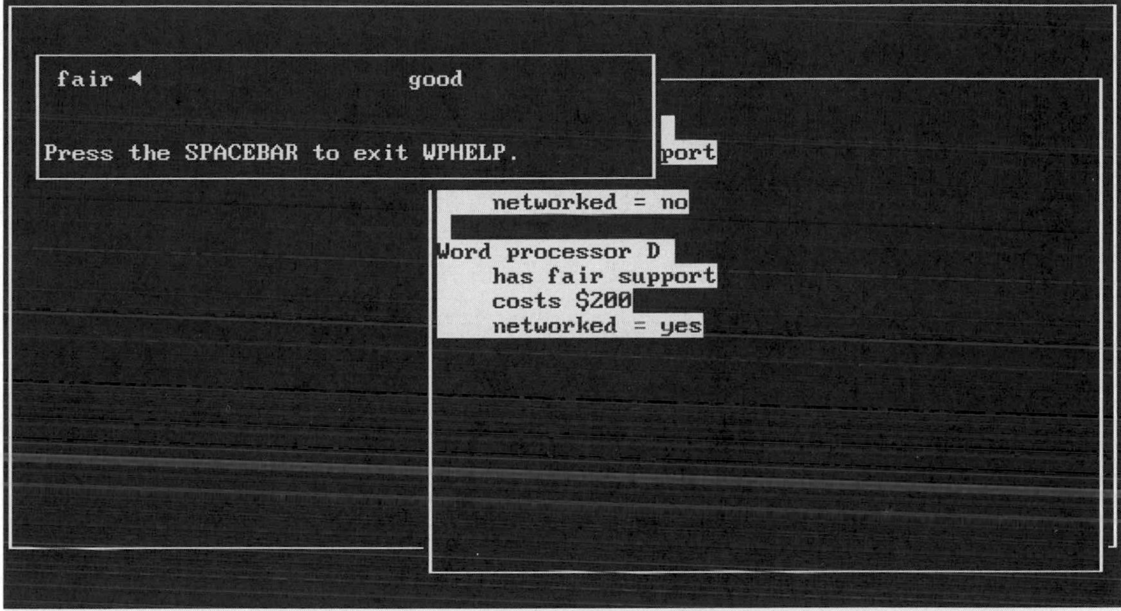

FIGURE M7.11 LAST CONSULTATION SCREEN FOR wp24.kbs

to the user are in the upper left and the results of the consultation on the right. We asked for word processors with good technical support, and we displayed the entire record of each in the right window. The user was also told to turn the printer on for a hard copy of the results.

When we examine the program code (Figure M7.12), we again see the opening of two windows and the switching between the two windows, as indicated in Comments 1 and 2. In C1, we open both windows. Window 1 has a background color of cyan (3) and Window 2 has a background color of white (7). In C2, Window 1 is made active with text in red (4), whereas Window 2 will have text in black (0). Note that there are 3 ACTIVE clauses so that text is written to Window 1, then Window 2, and then Window 1 again. This is why the last message appears in the first window. In C3, we find the EJECT clause, which causes your printer paper to move to the top of the next page. This feature allows you to tear off your report without having to realign the printer.

One more example is appropriate. If you have a color adapter, program wp25 will let you select the background color that you want for your consultation run. If you do not have a graphics adapter, you will see only black and white. Run the consultation and obtain a copy of the program code.

Program wp25 advises you on selecting a word processor according to Functionality and Cost. After the opening screen, however, the color numbers appear on the display, as shown in Figure M7.13.

```
!The file wp24.kbs will access information
!         from a database file called wpdb.dbf.
!         The field names in the database are:
!         Word_proce, Networked, Support, Cost.
!         Overlapping windows will be used.

          ENDOFF;
          RUNTIME;
ACTIONS
          DISPLAY "This expert system WPHELP will display
information from a database file.  Be sure that
the printer is on.

Press the Spacebar to start the consultation.~"
          CLS
          PRINTON
          WOPEN 1,2,2,5,44,3
          WOPEN 2,3,30,20,48,7

          ACTIVE 1
          COLOR = 4
          MENU List, ALL, wpdb, Support
          FIND List

          ACTIVE 2
          COLOR = 0
          WHILEKNOWN Support
               GET List = Support, wpdb, ALL
               RESET Screen
               FIND Screen
          END
          EJECT
          PRINTOFF
          ACTIVE 1
          DISPLAY "Press the Spacebar to exit WPHELP.~"
          CLS;

RULE 1
          IF Support <> UNKNOWN
          THEN Screen = Yes
               DISPLAY " "
               DISPLAY "Word processor {Word_proce}"
               DISPLAY "       has {Support} support"
               DISPLAY "       costs ${Cost}"
               DISPLAY "       networked = {Networked}"
          ELSE Screen = No;

ASK List: "What kind of technical support do you want for your
word processor?";
```

FIGURE M7.12 PROGRAM CODE WITH WINDOWS AND PRINTER OUTPUT (wp24.kbs)

The user is to select the number of the desired color. The screen shows white text (7) on a blue background (1). If the user chooses white for the background, then almost nothing will appear on the screen, except certain words and colors controlled in the text of the ASK statement by the inference engine. After the selection is made, the screen changes color, and the consultation proceeds with a screen as selected. The program code for this consultation run (Figure M7.14) shows how we control color in Text mode.

From C1, we can see that the foreground color is white (7) and the background color (from WOPEN) is blue (1). The search for the desired new background color is

```
What color do you want your background screen?
(The foreground color is set to white.)
Enter a number from the list:

          0 for Black

          1 for Blue

          2 for Green

          3 for Light Blue

          4 for Red

          5 for Magenta

          6 for Brown
```

FIGURE M7.13 CONSULTATION SCREEN FROM wp25.kbs

identified with (C2) FIND Screen. The inference engine takes over and sends the text of the corresponding ASK statement to the screen. This value of the background color number is stored in the variable called Screen. In C3, we read that @Screen holds the user's choice and is the last parameter in the WOPEN clause. The consultation then proceeds with the new background color for the display screen. The foreground, or writing color, is still white. You may note from C3 that we opened as large a window as possible (80 columns by 25 rows), another way to override VP-Expert's settings.

Specialized Graphical User Interfaces in VP-Expert

Let's turn our attention to building specialized user interfaces in VP-Expert. Built-in commands help us implement Dynamic Images and SmartForms as specialized front ends to our VP-Expert consultations. The next two sections will discuss these additional options for the user interface in VP-Expert.

Dynamic Images

VP-Expert defines a *Dynamic Image* as a predefined graphics object that is dynamically linked to or associated with a specified variable. When the value of the variable

```
!The file wp25.kbs lets the user set the background color
!          for a WPHELP consultation.

           ENDOFF;
           RUNTIME;
ACTIONS
           WOPEN 1,0,0,24,79,1 <-- (C1)
           ACTIVE 1
           COLOR = 7 <-- (C1)
           DISPLAY "This expert system WPHELP will advise you on
selecting a word processor for your IBM compatible microcomputer
in your business.

Press the Spacebar to start the consultation.~"
           CLS
   (C2) -> FIND Screen
           WCLOSE  1                (C3)
           WOPEN 2,0,0,24,79,@Screen
           ACTIVE 2
           FIND Word_processor
           DISPLAY "WPHELP advises {word_processor}.
"
           DISPLAY "       Press the Spacebar to exit WPHELP.~"
           WCLOSE 2
           CLS;

RULE 1
IF         Functionality = Simple AND
           Cost = Low
THEN       Word_processor = Product_A;

RULE 2
IF         Functionality = Simple AND
           Cost = High
THEN       Word_processor = Product_A;

RULE 3
IF         Functionality = Complex AND
           Cost = Low
THEN       Word_processor = none;

RULE 4
IF         Functionality = Complex AND
           Cost = High
THEN       Word_processor = Product_B;
ASK Functionality: "What type of document will you need to
produce with your word processor?";

CHOICES Functionality: Simple, Complex;

ASK Cost: "Estimate the amount of money that you can afford.";
      .
      .
      .
```

FIGURE M7.14 PARTIAL PROGRAM CODE FOR SELECTING BACKGROUND COLOR IN TEXT MODE (wp25.kbs)

changes during a consultation, the "appearance" of the graphics object on the screen is changed or updated to reflect this new value. Dynamic Images are used to communicate information in a graphical rather than text format. For example, a dangerously high temperature could be communicated by displaying a picture of a gauge showing the temperature in a red zone. A meter or scale could indicate how a user is doing while taking a test.

Dynamic Images can be used to create a control panel on the screen for the user. The user interacts with the control panel by manipulating a graphic image rather than selecting from a menu of choices or typing in a response. Clicking with a mouse on a rectangular figure or button labeled DONE to end a consultation would be one example.

A Dynamic Image is not displayed on the screen until the variable to which it is linked is assigned a specific value other than UNKNOWN. Once the image appears on the screen, it can be removed by using the RESET clause to change the variable to UNKNOWN. The user can change the value of the variable only by clicking the button or object with a mouse. The kinds of images that can be easily created are gauges, meters, buttons, formfields, and time-series graphs. The following is a summary of the specialized commands needed to implement Dynamic Images:

VARIABLE INTERACTION (USING A MOUSE)

BUTTON	Displays a button-like or oval image on the screen that is linked to a variable's Yes or No value.
LBUTTON	Displays a button-like screen image containing a text label that is linked to a variable's Yes or No value.
FORMFIELD	Displays a rectangular box or formfield containing the value of a linked variable.
HOTREGION	Displays an invisible rectangular "hot" or active region on the screen. When this area is clicked on with a mouse, a linked variable is toggled or switched between a variable's Yes or No value.

IMAGE DRAWINGS

METER	Displays a graphic meter or scale image linked to the value of a given variable.
HGAUGE	Displays a horizontal graphical slide scale linked to the value of a given variable.
VGAUGE	Displays the same as HGAUGE but in a horizontal format.

OTHER COMMANDS

HYPERTEXT	Displays external file data (hypertext) when a specified variable is assigned a hypertext value.
TRACK	Displays a graphical time-series graph linked to the last 50 values of a given variable.

The graphical results of executing BUTTON, LBUTTON, FORMFIELD, FORMFIELD/ CHOICES, HGAUGE, and METER commands are illustrated in Figure M7.15.

All of these dynamic image commands work only in Graphics mode with graphical coordinates determined by the pixel resolution of the GMODE setting. Two images, however, FORMFIELD and LBUTTON, are text images and require text row and column coordinates. For text images, the x (column coordinate) varies from 0 to 79 and the y (row coordinate) ranges from 0 to 24. The following code segment will help to unify these concepts.

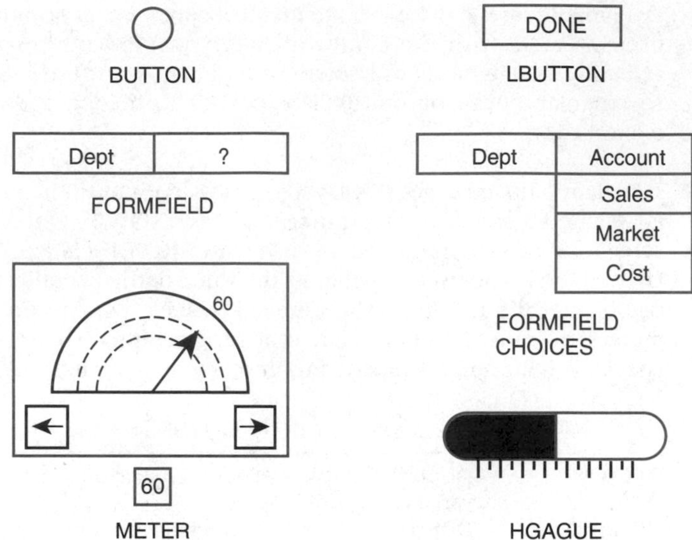

FIGURE M7.15 GRAPHICAL OUTPUT OF SELECTED COMMANDS

```
ACTIONS
    GMODE = 14
        :
    Speed = 10
    Done = No
    Gas = No
    Brake = No
    WHILETRUE Done = No
        THEN
        END
    TMODE;
HGAUGE Speed: 30,30,0,100;
LBUTTON  Gas: 5,7;
LBUTTON  Brake: 15,7;
LBUTTON  Done; 70,2
WHENEVER accelerate
IF       Gas = Yes
then       Speed = (Speed + 2)
        Gas = No;
WHENEVER decelerate
IF       Brake = Yes and Speed > 0
THEN       Brake = No
        Speed = (Speed - 2);
    :
    :
```

The GMODE setting of 14 places the screen in graphics modes with 640 columns and 200 rows. Sixteen colors are possible in this mode. The mouse commands needed to manipulate the values of the three label buttons are not shown. The three label or text buttons (LBUTTON) are displayed on the screen at the specified text coordinates. VP-Expert automatically sizes the text name of the variables (Gas, Brake, Done) inside a rectangle. Pressing a mouse button while pointing to a label button causes the value of the associated variable to switch from Yes to No, or vice versa.

The HGAUGE statement creates a horizontal slide gauge or scale. This image is linked to the variable Speed. The graphic image is displayed on the screen at graphical coordinates 30,30. The horizontal meter is scaled between a low value of 0 and a high value of 100. In addition, the HGAUGE statement links the image to the variable Speed. When the value of Speed is increased or decreased by a WHENEVER rule, the change is reflected by the gauge. When the user clicks on the Done button, the value of Done toggles to Yes. The condition Yes = No is, therefore, a false statement, and the program execution exits to the command after the End of the WHILETRUE loop. The TMODE command would need to be executed and the program consultation ends. (Many of the details of these commands can be found in the VP-Expert Help facility.)

SmartForms

Our next and last graphics topic is SmartForms. By using a host of advanced VP-Expert features, you can create user interfaces that resemble regular paper forms. These forms could resemble the forms you fill out when you prepare your taxes, apply for a loan, or visit a doctor's office. Unlike a traditional static form, a carefully designed VP-Expert SmartForm can be progressive in its display of form parts. Appropriate form fields pop up when needed and disappear when not needed. For example, a SmartForm application might cause a form field labeled "Do you have diabetes?" to appear. A user response of Yes would then cause additional form fields to appear. A No reply would bypass that part of the form.

SmartForms are useful in data-intensive consultations. They allow the user to view many questions at a time, respond to the displayed fields in any order, and change responses; they imitate the actual paper forms needed in some applications. The choice of a graphical image, the interview style, and the form field approach gives the expert system designer more flexibility in the selection of a user interface.

Building a SmartForm in VP-Expert contains three basic elements: the ACTIONS block, WHENEVER rules, and special statements. The ACTIONS block must include a GMODE setting for a graphics screen and a WHILETRUE-END loop to keep VP-Expert in a loop while the forms are being displayed and filled out by the user. The WHENEVER rules are rules that are tested whenever the value of the associated or linked variable changes value. Hence, WHENEVER rules can be used to do the following:

- Offer advice to the user if a certain response is entered in the form.
- Remove irrelevant fields from the screen display.
- Add new fields to the form display.
- Maintain consistency in the form layout.

The third part of SmartForm construction consists minimally of two Dynamic Image statements, FORMFIELD and LBUTTON. The FORMFIELD statement is used to create the form fields, and the LBUTTON statement is used to create a button called Exit or Done. Responses from the user are effected with the use of a mouse.

The FORMFIELD command appears on the screen as a rectangle with text to the right of the figure and information within the rectangle. Using the mouse, the user may change the value appearing inside the form field. Let's discuss the FORMFIELD statement in detail. The syntax for the command is

```
FORMFIELD variable-name, x,y, length, color-value.
```

The variable name or first parameter is the variable linked to the form field. The current value of this variable is displayed inside the form field. (The form field cannot be empty.) The form field, moreover, is located on the screen with its upper-left corner at text column x and row y and is sized according to the length specified in the FORMFIELD statement. The color-value (0–7) is the text background color. If no value is specified, the text background color defaults to black. A label to accompany the form field can be generated by an ASK statement that specifies the same variable. The ASK text is placed, right justified, against the left edge of the form field. If there is a CHOICES statement (or MENU clause) for the same variable, an "invisible" menu is created. The user can make the menu visible by pressing the Tab key or pressing the mouse button. A pull-down menu containing the value from the associated CHOICES statement and variable is displayed, as illustrated by the following code section:

```
ACTIONS
   GMODE 14
   Season = Needed
   WHILETRUE Season  <> Quit
         THEN
         END
   TMODE;
      :
      :
   FORMFIELD Season: 11,5,12,5
   CHOICES Season: Summer, Fall, Winter,
            Spring, Quit;
   ASK Season: "Season:"
```

From this example, FORMFIELD displays a form field of length 12 that contains the value "Needed" with the text season positioned to the left of the form field. If the Tab key or mouse button is pressed while pointing to the form field, a pull-down menu of

Summer, Fall, Winter, Spring, Quit will be displayed. If a new value is selected, the new value of the variable (Season) will be displayed in the form field. When the user selects Quit, the condition in the WHILETRUE-END loop is evaluated as false. Execution falls out of the loop to the TMODE statement and the consultation ends.

The LBUTTON statement can also be used to "Exit" a SmartForm. The format of this instruction is as follows:

```
LBUTTON variable-name x,y, color-value1, color-value2, text.
```

This statement creates a rectangular Dynamic Image called a "label button" linked to the Yes or No value of variable-name. The button is automatically sized to enclose the specified text (last parameter above) or the name of the variable if no text is given on the command line. The text has a maximum length of 40 characters. Pressing the mouse button while pointing to a label button causes the value of the variable to toggle between Yes and No. If the variable-name has no preassigned value, it defaults to No. The difference between BUTTON and LBUTTON is that BUTTON appears "round" without text or values and LBUTTON displays a rectangular figure with the name of the variable or other text within the figure. The following code excerpt illustrates these points:

```
ACTIONS
GMODE 14
Exit_button = NO
WHILETRUE Exit_button = No
   THEN
   END
TMODE
DISPLAY "Press any key to exit the
     consultation.";
   :
   :
LBUTTON Exit_button: 5,7,1,2,EXIT
```

In this example, the LBUTTON displays a label button linked to the variable Exit_button. The label button box contains the text characters "EXIT" from the last parameter of the LBUTTON statement. The value of the variable Exit_button is No. When the user clicks the mouse button while pointing to the EXIT button, the value of Exit_button changes to Yes. Program control then goes out of the loop (false condition) and passes to TMODE and DISPLAY. The EXIT button is positioned with its left corner at text column 5 and text row 7. When the value of the variable Exit_button is No, the interior of the EXIT button is color 1 (blue). The interior color changes to color 2 (green) when the Exit_button variable has the Yes value. The program code could contain many other label buttons and form fields for the complete user interface.

Two additional commands, HYPERTEXT and HOTREGION, can add more features to Dynamic Images or SmartForms. Upon execution of the HYPERTEXT instruction, VP-Expert scans an external file of "text" and retrieves the text block associated with the current value of an associated variable. This feature is useful if more textual information is needed about a given "word" that appears on the screen. It is functionally equivalent to having an on-line dictionary available for selected "key words." Hypertext words appear on the screen in uppercase and can be accessed by pointing with a mouse. Hypertext words can also be nested.

The other command, HOTREGION, can be used to place an invisible graphic trap on the screen for use by a mouse. This trap, or "hot region," covers a rectangular region on the graphics screen. When the mouse button is pressed while pointing to an area on the screen covered by a HOTREGION, the value of the linked or associated variable changes from Yes to No, or vice versa.

Examples of Advanced Interfaces

Examples of these advanced user interfaces are appropriate. The following chart summarizes four programs (from the Assignment section of this module) and selected commands:

PROGRAM NAME	ASSIGNMENT NUMBER	COMMANDS USED
carpict.kbs	7	Graphics primitives HOTREGION LBUTTON
formdemo.kbs	12	BUTTON FORMFIELD CHOICES METER
fromage.kbs	13	LBUTTON FORMFIELD
tempsim.kbs	14	BUTTON LBUTTON FORMFIELD METER VGAUGE
goodies.kbs	15	BUTTON LBUTTON HOTREGION VGAUGE TRACK HOTREGION External text files

The user interface of carpict.kbs shows a graphic of a car surrounded by label buttons, as shown in Figure M7.16.

Another program example, formdemo.kbs, shows form fields, meters, and a pull-down menu from the CHOICES command. This interface is shown in Figure M7.17.

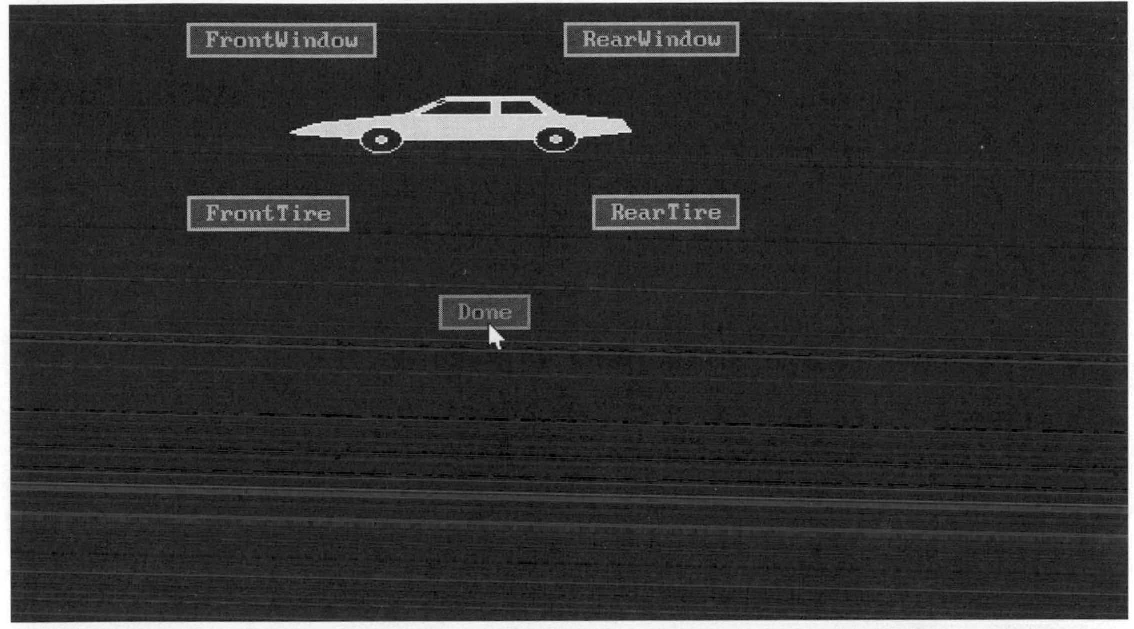

FIGURE M7.16 GRAPHICAL USER INTERFACE FOR carpict.kbs

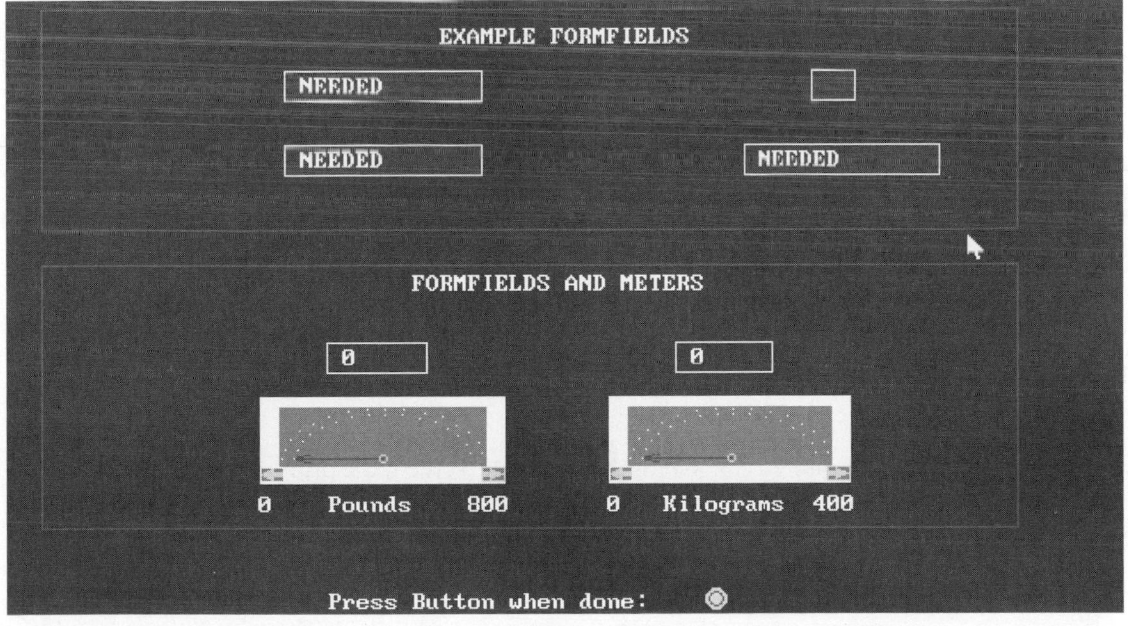

FIGURE M7.17 GRAPHICAL USER INTERFACE FOR formdemo.kbs

Another example of a graphical user interface in found in the program fromage.kbs, illustrated in Figure M7.18.

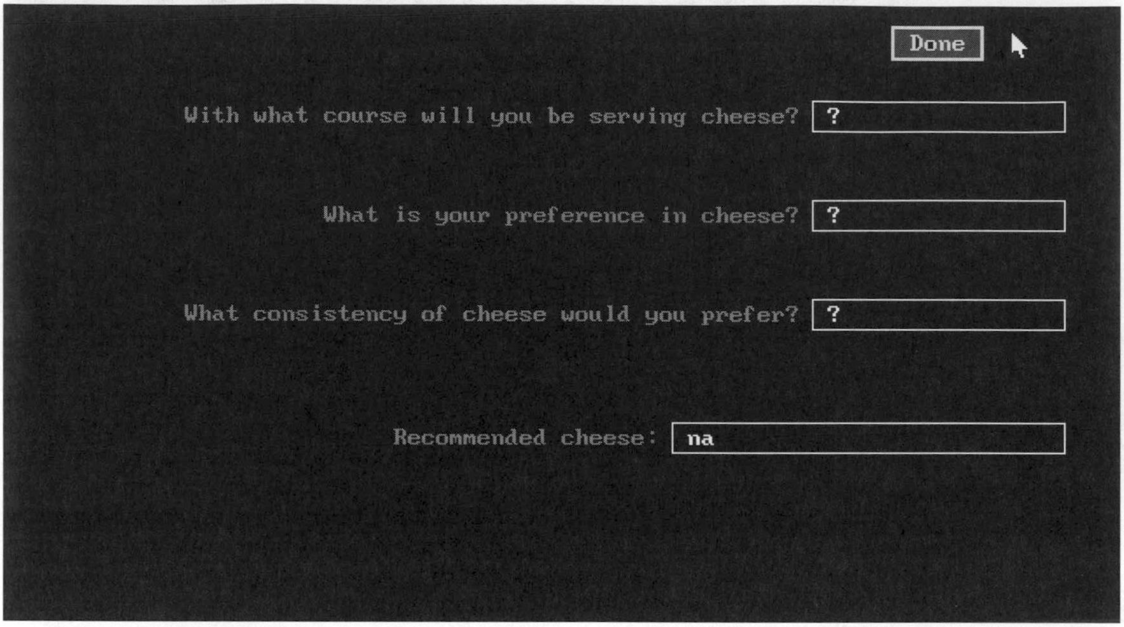

Done

With what course will you be serving cheese? ?

What is your preference in cheese? ?

What consistency of cheese would you prefer? ?

Recommended cheese: na

FIGURE M7.18 GRAPHICAL USER INTERFACE FOR fromage.kbs

A control panel interface consisting of meters, gauges, buttons, label buttons, and form fields is shown in the user interface for temperature simulation, tempsim.kbs, in Figure M7.19.

All of these advanced features of graphical user interfaces work best with a mouse. Designing and selecting the most appropriate interface for the expert system consultation is a real challenge.

Summary

We have explored commands to control the screen, the printer, the run-time environment, color, graphics, and a mouse. We ran sample programs to illustrate the methods by which VP-Expert allows you to implement the user interface. Advanced graphical user interfaces can be implemented with Dynamic Images and SmartForms. If you have not read the guidelines in Chapter 7 yet, we suggest that you do.

In these seven modules, we have discussed in detail the implementation of the three major components of an expert system, the knowledge base, the inference engine, and the user interface. None of the three parts can be neglected if you are to design

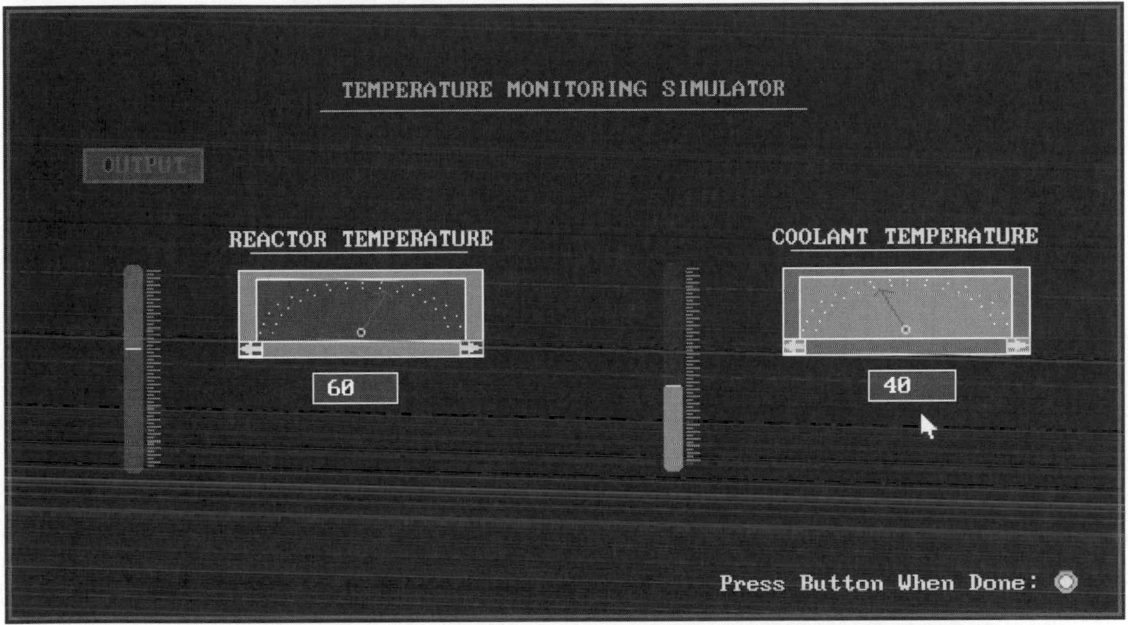

FIGURE M7.19 GRAPHICAL USER INTERFACE FOR tempsim.kbs

and implement a workable and usable expert system. In the next and last module, we will discuss additional commands that you can use to implement an entire system.

Key Points

- The user interface is one of the three major components of an expert system.
- The user interface is the total communications interface between the user and the expert system. The user interface is concerned with screen design, text, color, graphics, printer output, and alternative input or output devices.
- Text can be output to the screen with the DISPLAY clause. For displaying variable values, various parameters in this clause can format the size of the field or output the CNF associated with the value.
- For IBM microcomputers, there is a one-to-one association between symbols and numbers (0–255). This correspondence is called the American Standard Code for Information Interchange (ASCII); it can be displayed using the CHR command.
- Seven clauses help manipulate characters and strings (lists of characters):

CHR	Places the ASCII character of a given integer in a variable
ASC	Stores the numeric value of an ASCII character in a variable
UPPER	Converts a string to uppercase letters
LOWER	Converts a string to lowercase letters
CONCAT	Places a string at the end of another string
POS	Finds the beginning position of a substring within another string
SUBSTR	Finds a substring within a string and places it in a variable

- Windows are rectangular regions of the display screen. Various arrangements of multiple windows on the video display can aid in the organization of information during a consultation.
- Windows can be created with the WOPEN clause. Text can be written within a given window after an ACTIVE clause is executed, followed by LOCATE and DISPLAY commands.
- If a microcomputer system has a graphics board, color can be utilized in the user interface. Numbers are associated with the background color and the foreground color. The BKCOLOR and COLOR commands are used to set these colors (for Text modes).
- Graphics can be utilized in the user interface. Graphics consist of circles, lines, polygons, and so on, in addition to text. To utilize graphics, use the GMODE command and the appropriate command to draw the graphic. Text mode settings are 0 to 3 or 7; Graphics mode settings range from 4 to 6 or 8 to 19. The "size" of the graphics screen or resolution changes in the various Graphics modes.
- Three commands, GLOCATE, GDISPLAY, and GETCH, support the display of text and handling of text in Graphics mode.
- The PCX instruction is used to load and display preexisting picture files.
- Common graphics boards include the Color Graphics Adapter (CGA), the Enhanced Graphics Adapter (EGA), the Video Graphics Adapter (VGA), Super VGA, and Enhanced Super VGA, and the Hercules Graphics Adapter (HGA).
- Printing in VP-Expert can be controlled with the PRINTON, PRINTOFF, PDISPLAY, and EJECT commands.
- Five commands, EXECUTE, RUNTIME, ENDOFF, BLINEON, and BLINEOFF, can improve the run-time environment for the user of an expert system.
- A group of commands within VP-Expert supports a mouse for user input to an expert system. Mouse commands work in Graphics mode, as opposed to Text mode.
- VP-Graphics is an add-on library to support presentation graphics in the user interface for VP-Expert applications.
- Dynamic Images—gauges, buttons, formfields, meters, and time-series graphs—can be drawn with special commands. Dynamic Images help to communicate information in a graphical context useful in scientific, engineering, or business applications.
- SmartForms allow the user interface to be a form-like questionnaire instead of an interview. A WHILETRUE-END loop is used to keep VP-Expert in a loop as the user fills out the form.

Exercises

1. Indicate what is displayed on the screen for each of the following:
   Cost = 324
   a. DISPLAY "The Cost is {Cost}."
   b. DISPLAY "{Cost} is the cost."
   c. DISPLAY "The Cost is {10 Cost}."
   d. DISPLAY "The Cost is ${10 Cost}."
   e. DISPLAY "The Cost is {# cost}."

2. Using the ASCII table in Appendix B, what is the number associated with each of the following symbols?
   a. ?
   b. Q
   c. g
   d. 3
3. Using the ASCII table in Appendix B, what is the symbol associated with each of these numbers?
   a. 27
   b. 77
   c. 127
   d. 177
4. In the following clause
   WOPEN 1,2,3,4,5,6
   a. Which number identifies the background color?
   b. Which two numbers give the row and column coordinates of the upper-left corner of the window?
   c. Which number identifies the window number?
   d. Which two numbers determine the width (rows) and length (columns) for the window?
5. Which of the following commands opens a window across the bottom half of your screen?
   a. WOPEN 3, 0, 40, 12, 7
   b. WOPEN 3, 10, 10, 40 12, 7
   c. WOPEN 3, 39, 12, 12, 40, 7
   d. WOPEN 3, 10, 0, 10, 40, 7
6. Given the clause
   WOPEN 9,3,4,12,20,8
   what are the coordinates of
   a. the upper-left corner?
   b. the lower-left corner?
   c. the upper-right corner?
   d. the lower-right corner?
7. Given the clause
   WOPEN 3,7,6,4,9,7
   enter the actual value for row and column for the following:
   a. LOCATE 0,0
   b. LOCATE 1,1
   c. LOCATE 2,1
   d. LOCATE 3,2
8. Given the code
   WOPEN 7,0,4,9,20,7
   WOPEN 9,4,7,9,20,9
   ACTIVE 7
   ACTIVE 9
   LOCATE (4,7)
   CHR 70,X
   DISPLAY "(X)"

a. How many windows are created?
b. Which window will be written to?
c. Where will the output be written?
d. What is the output?

9. In regular Text mode, the VP-Expert consultation screen has
   a. 25 rows and 25 columns.
   b. 80 rows and 25 columns.
   c. 25 rows and 80 columns.
   d. 24 rows and 79 columns.

10. For each of the following, write two Text mode commands that will allow you to write:
    a. Red text on a blue screen
    b. White text on a red screen
    c. Green text on a white screen
    d. Yellow text on a yellow screen

11. To utilize graphics, what command must be executed?
    a. GRAPHICS
    b. GMODE number
    c. DRAW shape
    d. CGA

12. What are the printer output and screen output for the following sequence of code?
    DISPLAY "X"
    PRINTON
    DISPLAY "Y"
    DISPLAY "Z"
    PDISPLAY "Q"
    PRINTOFF
    PDISPLAY "R"

13. What is the output to the screen from the following code sequence?
    CHR 36, dollar_sign_character
    ASC @dollar_sign_character, dollar_sign_value
    DISPLAY "The ASCII {character} {dollar_sign_character} has a numeric value of {dollar_sign_value}.";

14. What is the screen output from the following code segment?
    product_name = VP-Expert
    type = expert
    POS @topic, @product_name, offset
    DISPLAY "The substring {topic} starts in position {offset} of the string {product_name}.";

15. What is the output to the screen?
    product_name = VP-Expert
    SUBSTR @product_name, type, 3, 6
    DISPLAY "The variable called topic which has the value {topic} is a substring of

the string {product_name} which is stored in the variable called product_name. The substring {topics} starts in position 3 and has a length of 6 characters.";

16. Which of the following commands will not work in Graphics mode?
    a. GBCOLOR
    b. RECTANGLE
    c. ASK
    d. ELLIPSE
    e. PDISPLAY
17. In Graphics mode, which of the following is false?
    a. Color is available.
    b. Text can be displayed.
    c. Polygons can be drawn.
    d. The number of columns and rows is constant.
    e. The number of available colors varies.
18. To get into a Graphics mode, set the GMODE number to an integer
    a. Greater than 0.
    b. 4 to 6 or 8 to 19.
    c. 0 to 19.
    d. Greater than 10.
    e. 5 to 8 or 9 to 19.
19. Which VP-Expert command transfers control of a consultation to DOS?
    a. SYSTEM
    b. DOS
    c. EXIT
    d. QUIT
    e. TRANSFER
20. Which of the following is not valid for Graphics mode?
    a. mouse commands
    b. SmartForms
    c. Dynamic Images
    d. HYPERTEXT
    e. WOPEN

Assignments

1. Make a copy of wp21.kbs. At the c:\VPX_samp prompt, type:
   COPY wp21.kbs my21.kbs.
   Use the editor to change the numbers in the FOR loop in wp21.kbs from
   FOR X = 50 to 65
   to
   FOR X = 41 to 54
   a. Run a consultation and obtain a printout of the results with Shift-Print Screen.
   b. Insert the PRINTON and PRINTOFF commands. What happens when you run the consultation now?

2. Make a copy of wp22.kbs under the name my22.kbs.
   a. Use the editor to delete the tilde (~) from two DISPLAY statements. Run a consultation and record the results.
   b. Use the editor to delete the clause
      COLOR = 12
      Replace this clause with the clause
      ACTIVE 1
      Run a consultation and record the results.
3. Make a copy of wp23.kbs and call it my23.kbs
   a. Use the editor to change COLOR = 7 to COLOR = 5. Also insert EXECUTE; above the ENDOFF statement.
   b. Run a consultation and record the results.
4. Make a copy of wp25.kbs and call it my25.kbs. Rewrite the code so that:
   a. The opening message appears in Window 1.
   b. The advice message appears in Window 2.
   c. The ending message appears in Window 3.
   Enter your revisions and test the results.
5. Make another copy of wp25.kbs and call it my25a.kbs.
   a. Write the code to let the user select the foreground color in addition to the background color.
   b. Using the editor, add the new code to my25.kbs.
   c. Run consultations with your new version of my25.kbs.
   d. Does your new program run correctly?
6. Make a copy of wp24.kbs and call it my24.kbs.
   a. The PDISPLAY command sends the output of the DISPLAY statements to the printer but not the display screen. Write the code to modify my24.kbs so that you can send output to the printer or the screen, but not both.
   b. Using the editor, enter your changes and test the results.
7. If you have an EGA or VGA graphics adapter and mouse, run carpict.kbs from the Sample File directory.
   a. Record the results.
   b. Obtain a printout of the code.
   c. Circle the graphics commands.
8. Use the VP-Expert Help Facility to obtain a printout of the details of the following commands:
   a. Window commands
      WOPEN
      LOCATE
   b. RUN-TIME commands
      EXECUTE
      RUNTIME
      ENDOFF
      BLINEON
      BLINEOFF
   c. Printer Commands
      PDISPLAY
      EJECT

9. Use the VP-Expert Help Facility to obtain a printout of the details of the following graphics commands:

MANAGING THE SCREEN	DRAWING THE GRAPHICS
GBCOLOR	ELLIPSE
GCLS	RECTANGLE
GCOLOR	LINETO
GLOCATE	PSET
GDISPLAY	MOVETO
GETCH	PCX

10. If you have a mouse on your system, use the VP-Expert Help Facility to obtain printed information on each of the following:
    a. BUTTON
    b. LBUTTON
    c. MOUSEON
    d. MOUSEOFF
    e. GETMOUSE
    f. MOUSEX
    g. MOUSEY
11. Make a copy of wp22.kbs and call it graph.kbs.
    a. Write the changes needed to convert this program to Graphics mode. Select any mode available on your system. Use the RECTANGLE command to draw a rectangle on the screen.
    b. Using the editor, make these changes.
    c. Run the consultation. Is there any difference in the screen from the user's perspective?
    d. Summarize the difference from a programming prospective.
12. If you have a mouse and a VGA or EGA graphics adapter, run a consultation with formdemo.kbs from the subdirectory VPX_samp.
    a. What does the consultation add to the "user interface"?
    b. To examine the source code, obtain a printout of formdemo.kbs.
    c. Is this program in Graphics or Text mode?
    d. Why are WHENEVER rules used?
13. If you have a mouse and a VGA or EGA adapter, run a consultation with fromage.kbs.
    a. Describe the consultation.
    b. Obtain a printout of fromage.kbs.
    c. Does this consultation run in Text or Graphics mode?
    d. Explain the instruction "WHILETRUE DONE = NO THEN END."
    e. What commands are used to implement the mouse?
14. If you have a mouse and a VGA or EGA adapter, run a consultation with tempsim.kbs.
    a. Describe the consultation.
    b. Obtain a printout of tempsim.kbs.
    c. Does this consultation run in Graphics or Text mode?
    d. What commands are used to create the "graphical display"?
    e. How are the "scales" drawn on the screen?

15. If you have a mouse and a VGA or EGA adapter, run a consultation with goodies.kbs. Be sure to use the mouse to reveal the pull-down menus.
    a. Describe the consultation.
    b. Obtain a printout of goodies.kbs.
    c. Is this consultation in Graphics or Text mode?
    d. Explain what the HYPERTEXT and HOTREGION commands do.
    e. Why is the FDISPLAY command used?
16. Use the VP-Expert Help facility to obtain printed documentation on the following advanced graphics features:
    Dynamic Images
    BUTTON
    LBUTTON
    FORMFIELD
    DBFORM
    METER
    HGAUGE
    VGAUGE
    TRACK
    HYPERTEXT
    HOTREGION
17. Use the VP-Expert Help Facility to obtain printed information on control measures needed to implement "SmartForms."
    WHENEVER RULES
    WHILETRUE-END loop

Module 8

Systems Development with VP-Expert

This last module is designed to give you an overview of the power and flexibility available within VP-Expert. As you are designing and developing an expert system application, it is helpful if one of the team members is familiar with the features of the expert system shell. This familiarity will enhance the entire software engineering cycle.

This module is divided into two basic parts. The first section will take you through a short review and summary of the VP-Expert commands by category. We will summarize commands relating to the inference engine, confidence factors, loops, screen display and painter, database, and worksheets. New commands will be presented in more detail than commands previously covered. The categories are organized by commonality rather than application. The second section will give you an overview of some additional advanced features available for the software engineer. Included in each section are summaries of commands not discussed elsewhere in the text. Commands added to version 3.0 of VP-EXPERT are also summarized.

Summary of Commands by Category

Inference Engine

The power of the **inference engine** is invoked when a FIND command has a variable with an UNKNOWN value. If the variable has a known value, the FIND variable will be skipped and control will pass to the next command. When the FIND variable is activated, the inference engine will search the Rules, which are expressed in the IF-THEN or IF-THEN-ELSE format. If a value of the variable cannot be found there, the search for the goal variable is extended to ASK variable commands. If the ASK statement is followed by a CHOICES statement, then the user will have a selection to make; otherwise, the user types in an appropriate response. This chaining of rules is altered somewhat if the Rule is preceded by a WHENEVER value; if the value changes, the Rule is automatically tested. The RESET command will change the value of a variable to UNKNOWN. Hence, a subsequent FIND variable command will initiate a search for the goal variable.

If the inference engine is attempting to FIND values for a PLURAL variable, there are several useful commands that apply to PLURAL variables only: POP, COUNT, and SORT. The PLURAL clause allows the value of a variable to have multiple values. Hence, the inference engine continues its search even after it has found one value.

The POP command can be used to manipulate the individual values of the PLURAL variable. The syntax is

```
POP plural_variable, non_plural_variable.
```

When this clause is executed, the most recently found value of the plural_variable is moved or stored in the non_plural_variable. The next POP clause will move the second value to the destination variable, and so on. Placing the POP clause inside a WHILEKNOWN loop can display the multiple values of a PLURAL variable. For example,

```
ACTIONS
    FIND Function
    WHILEKNOWN One_function
        RESET One_function
        POP Function, One_function
        DISPLAY "{one_function} is a
            desired feature."
    END;

    ASK Function: "Select five features you
        want in your word processing
        package.~"
    CHOICES Function: ....
    PLURAL: Function;
```

In this portion of code, the user selects five features desirable for word processing. The WHILEKNOWN loop displays these choices to the screen by using the POP clause. The last value selected is the first function displayed. The fourth value selected by the user is the second value displayed. The stack of values, hence, is displayed from last selected to first selected.

Another clause used to implemented the PLURAL variable is COUNT. This instruction counts the number of values of a PLURAL variable. The syntax is

```
COUNT plural_variable_name, counter_variable.
```

The COUNT clause will place the number of values for the plural_variable_name in the counter_variable. In our POP word processing example, the clause

COUNT Function, Counter

would place the number 5 in the variable called Counter.

Another useful command related to a PLURAL variable is SORT. The clause SORT plural_variable will sort the various values of the plural_variable by confidence factors (CNFs) in descending order. Hence, POP, COUNT, and SORT are clauses used to manipulate PLURAL variables.

Loops

Loops are clauses that cause a repetition of a given set of instructions. VP-Expert has three repetitive structures:

```
WHILEKNOWN variable_name
END

WHILETRUE condition
END

FOR variable = start_value to end_value
END
```

These three constructs, WHILEKNOWN-END, WHILETRUE-END, and FOR-END, allow all the instructions between the stated keyword and END to be repeated. In the WHILEKNOWN loop, the instructions are repeated until the END is reached and the value of the variable is UNKNOWN. As long as the variable has a KNOWN value, the instructions within the loop are repeated.

The WHILETRUE loop works in the same manner except that there is a condition after the keyword instead of a variable name. The condition may be a simple statement or a compound one using AND and OR connectives. As long as the condition has the value TRUE, the loop is executed. At the start of the loop, the system tests the condition. If the condition has a FALSE value, then control passes to the instruction after the END clause. If the first test of the condition is FALSE, then the loop is skipped entirely.

The FOR-END loop works on an automatic counter. The start_value and the end_value must represent numbers. At the END of the loop, the loop variable is automatically incremented by one. A test is made to determine if the variable is past the end_value. If it is, the loop is skipped; otherwise, the loop is repeated. It is possible, however, to use instructions within the loop to increase or decrease the control variable using assignment statements. These three loop constructs—WHILEKNOWN_END, WHILETRUE_END, and FOR_END—are very useful for system development work.

Confidence Factors

Expert system development is concerned with the degree of certainty or confidence for conditions and conclusions. Hence, handling the *confidence factors* or CNFs is part of the system development work. Three clauses—CNF, TRUTHTHRESH, and SORT—assist with this task. The statement CNF = number can be inserted at the end of a conclusion in a rule of the knowledge base. As the inference engine does its work, it uses these numbers to calculate the certainty of its final conclusion(s), the value(s) of the goal variable, or the FIND variable.

The TRUTHTHRESH is a clause that affects the firing of rules. If the condition of a rule is true and the CNF of the condition is greater than or equal to the TRUTHTHRESH value, then the rule fires. However, if the condition is true and the CNF value for the condition is below the TRUTHTHRESH value, the condition is reevaluated to be false. This clause may occur in the ACTIONS block and affects all rules, or within the conclusion of a particular rule, in which case it affects all actions by the inference engine after the rule is fired. If there is no TRUTHTHRESH value in a program, VP-Expert assumes a TRUTHTHRESH of 40. Another command for certainty factors, the SORT clause, sorts the CNF numbers associated with the values of a PLURAL variable in descending order. In summary, CNF, TRUTHTHRESH, and SORT are instructions relating to certainty factors of conditions or conclusions. (Refer to Module 6 for further explanation of these commands.)

VP-Expert has added two clauses, GETCNF and MODCNF, that allow factors assigned before a consultation begins (in the conclusion part of a rule) to be altered upward or downward during the consultation. This feature can be very helpful as new information becomes available during the running of the consultation. It can also be useful because, after firing a series of rules, the confidence factor may be lessened because of the mathematical operations used to calculate confidence factors.

The syntax for GETCNF is

```
GETCNF variable1, variable2.
```

Variable1 is the name of the variable that the CNF is associated with. Variable2 is the name of the variable where the value of the CNF is to be stored.

The syntax for MODCNF is

```
MODCNF variable, value.
```

Variable is the name of the variable associated with the CNF (variable1 from above). Value is a number between 1 and 100 that represents the new CNF (variable2 from above).

Screen Display and Printer

We will present a brief overview of the commands for controlling the *display screen and printer*. Text can be displayed with the DISPLAY statement. The FORMAT command can be used to format the display of numeric variables. The CLS instruction will clear the video display screen. The color of the background and foreground can be controlled with the BKCOLOR and COLOR clauses, respectively. Windows on the screen can be opened with the WOPEN clause and accessed via the ACTIVE and LOCATE instructions. The WCLOSE command will close a window.

One useful command relating to the screen and text is GETCH. This clause will cause the system to wait until the user enters a character at the keyboard. Its syntax is

GETCH variable_name.

VP-Expert waits until the user presses a key. When a key is pressed, its corresponding ASCII value is placed in variable_name. The difference between the GETCH clause and the tilde within a DISPLAY statement is that the GETCH clause captures the ASCII code of the key pressed. GETCH can be used as a control feature in VP-Expert programs.

The printer can be controlled with the PRINTON, PRINTOFF, PDISPLAY, and EJECT instructions.

Database

Access to certain database packages is a powerful tool for expert systems development. VP-Expert has a variety of commands to support this feature. The MENU clause can be used to help generate choices from a database column. For large MENUS, the MRESET clause is used to release the memory used in the MENU clause. The GET clause is used to get a record from the database file. RECORD_NUM is a system variable that indicates the current number of the record from the database. The PUT clause will allow you to place new values back into the database file, whereas APPEND will allow you to add a new record to the database file. The CLOSE command will close a database file.

One helpful command that we have not yet used is the INDEX clause. To speed access to a database file, the database management system may allow the creation of an index to a field or attribute of the database. These **indices**, or **pointers**, are stored in an INDEX file. If the database file to be accessed has an existing INDEX file, then VP-Expert can utilize this index. The clause's syntax is

INDEX index_file_name, database_filename_for_index.

The first parameter is the name of the existing index file. The physical index file has a .ndx extension. The second parameter is the database name of the field upon which the index is constructed. The advantage of using indexed database files is increased speed of retrieval; in addition, the index will list the values of the indexed field in order. The INDEX clause must be executed before a GET clause.

The SETPOS command is used to set the access point in a database file. This feature allows random access to any record in the file. Its syntax is

SETPOS file_name, value.

The value parameter represents the record number to be accessed.

Worksheets

VP-Expert allows access to *worksheet information or spreadsheet information* during a consultation. Like database access, this is a powerful tool for the software engineer in expert systems. A worksheet can be read with the WKS instruction. The entire worksheet information is read into arrays in the microcomputer's memory. Hence, it may take a few seconds to load the information and more memory to hold the information. Worksheet information can be accessed by COLUMN, ROW, or NAMED ranges. Updated information can be written to the worksheet with the PWKS clause. The numeric format of the worksheet cell can be altered with the WFORMAT instruction.

Two instructions, WORKON and WORKSHEET, were not used in the earlier modules. The WORKSHEET clause is used before a WORKON instruction. The syntax for WORKSHEET is

WORKSHEET VP

OR

WORKSHEET VPP.

VP is a setting for VP-Planner; VPP is a program setting for VP-Planner PLUS. The WORKSHEET clause indicates which spreadsheet program will be used, and the subsequent WORKON clause identifies the worksheet file for the appropriate spreadsheet program.

The syntax of WORKON is

WORKON path, worksheet_file_name.

The path is the name of the drive where the worksheet program is to be found. After this clause, the instruction takes the user immediately to the opening screens of VP-Planner or VP-Planner PLUS. Control passes from VP-Expert to either of the two spreadsheet programs where normal worksheet operations can be performed.

When you exit the spreadsheet software, you are returned to VP-Expert. If WORKSHEET is omitted, then VP-Expert assumes WORKON for VP-Planner. One WORKSHEET clause affects all subsequent WORKON commands until a new WORKSHEET instruction is encountered. (To use Lotus 1-2-3 or Symphony, use a CCALL, which we will discuss later.)

An example is appropriate here. In a large-scale expert system, we may want to update a special worksheet related to Chinese word processors. The following code accomplishes this:

```
RULE 12
IF Functionality = Chinese
THEN
   DISPLAY "Press any key to update
      information on Chinese
      word processors."
   WORKSHEET VPP
   WORKON C:\VPP, ChineseWP;
```

If the Functionality is Chinese, then the rule displays a message to the user and passes control to VP-Planner PLUS to WORKON, the worksheet ChineseWP. VP-Planner PLUS is found in the vpp directory on drive C. The advantage of this interface being used for two worksheet programs (not just work files) is apparent.

We have reviewed some VP-Expert commands by related category rather than by applications. We added a few new related instructions to our list. In the next section, we will discuss some advanced features that the software engineer can utilize.

General Files

Several clauses—EXISTS, DELETE, and RENAME—are useful for file maintenance. The EXISTS clause allows you to ask if a particular file exists and then place the response (either Yes or No) in a variable. Its syntax is

```
EXISTS file_name, variable.
```

You must be careful, however, because file names with DOS extensions can cause problems. The period in the file name is not allowed as a character in a VP-EXPERT variable. Use the CONCAT command to overcome this problem.

```
file_name = @name
CONCAT ., file_name
CONCAT @extension, file_name
```

The DELETE clause works just as you would expect. It deletes the specified file. Its syntax is

```
DELETE file_name.
```

Of course, the extension must be specified with the file name.

The RENAME clause is used to rename an existing file. Its syntax is

```
RENAME old_file_name, new_file_name.
```

The same comments apply here as made with the EXISTS clause relative to file name extensions.

System Control and Data

VP-EXPERT includes three commands that allow you to interact with DOS. The SYSTEM command transfers control to DOS and allows the use of any DOS command. To return to VP-Expert, type EXIT at the DOS prompt. The DATE and TIME commands will return the system date and time. These two commands could be used to record when a particular application was run. This can be done by including the following code:

```
DATE cur_date
TIME cur_time
DISPLAY "This application was last run on {cur_date} at {cur_time}."
```

Advanced Features

VP-Expert has a variety of advanced features to aid the software engineer in implementing expert systems. We will present an overview of these advanced topics under the following categories: chaining, text files, external programs, graphics, Dynamic Images, Smartforms, and HYPERTEXT. All of these features can add to the expert system implementation.

Chaining Knowledge-based Files

VP-Expert includes instructions that allow the results of one consultation to be passed as input to another, but related, knowledge base. This is an important feature because of the memory required for large expert systems on microcomputers.

With the RULECHAIN clause, knowledge bases can share facts with other .kbs files. In earlier versions of VP-EXPERT, you were required to save facts to a file. Now, facts obtained during the consultation are automatically stored in memory.

The syntax of the clause is

```
RULECHAIN file name
```

where file name is the name of the .kbs file that is to be chained to the current knowledge base. As a final note, RULECHAIN replaces three commands (CHAIN, SAVEFACTS, and LOADFACTS) used in earlier versions of VP-Expert.

Text Files

VP-Expert has several instructions to enhance the transfer of *external text file information* to and from consultations. The clauses—RECEIVE, SHIP, FDISPLAY, SHOWTEXT, CLOSE, RECEIVE, and SHIPPOS—provide for the transition of text material to and from a consultation.

The RECEIVE clause transports a line of text from an external file into a VP-Expert variable. The syntax of the instruction is

```
RECEIVE text_file_name, variable.
```

The current line of text found in the external text_file_name is placed in the VP-Expert variable. The external file is an ASCII text file, which may or may not have a drive specification. When the first RECEIVE clause is executed, the first line from the file is read into the receiving variable. The receiving variable after the RECEIVE command can hold lines of up to 1,000 characters. This is an advantage over the usual 20-character limit on values assigned to VP-Expert variables. Subsequent RECEIVE clauses extract the second line, the third line, and so on. When RECEIVE and DISPLAY commands are placed inside a WHILEKNOWN-END loop, the entire contents of the external file will be displayed to the screen. For example, with this code

```
WHILEKNOWN Description
RECEIVE wpdata, Description
DISPLAY "{Description}"
END;
```

the contents of the file wpdata will be transferred to the variable Description and then displayed on the screen.

The opposite action is implemented with the SHIP clause. The command

```
SHIP text_file_name, variable
```

will transfer the value of the variable to the external text file. If the text_file_name does not exist, the system will create it and place the contents of the variable as the first information on the file. If the file already exists, the information will be written at the end of the file. Placing SHIP inside a WHILEKNOWN-END loop will write many lines of text to the external file.

A more powerful version of SHIP is the FDISPLAY clause. Its syntax is

```
FDISPLAY text_file_name, "text".
```

Instead of a variable to hold the information, the information inside the double quotes is written to the file indicated. The value of variables can be embedded within this text. For example,

```
WHENEVER Change
IF  Cost < (New_cost)
THEN FDISPLAY wpdata, "{Word_proc} has a new
    cost of {Newcost}.";
```

Both the text and the values of the two variables, Word_proc and New_cost, will be written to the text file wpdata. One interesting use of the FDISPLAY clause is to record the results of a consultation. If an FDISPLAY is placed after each DISPLAY statement, the results of the consultation will be saved to disk.

Another clause that handles text is SHOWTEXT, which displays a page of text from a special indexed text file. The syntax is

```
SHOWTEXT text_file_name, variable.
```

The text_file_name must have a .txt extension and must be created with the BASIC program indexer.bas. The file is created with a special index. When the clause SHOWTEXT is executed, a page of text from the external file with the heading listed in the variable is displayed to the screen. For example, the code

```
DISPLAY "Information on Package_A"
PRINTON
Package = (Package_A)
SHOWTEXT wpdata, Package
PRINTOFF
```

would find, display, and print the page of text in the file wpdata whose heading is Package_A. This is a useful technique for handling large amounts of textual information.

External Programs

Another advanced feature of VP-Expert is its ability to call *external DOS programs*. In addition, VP-Expert allows a consultation to call other programs with the CCALL, CALL, and BCALL clauses.

The CCALL instruction allows the consultation to call an external DOS file that has a .com extension. Files with a .COM extension are limited to 64K. The syntax of the clause is

```
CCALL file_name, "parameters".
```

The file name must have a .com extension, but it is not used in the CCALL clause. The command line parameters are optional. After the new program is executed, control returns to the consultation.

The CALL command is almost identical to CCALL, except that it calls .exe files. .Exe files can be of any size and are relocatable object files. Hence, for this reason, we can say that CALL is more powerful than CCALL. The syntax of the clause is the same as

CCALL. Because of the additional memory, be careful not to conflict with the memory requirements of VP-Expert. These clauses allow you to integrate additional programs, such as telecommunications, graphics, and programs to do mathematical calculations into a consultation. For example, if Rule 25 were fired,

```
RULE 25
IF word_processor = technical
THEN CALL C:\symbols, " "
```

the system would call, load, and execute the file named symbols.exe, which would display the technical symbols available in the designated word processor. When symbols.exe finishes executing, you would be back in the calling consultation. Symbols.exe could be a program written in another language. The ability to incorporate these external DOS programs enhances both the expert system and the development time and costs.

A third clause, BCALL, turns control over to a DOS batch file. The file must have a .bat extension. Batch files are text files that contain commands for the operating system. They are useful in starting other programs without the user having to type in several lines of commands. The BCALL command functions the same as CCALL and CALL and has the same syntax.

Two additional commands, CLROFF and CLRON, affect the screen for CCALL and CALL. Following a CCALL or CALL clause, the CLROFF instruction cancels the normal clearing of the Consultation screen. The user still sees the same Consultation screen, as the external program executes transparently. In other words, background processing is allowed. The CLRON clause is used to turn off the effect of the CLROFF clause. Consider this example:

```
CLROFF
DISPLAY "Wait please."
CCALL C:\forecast,
CLRON
CALL C:\graphics," "
```

When CLROFF is executed, the Consultation screen remains, a wait message is displayed, and the program forecast.com is run in the background. When the forecasting program to project the savings of the word processor has finished, the CLRON command allows the screen to be cleared when the CCALL clause activates the graphics.exe file, which displays a bar chart of the five-year forecast of savings. Indeed, these five clauses—CCALL, CALL, CLRON, CLROFF, and BCALL—are power tools for the software engineer in expert systems.

The CLOSE clause closes files that are open; it may be used with either database (.dbf) or text files. Because of this duality, the file extension must be specified in the parameter. The syntax of the clause is

```
CLOSE file_name.ext.
```

The purpose of CLOSE is to reset the pointer to the beginning of the file. The close command should be used before issuing the SHIP, RECEIVE, DISPLAY, GET, and PUT commands.

The SHIPPOS clause is designed to increase VP-Expert's ability to process documents and forms. It does so by allowing you to specify an ASCII (text) file name, the name of the variable to be inserted at the specified position in the document or form, and the line and column numbers of the position. Its syntax is

SHIPPOS file_name, variable, line number, column number.

Line and column numbers start at 0. SHIPPOS works in "overwrite" mode, so you must be careful that there is enough space in the file to allow what is to be written to fit. Field size must be constantly considered. The alternative is to rewrite the remainder of the file.

The RECEIVEPOS clause is used to import from a text file and store the text in a variable. As with SHIPPOS, you must specify the ASCII file name, the variable to receive the text, the line number that contains the text to be imported, and the column number on the line that begins the text to be imported. Its syntax is

RECEIVE file_name, variable, line_number, column_number.

The clause imports all text on the specified line number beginning with the specified column number. As with SHIPPOS, line and column numbers begin with 0.

Graphics

VP-Expert contains a set of tools for creating drawings on the Consultation screen. These *graphics primitives* cannot be used unless the microcomputer system has a graphics adapter. To utilize graphics drawings, you must first set the proper setting code with the GMODE clause. Then a background color, GBCOLOR, and a drawing color, GCOLOR, must be selected.

Once these clauses have been executed, lines, polygons, circles, and ellipses can be drawn. The MOVETO command moves the drawing pen to a new x,y position on the screen. The PSET clause sets a particular pixel to the given drawing color. The LINETO clause draws a line (in the GCOLOR color) from the current cursor location to the indicated new point. A series of LINETO commands can generate a polygon. The RECTANGLE clause draws a rectangle on the screen, and the ELLIPSE instruction draws an ellipse or circle on the display. The FILL command can be used to fill the interior of a polygon, rectangle, ellipse, or circle with the preselected drawing color.

Text can also be incorporated into these drawings. The GLOCATE clause will locate the x,y position where the GDISPLAY clause will display text. When a graphics program is finished, you can return VP-Expert to its default screen setting (text and color only) with TMODE.

Dynamic Images

As we learned in Module 7, VP-Expert offers an additional feature for implementing user interfaces called *Dynamic Images*. A Dynamic Image is a predefined graphic object (drawing) that is dynamically linked to a value of a specific variable. When the value of the variable changes, the linked Dynamic Image changes. These Dynamic Images include gauges, buttons, form fields, meters, and time-series graphs. For example, a gauge could be used to indicate the rating of the performance of a word processing package. A meter could indicate the expected productivity gains from using the specialized word processor. Combinations of Dynamic Images could be used to create a control panel where the user selects symbols instead of word items from menus.

All Dynamic Images work in Graphics mode. Two of the images, FORMFIELD and LBUTTON, are text images, whereas other images, such as METER and BUTTON, are graphics images. Both types of images require x,y coordinates to determine the upper-left corner of the enclosing rectangle; however, the values of the coordinates vary in different Graphics modes. The text images range from 0 to 79 for x (column) and 0 to 24 for y (row).

A Dynamic Image is not displayed on the screen until the value of the variable to which it is linked (associated) has a known value. As long as the associated variable has an UNKNOWN value, the image is not displayed. Once the image is shown on the screen, however, it can be erased by issuing the RESET command with the linked variable.

The Dynamic Image instructions are statements. The BUTTON statement displays a button image that is linked to a Yes or No variable. The LBUTTON statement is the same except the displayed button contains a Text label. FORMFIELD displays a rectangular box, or form field, containing the value of the linked variable, whereas DBFORM displays a vertical array of rectangles or form fields corresponding to the fields of a .dbf file. The METER statement displays a meter image; the HGAUGE and VGAUGE commands display horizontal and vertical slide gauges, respectively. The TRACK instruction displays a time-series graph linked to the last 50 values of a given variable.

It is also possible to create your own Dynamic Images for use within a consultation by combining a series of graphics primitives with the HOTREGION statement. The HOTREGION statement creates an invisible hot, or active, region. When a mouse clicks on this region, the associated variable toggles between Yes and No. Now your custom-made image is linked to a variable value to control its display on the screen. Dynamic Image statements can add creativity and functionality to the expert system's application user interface.

SmartForms

Another advanced feature of VP-Expert is the ability to create *SmartForms* of screen displays, which resemble regular paper forms. Unlike the traditional paper forms used in business, the VP-Expert SmartForm can adapt to the user's responses. As the user fills out a form, the value of the response determines which form field will pop up next. For example, in a SmartForm application in which the user is filling out an upgrade agreement with the vendor of a word processing package, the reply to the question, "Are you networked?" would determine what part of the form would appear next. Comparing the SmartForms approach to traditional questioning (interview style), the SmartForms approach presents the user with a group of questions, allows him or her to change answers on the screen, and simulates the actual paper form used in the business application.

Creating a SmartForm involves several steps. First of all, the ACTIONS block must have a GMODE setting, a WHILETRUE-END loop to keep VP-Expert in a loop until the user has completed the form, and a TMODE clause to resume Text mode. WHENEVER rules are used to activate actions depending on the replies from the user in the SmartForm. These WHENEVER rules can offer advice if certain information is entered into a field, remove irrelevant fields, add new fields to the form, and maintain consistency between the fields in the forms. The FORMFIELD statement with the LBUTTON command can be used to create the form fields and link the reply to the click of a mouse when finished. Obviously, this SmartForm feature can enhance those expert systems involving extensive data entry with forms.

HYPERTEXT

Another useful feature available for VP-Expert applications is *HYPERTEXT*. HYPERTEXT is similar to SHOWTEXT and RECEIVE text in that it reads externally stored textual information into a consultation. HYPERTEXT, however, works only in Graphics mode, is activated whenever a variable is initialized, and appears inside a window. The interesting feature of HYPERTEXT is that the textual information is linked to other HYPERTEXT information. When the displayed information contains uppercase HYPERTEXT words, the user can click on these words with a mouse and the system will replace them with additional text related to the HYPERTEXT word. This feature allows the user to browse through information and find additional information on keywords.

HYPERTEXT is a special Dynamic Image. When it is not being used in a consultation, it must be removed from the screen. The HYPERTEXT window disappears when the linked variable is UNKNOWN. The HYPERTEXT feature can add flexibility to the expert system application.

Conclusion

The many features of VP-Expert can enhance the applications development of an expert system. Combining the graphics, Dynamic Images, and HYPERTEXT features is limited only by the design and imagination of the software engineer. The total development of an expert system application rests on the accuracy of the knowledge base and the consistent, creative, and meaningful implementation of the user interface. VP-Expert has all the power tools for the creation and implementation of useful expert systems in the business world.

Key Points

- Commands related to the inference engine are
  - FIND
  - ASK, CHOICES
  - RESET
  - RULES
- Commands involving a PLURAL variable of a FIND clause are
  - POP
  - COUNT
  - SORT
- Three loop structures for VP-Expert programs are
  - WHILEKNOWN variable
  - WHILETRUE condition
  - FOR variable = start_number to end_number
- Confidence factors can be implemented and utilized by
  - CNF
  - GETCNF
  - MODCNF
  - TRUTHTHRESH
  - SORT (PLURAL variables only)
- Commands for controlling the screen are
  - CLS
  - DISPLAY, FORMAT
  - COLOR
  - BLINEOFF
  - BLINEON
  - BKCOLOR
- Commands to process text characters are
  - CHR
  - GETCH
- Commands to utilize windows are
  - WOPEN
  - ACTIVE
  - LOCATE
- Printer commands include
  - PRINTON
  - PRINTOFF
  - PDISPLAY
  - EJECT
- Database files can be utilized with
  - MENU
  - GET
  - INDEX
  - MRESET
  - PUT
  - APPEND
  - CLOSE

RECORD-NUM
SETPOS

- Worksheet- (spreadsheet-) related instructions include
  WKS
  PWKS, WFORMAT
  WORKSHEET
  WORKON
- The RULECHAIN command allows facts generated by one knowledge base to be made available to other .kbs files.
- Commands to incorporate text files into a consultation are
  RECEIVE
  SHIP
  SHOWTEXT
  FDISPLAY
- Commands that will call specific external DOS files are
  CCALL—calls .com files
  CALL—calls .exe files
  BCALL—calls .bat files
- The CLROFF and CLRON commands toggle the clearing of the screen before the execution of CCALL or CALL.
- The following instructions implement chaining:
  CHAIN
  LOADFACTS
  SAVEFACTS
  EXECUTE
- Commands to set the graphics screen are
  GMODE
  GBCOLOR
  GCOLOR
- The PCX command will display a .pcx file.
- Commands to draw graphics include
  PSET
  MOVETO
  LINETO
  RECTANGLE
  ELLIPSE
  FILL
- Commands to control text on a graphics screen are
  GLOCATE
  GDISPLAY
  TMODE
- Mouse input can be implemented with
  GETMOUSE
  MOUSEX
  MOUSEY
- Dynamic Images can be implemented in graphics modes with
  BUTTON
  LBUTTON
  HGAUGE

VGAUGE
METER
TRACK

- Dynamic Images involving text are
  DBFORM
  FORMFIELD
- Commands for string manipulation are
  ASC
  CONCAT
  LOWER
  POS
  SUBSTR
  UPPER
- Commands to communicate with DOS are
  DATE
  TIME
  SYSTEM
- Commands for general file manipulation are
  DELETE
  EXISTS
  RENAME
- Custom Dynamic Images can be created with graphics primitives and HOTREGION.
- SMARTFORM allows the creation of dynamic data entry forms.
- HYPERTEXT is textual information that allows the user to obtain related information on keywords.

Exercises

1. Indicate the direction of movement in the following parameters by means of an arrow:
   a. POP X,Y
   b. SHIP X,Y
   c. RECEIVE X,Y
   d. COUNT X,Y
   e. SHOWTEXT X,Y
2. Which of the following commands are file names and which are variable names?
   a. CALL X
   b. FDISPLAY X
   c. RECEIVE X,Y
   d. POP X,Y
   e. INDEX X,Y
3. Which of the following instructions do not require any parameters?
   a. GETCH
   b. POP
   c. CLROFF
   d. EXECUTE
   e. CHAIN

4. Which of the following commands could execute Lotus 1-2-3 during a consultation?
   a. RULECHAIN
   b. CCALL
   c. WORKON
   d. WKS
   e. LOADFACTS
5. Given the code below, what can you assume about file x, in each instruction?
   a. CCALL X
   b. BCALL X,
   c. SHOWTEXT X, "Jones"
   d. INDEX X,Y
   e. WORKSHEET X
6. Which of the following commands apply only to a PLURAL variable?

POP	TRUTHTHRESH
CALL	COUNT
SORT	FIND
CNF	WHENEVER
GETCH	CLROFF

7. Which of the following commands work in Graphics mode only?

ELLIPSE	PUT
MOUSEX	PSET
LINETO	LOCATE
GETCH	WINDOW
RECTANGLE	GCOLOR

8. Answer the following questions about the code

```
FOR X = 2 to 5
DISPLAY (X)
END
```

   a. What is the start value?
   b. What is the end value?
   c. How many times is DISPLAY executed?
   d. What is the value of X when control passes to a command after the END?
   e. What is displayed on the screen?
9. Describe the functions of SHIPPOS and RECEIVEPOS. Why are they useful?
10. Why might DOS files be a problem for VP-Expert?
11. Why are the GETCNF and MODCNF commands useful?
12. What commands did RULECHAIN replace in version 3.0 of VP-Expert? Why is it a useful command?

Assignments

1. Readme.kbs is a file that demonstrates how to start a graphics screen and use the LBUTTON and BUTTON commands. It is in the Sample Files directory. You must first change the GMODE setting to match your adapter.

a. Copy readme.kbs to your Data Disk.
b. Run a consultation.
c. What does the expert system do?
d. Obtain a printout of the code.
e. Circle the graphics primitives and new commands.

2. Formdemo.kbs is another file in the Sample Files directory. It demonstrates use of the FORMFIELD command. Repeat the instructions for Assignment 1.
3. Fromage.kbs is a file that demonstrates use of the FORMFIELD statement and WHENEVER rules. It is a cheese advisor system in the Sample Files directory. Check the GMODE setting before running it. Repeat the instructions for Assignment 1.
4. Tempsim.kbs is a demonstration file of graphics primitives in the Sample Files directory. It is combined with Dynamic Images to simulate data acquisition of the current temperature. Repeat the instructions from Assignment 1.
5. Transfer wp25.kbs to your Data Disk as my25.kbs.
   a. Redesign the program to use the GETCH instruction.
   b. Using the editor, enter your changes.
   c. Run a consultation.
   d. Record the results.
6. Transfer wp9.kbs to your Data Disk as my9.kbs.
   a. Rewrite the program using a WHILETRUE-END loop.
   b. Using the editor, enter your revisions to the code.
   c. Run a consultation with your revised code.
   d. Record the results.
7. Try using the RULECHAIN command. Rewrite wp4.kbs and wp1.kbs so that if wp4.kbs advises that you need a word processor, you are automatically taken to a consultation with wp1.kbs. Copy both files to your Data Disk using different file names.
8. Use the VP-Expert Help Facility to obtain details on Dynamic Images:
   BUTTON
   LBUTTON
   HGAUGE
   VGAUGE
   METER
   TRACK
9. Use the VP-Expert Help Facility to obtain details of two Dynamic Images with text:
   DBFORM
   FORMFIELD
10. Use the VP-Expert Help Facility to obtain details about HYPERTEXT.
11. The readme.txt file in the Sample Files directory contains useful information about the commands and latest improvements to VP-Expert. Obtain a printout by entering
    A> COPY C:README.TXT LPT1
12. File readme.kbs in the Sample Files directory demonstrates the use of HYPERTEXT. The expert system file reads the HYPERTEXT readme.txt file. Run a consultation in VP-Expert with readme.kbs.

13. Throughout the text, we have used a knowledge base for choosing a word processor to demonstrate characteristics and features of VP-Expert. Modify the knowledge base to meet your own set of criteria for choosing a word processor. Make use of windows.
14. Create a knowledge base that could be used for advising first-year students at your school.
15. Create a knowledge base that will help a person choose his or her next vacation.
16. Create a knowledge base that will help a person purchase a car.

VP-Expert Command Hierarchy Chart

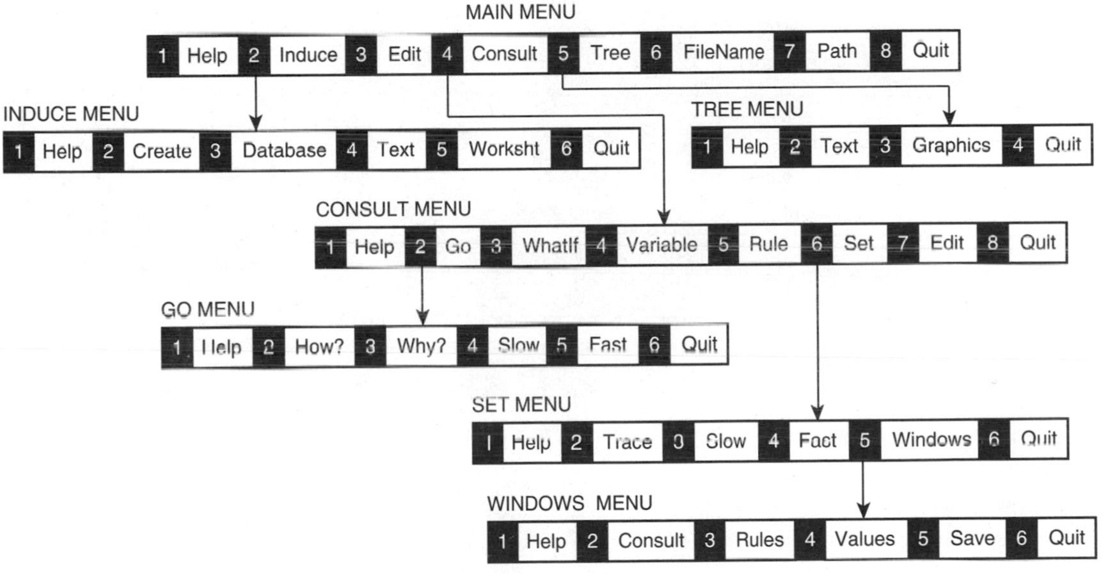

B

Table of ASCII Characters

ASCII CODE	DISPLAY	CONTROL SEQUENCE	CONTROL CHARACTER	ASCII CODE	DISPLAY	CONTROL SEQUENCE	CONTROL CHARACTER	ASCII CODE	DISPLAY	ASCII CODE	DISPLAY	ASCII CODE	DISPLAY	ASCII CODE	DISPLAY
0		Ctrl 2	NUL	16		Ctrl P	DLE	32		48	0	64	@	80	P
1		Ctrl A	SOH	17		Ctrl Q	DC1	33	!	49	1	65	A	81	Q
2		Ctrl B	STX	18		Ctrl R	DC2	34	“	50	2	66	B	82	R
3		Ctrl C	ETX	19		Ctrl S	DC3	35	#	51	3	67	C	83	S
4		Ctrl D	EOT	20		Ctrl T	DC4	36	$	52	4	68	D	84	T
5		Ctrl E	ENQ	21		Ctrl U	NAK	37	%	53	5	69	E	85	U
6		Ctrl F	ACK	22		Ctrl V	SYN	38	&	54	6	70	F	86	V
7		Ctrl G	BEL	23		Ctrl W	ETB	39	‘	55	7	71	G	87	W
* 8		Ctrl H	BS	24		Ctrl X	CAN	40	(	56	8	72	H	88	X
* 9		Ctrl I	HT	25		Ctrl Y	EM	41	)	57	9	73	I	89	Y
10		Ctrl J	LF	26		Ctrl Z	SUB	42	*	58	:	74	J	90	Z
11		Ctrl K	VT	*27		Ctrl [	ESC	43	+	59	;	75	K	91	[
12		Ctrl L	FF	28		Ctrl \	FS	44	,	60	<	76	L	92	\
*13		Ctrl M	CR	29		Ctrl]	GS	45	-	61	=	77	M	93	]
14		Ctrl N	SO	30		Ctrl ^	RS	46	.	62	>	78	N	94	^
15		Ctrl O	SI	31		Ctrl _	US	47	/	63	?	79	O	95	_

*These characters cannot be entered in LABEL mode. Characters 0 through 31 (ASCII) cannot be entered in EDIT mode.

ASCII CODE	DISPLAY	ASCII CODE	DISPLAY	ASCII CODE	DISPLAY	ASCII CODE	DISPLAY	ASCII CODE	DISPLAY	ASCII CODE	DISPLAY	ASCII CODE	DISPLAY	ASCII CODE	DISPLAY
96	‘	116	t	136	ê	156	£	176	░	196	─	216	╪	236	∞
97	a	117	u	137	ë	156	¥	177	▒	197	┼	217	┘	237	φ
98	b	118	v	138	è	158	₧	178	▓	198	╞	218	┌	238	ε
99	c	119	w	139	ï	159	ƒ	179	│	199	╟	219	█	239	∩
100	d	120	x	140	î	160	á	180	┤	200	╚	220	▄	240	≡
101	e	121	y	141	ì	161	í	181	╡	201	╔	221	▌	241	±
102	f	122	z	142	Ä	162	ó	182	╢	202	╩	222	▐	242	≥
103	g	123	{	143	Å	163	ú	183	╖	203	╦	223	▀	243	≤
104	h	124	¦	144	É	164	ñ	184	╕	204	╠	224	α	244	⌠
105	i	125	}	145	æ	165	Ñ	185	╣	205	═	225	β	245	⌡
106	j	126	~	146	Æ	166	ª	186	║	206	╬	226	Γ	246	÷
107	k	127	Δ	147	ô	167	º	187	╗	207	╧	227	π	247	≈
108	l	128	Ç	148	ö	168	¿	188	╝	208	╨	228	Σ	248	°
109	m	129	ü	149	ò	169	⌐	189	╜	209	╤	229	σ	249	·
110	n	130	é	150	û	170	¬	190	╛	210	╥	230	µ	250	·
111	o	131	â	151	ù	171	½	191	┐	211	╙	231	τ	251	√
112	p	132	ä	152	ÿ	172	¼	192	└	212	╘	232	Φ	252	ⁿ
113	1	133	à	153	Ö	173	¡	193	┴	213	╒	233	Θ	253	²
114	r	134	å	154	Ü	174	«	194	┬	214	╓	234	Ω	254	■
115	s	135	ç	155	¢	175	»	195	├	215	╫	235	δ	255	

C

Answers to Odd-Numbered Exercises

Chapter 1

1. According to Webster's *New World Dictionary*, Second Edition, intelligence is "the ability to learn or understand from experience; the ability to acquire and retain knowledge; the ability to respond quickly and successfully to new situations; or the use of the facility of reason in solving problems, directing conduct, etc., effectively."
3. Answers will depend on your particular situation.
5. *Advantages*
   1. Solves complex problems.
   2. Increases productivity.
   3. Prevents loss of human expertise.
   4. Removes people from working in selected dangerous environments.

   Limitations (disadvantages)
   1. Large overhead in time, money, and resources.
   2. Time required to learn new computer languages.
   3. Lack of inexpensive, usable, and reliable AI development tools to implement needed systems.
   4. Lack of qualified professionals to work on teams to implement AI systems.
7. Data processing systems (DPS) use predetermined procedures to manipulate huge volumes of data. Management information systems (MIS) summarize and select data from the massive DP files so as to give meaning to the computer output. Decision support systems (DSS) are designed to help managers make more effective and efficient decisions. They can be used when problems become too large or complex for an MIS to be adequate, especially when the individual does not have any relevant experience in solving the problem. Expert systems (ES) are systems designed to simulate the behavior of a human expert as he or she attempts to solve a complex problem in a particular domain. DSS and ES evolved from similar disciplines, but the general goal of an ES is to replace human knowledge while the general goal of a DSS is to assist, rather than replace, the expert.

Chapter 2

1. An expert system is a computer program that emulates the behavior of human experts who are solving real-world problems associated with a particular domain of knowledge. The most basic difference between an expert system and a conventional computer program is that expert systems manipulate knowledge while conventional computer programs manipulate only data. Conventional programs are algorithmic in nature and therefore guarantee a solution to a problem, while expert systems use heuristic (rules of thumb) and inferential (reasoning) processes and therefore do not guarantee a solution to a problem. Expert systems and conventional computer programs are similar in that they are both computer software designed to accomplish a specific task.

3. The early efforts at expert system development began in the 1950s with the work of Newell, Shaw, and Simon, who created a system called the General Problem Solver (GPS). Other groups, such as Minsky and McCarthy, and Ledburg, Buchanan, and Feigenbaum, also did significant work. In the 1970s, researchers realized that expert systems should be built with narrow-scoped problem domains; they also realized that the quality of the knowledge possessed by the expert system was at least as important as the inference techniques used by the system. These two realizations, along with advances in microcomputer technology, were the real catalysts to progress in the 1970s and 1980s. Expert systems had begun to move from the research to the commercial environment.

5. The two types are (1) knowledge engineering languages/systems (shells) and (2) high-level programming languages. Shells are easier to use for inexperienced expert system builders, but they are not as flexible relative to the range of potential problems to be solved because the inference engine is preconstructed. High-level programming languages offer greater flexibility in the range of problems to be solved, but they are more difficult because the inference engine must also be constructed and decisions about how knowledge is to be represented must be made.

7. Advantages:
   1. The knowledge base is (usually) permanent and easy to maintain.
   2. The system is relatively inexpensive to maintain and can be used in several locations at the same time.
   3. The system is easy to document and produces consistent results.

   Limitations:
   1. The system cannot learn from experience.
   2. The knowledge-acquisition process is not well understood.
   3. Not all problems are candidates for expert systems.
   4. All knowledge put into the expert system must be represented in the same form.

9. A word processor is a software program that manipulates text.

Chapter 3

1. There are four models for knowledge representation.

3. Antecedent = If today is Tuesday and the temperature is above 60 degrees
   Consequent = I will go fishing
   A = today is Tuesday
   B = the temperature is above 60 degrees
   C = I will go fishing
   Example
   A = T
   B = F
   A and B = F
   C = F
   IF (A and B), THEN C = T
   Answers to the last part of the question will vary, depending on your situation.

5. Example
   1. Monte Carlo
   2. 789321
   3. 1985
   4. Black
   5. 305
   6. Chevrolet

   Answers will vary.

7. | | |
   |---|---|
   | Frame name | Monte Carlo |
   | License plate number | 789321 |
   | Model year | 1985 |
   | Color | Black |
   | Type of Engine | 305 |

9. | A | B | NOT A | NOT A AND B | A OR NOT B | B → A |
   |---|---|---|---|---|---|
   | T | T | F | F | T | T |
   | T | F | F | F | T | T |
   | F | T | T | T | F | F |
   | F | F | T | F | T | T |

11. IF functionality is simple and cost is low,
    THEN select a word processor equal to A.

    IF functionality is simple and cost is high,
    THEN select a word processor equal to A.

Chapter 4

1. Backward Control. We know what the goal state is in advance (the light bulb is on). We know that to turn the light bulb on we need to flip the light switch.

3. C
   B, E
   H, I
   D, F, A, H, I
   D
   G

None
C, B, D, E, F, A, G, H, I
C, B, E, D, F, A, G, H, I
Both methods would be the same.

5. If A and B then F
   If C and D then G
   If F and G then H
   If E and H then I
7. For example:
   Node 5 = 100(0) + 10(1) + 0 − (100(0) + 10(1) + 1) = 1
   Node 8 = 100(0) + 10(2) + 0 − (100(0) + 10(2) + 0) = 0
   Formula: 100A + 10B + C − (100D + 10E + F)
   Where
   A = number of rows, columns, diag. with 3 Xs
   B = number of rows, columns, diag. with 2 Xs and 0 Os
   C = number of rows, columns, diag. with 1 X and 0 Os
   D = number of rows, columns, diag. with 3 Os
   E = number of rows, columns, diag. with 2 Os and 0 Xs
   F = number of rows, columns, diag. with 1 O and 0 Xs
9. a. Blue
   b. 14
   c. 55

Chapter 5

1. Waterman [1986] suggests that three questions need to be answered.
   1. Is expert systems development possible for the problem solution?
   2. Is expert systems development justified for the problem solution?
   3. Is expert systems development appropriate for the problem solution?
3. It is often difficult for an expert to express thoughts pertinent to subject domain. Because an expert has a deep knowledge about the subject, he or she makes assumptions about the qualifications of the listener (knowledge engineer). If the listener is unprepared for the subject, what the expert says is likely to be meaningless, causing frustration during the communication process and possibly wasting time as each tries to understand the other.
5. Introspective modeling involves examining portions of the expert's previous experience. It aids in the interview process because the interviewer is better able to understand the point of view and experience of the expert.
7. The art of listening is the ability to fully understand what we hear. This is important to a knowledge engineer because during an interview, the knowledge engineer must be continually aware of the potential imprecision of language in verbal communications with the expert and must gain an understanding of the expert's thinking process.
9. If a high-level programming language is going to be used, any of the four models of knowledge representation discussed in Chapter 3 could be used. Most expert system shells use the rule as the form of knowledge representation.

11. The knowledge engineer queries the expert to try to capture his or her knowledge and to determine what strategies he or she uses to solve a given problem. The query/response process, the key to the knowledge-acquisition process, ideally should be a flow of the expert's knowledge and strategies with minimal or nonexistent gaps.

Chapter 6

1. Uncertainty can be considered as the lack of adequate information needed to make a decision.
3. 1. Reasoning errors can occur in the rule as a whole.
   2. Uncertainty is involved in determining how the knowledge should be combined in the antecedent. Giarratano and Riley [1989] note that there are three basic areas of uncertainty pertaining to rules—individual rules, conflict resolution rules, and rule compatibility.
5. Because most expert systems are rule-based, and even if each individual rule can be verified, there is no guarantee that the inference chain will give valid results. Uncertainties exist as inference chains are formulated. Conflict resolution is used to determine the order in which rules are fired (executed).
7. Rule incompatibility occurs when two rules have the same antecedent but have conflicting consequences, which causes a lack of adequate information to make a decision.
9. The original definition of the certainty factor for a given hypothesis H and the given evidence used the difference of the measure of belief (mB) and the measure of disbelief (mD). mB and mD were probably numbers assigned by the physicians (experts).
11. VP-Expert uses the term *confidence factor* instead of *certainty factor*. Its calculations are similar to those of MYCIN.
13. a. 70
    b. 40
    c. 74
    d. 68
    e. 84
    f. Not possible using the VP-Expert method

Chapter 7

1. The user interface is the total communications medium between the participant and the expert system.
3. Keyboard—cursor keys
5. Checking
   1. Cost of Checks
   2. Monthly Service Charge
   3. Rate of Interest
   4. Rate of Interest

   Savings
   1. Money Market
   2. Certificates of Deposit
   3. Closing Costs

Mortgage
1. Escrow
2. Principal
3. Points
4. Fixed
5. Adjustable Rates
6. Closing Costs
7. Insurance
8. Rate of Interest

7. * Main Menu
1Disk 2Create 3Edit
* Disk Submenus
1Save 2GetFile 3Quit
* Create Submenu
1Document 2Spreadsheet 3Database 4Outline
* Edit Submenu
1Typeover 2Insert 3Delete

Chapter 8

1. An expert system is a piece of software; therefore, its development process can use principles of software engineering.
3. The expert system life cycle includes the following phases: development, use/integration, and updating/maintenance. In the development phase, the expert system is actually created. In the use phase, the expert system solves problems related to the domain area. The updating/maintenance phase is where the knowledge base is updated or modified and where errors are corrected.
5. The development phase of the expert system life cycle is analogous to the development phase of the conventional software system life cycle in that the same basic steps are used. However, there are some differences. For example, a conventional software system is designed to deal with a single application or problem, and its specifications are developed in conjunction with the user who has a task to be accomplished. On the other hand, an expert system is intended to solve a range of applications or problems in a narrow domain of knowledge. Knowledge must be put into a form that the expert system can understand. Formal specifications are not developed with a conventional software system, but the domain of the knowledge in an expert system must be clearly defined. The knowledge engineer works with the human expert to acquire true knowledge, but does not work with the user.
7. 1. Validate the choice and definition of the problem to be solved.
   2. Make an initial determination of the appropriateness of the chosen development tool.
   3. Assess the initial completeness of the knowledge acquisition process.
   4. Initially evaluate the design of the expert system.
   5. Provide an initial test of the reactions of potential users.
   6. Allow management to see initial results of the expert system development process.

9. The four major areas are:
   1. Scarce Resources—It is very difficult to free resources for expert systems development within a company or organization because expert systems is a new technology, relatively unfamiliar to many computer specialists.
   2. Development Time—The knowledge acquisition process cannot be rushed. Also, many organizations increase the size of the team working on the development. This takes extra time and can be counterproductive if the number in the group exceeds the optimum team size.
   3. Sources of Errors—The sources of errors in an expert system include the human expert, the knowledge engineer, the knowledge base, the inference engine, and the user interface. Most of these errors should be corrected during final system evaluation, although the total elimination of errors is very difficult.
   4. Inherent Limitations—The current state of expert systems technology has some inherent limitations. Expert systems are not capable of acquiring knowledge directly without the knowledge engineer. They are not capable of dealing with knowledge that has a mixed representation. They cannot refine their own knowledge bases and cannot learn from experience.

Chapter 9

1. Expert systems of the future will likely be able to deal with more than one knowledge-representation scheme at the same time. New knowledge representation schemes will be developed that will be more powerful and easier to use and will allow the use of single time dimension.
3. 1. Decrease the amount of time required to acquire the knowledge.
   2. Decrease the skill level required of the person acquiring the knowledge.
   3. Automate some aspects of the knowledge-acquisition process.
   4. Enable users and human experts to add to the knowledge.
5. They are used to locate duplicate or contradictory rules in the knowledge base.
7. System development tools will be more powerful and easier to use with the appearance of new shells. They will include graphics and more natural-language interfaces. Intelligent design tools will aid the knowledge engineer by determining problem domain characteristics, ensuring that all important problem attributes are considered, determining which knowledge representation scheme(s) and inferencing mechanism(s) are used, assessing system user capabilities and selecting appropriate user interface design schemes, and helping detect appropriate shells for given applications.
9. The use of natural language is an attempt to make the human/machine interface as human-like as possible. The current status of natural-language technology shows that there is much progress to be made. The problem of natural-language processing is very complex, and solutions will not be easily found.
11. User interfaces in expert systems of the future will be verbal and "user-friendly" with increased attention by the expert system developer to the quality of the interface.
13. Cost will always be a problem; while hardware costs tend to decrease, software development costs tend to increase. The development of automated tools will help decrease costs in the future.

Module 1

1. a. 1. The knowledge base.
   2. The inference engine.
   3. The user interface.
   b. 1. They can give advice on investments.
   2. They can determine the acceptability of a credit application.
   3. They can predict backorders.
   4. They can control inventory.
   5. They can design investment portfolios.
3. We need to protect ourselves against the accidental loss of data and damage to programs.
5. a. 1. DISKCOPY
   d. 2. VER
   c. 3. FORMAT
   b. 4. COPY
7. d

Module 2

1. Quit, Edit, Path, Help
3. Go, Variable, Rule, Edit, Help
5. b
7. Help, How?, Why?, Slow, Fast, Quit
9. e

Module 3

1. All variables are valid except for COST_7 and COST/.
3. RULE
   BECAUSE "Product X has tutorial and telephone help."
   RULE 1
   BECAUSE "Product X has tutorial but does not have telephone help."
   RULE 2
   BECAUSE "Product X does not have tutorial but does have telephone help."
   RULE 3
   BECAUSE "Product X does not have tutorial or telephone help."
5. See Figure C.1.
7. a. 3
   b. Advice
   c. 3
   d. 3
   e. IF Name = I
   AND Type = J
   AND Model = K
   THEN Advice = L

```
!The file wp3.kbs has one compound AND
!     IF-THEN-ELSE rule.

!ACTIONS BLOCK

ACTIONS
      Display "This expert system WPHELP will advise you on whether you
need to buy a word processor soon or wait.
Press the Spacebar to start the consultation."
      CLS
      FIND Advice
      DISPLAY "concerning the purchase of a word processor, WPHELP
advises you to {Advice}.
      Press the SPACEBAR to exit WPHELP."
      CLS
!RULES Block

Rule 1
IF    Afford = Yes
      AND Help = Yes
      AND Time = Yes
      AND Need = Yes
THEN Advice = proceed CNF 90
ELSE Advice = wait CNF 70;
!STATEMENTS Block
ASK Afford: "Can you afford the money to buy a word processor at
this time?";
CHOICES Afford: Yes, No;
ASK Help: "Do you have some word processing help--people,
tutorials, experience, etc.--available to you?";
CHOICES Help: Yes, No;
ASK Time: "Do you have time to search for, install, and learn a
word processor?";
CHOICES Time: Yes, No;
ASK Need: "Do you need a word processor?";
CHOICES Need: Yes, No;
```


FIGURE C.1 PROGRAM CODE FOR WP3.KBS

9.

Cost	Warranty	Advice
400	Yes	OK
400	No	No
800	Yes	Maybe
800	No	No

11. a. Amount = (3 * 4)
    b. Amount = (3 * Increment)
    c. Amount = @(tan(x))
    d. Amount = @(sqrt(x))
    e. Correct syntax

Module 4

1. a. Trace, d. Fast
3. Path, Filename, Consult, Set, Trace, Go, Tree, Text
5. Document = Type4
   Functionality = Complex
   Cost = High
   Word_processor = Product_B

7. RUNTIME—statement
   ASK—statement
   CHOICES—statement
   FIND—clause
   DISPLAY—clause
9. c. 0 and 100
11. a. T b. F c.T d.T e.F
13. a. T b. F c. F d. F e.T

Module 5

1. b. GET
3. a. MENU Reply, ALL, wpdb, Word_Proce
   b. MENU Reply, ALL, wpdb, Networked
   c. MENU Reply, ALL, wpdb, Support
   d. MENU Reply, ALL, wpdb, Cost
5. a. Answer is: The entire table
   b. A B C D E
   c. fair good
   d. 200 500
   e. no yes
7. a. A B C D
   b. D E
   c. A D
   d. E
9. a. F b. T c. T d. T
11. a. T b. F c. F d. F

Module 6

1. Confidence factors are the same as certainty factors. They are a measure of belief/disbelief in an assertion.
3. It uses the minimum value of the set of condition CNFs.
5. Threshold values represent a minimum acceptable CNF for a rule condition to be considered true.
7. The final CNF for a rule is the product of the CNF for the rule condition and the CNF for the rule conclusion.
9. Confidence factors for rule conclusions are entered by the knowledge base developer at the time the rule is entered into the knowledge base.
11. First, calculate CNF for J or K.
    60 + 70 - (60 * 70)/100
    = 88
    Using this value (88) with "or L",

88 + 90 - (88 * 90)/100
= 99
which is the CNF of the rule condition. The final rule CNF is
99 * 40/100 = 40

13. 73

Module 7

1. a. The cost is 324.
   b. 324 is the cost.
   c. The cost is 324.
   d. The cost is $324.
   e. The cost is 324. CNF 95.
   (NOTE: e. assumes that the confidence factor of Cost is 95. The command will print whatever confidence factor is associated with COST at that time.)
3. a. ESC
   b. M
   c. DEL
   d. ?
5. d.
7. a. Row = 7 Column = 6
   b. Row = 8 Column = 7
   c. Row = 9 Column = 7
   d. Row = 10 Column = 8
9. c
11. b

Module 8

1. a. The most recently found value of the plural_variable X is moved or stored in the non_plural_variable Y.
   b. The value of the variable Y is transferred to the external text file X. X is created if it does not already exist.
   c. The current line of text found in the external text file X is placed in the VP-Expert variable Y.
   d. The number of values for the plural_variable X are placed in the counter-variable Y.
   e. A page of text from the external file X with the heading listed Y is displayed to the screen.
3. c. CLROFF
   d. EXECUTE
5. a. X is an external DOS file that has a .com extension not used in the CCALL clause. This file X is limited to 64k.

b. X is a DOS batch file with a .bat extension. Batch files are text files that contain commands for the operating systems.
c. X is a text file with a .xxt extension that was created with the BASIC program indexer.bas. X was created with a special index.
d. X is an existing index file with a .ndx extension that contains indices or pointers to a field or attribute of the database named in Y.
e. X is VP or VPP, which indicates which spreadsheet program, VP-Planner or VP-Planner PLUS, will be used. X is not a file.

7. ELLIPSE
LINETO
RECTANGLE
PSET
GCOLOR

9. SHIPPOS allows you to put text at a specified location in a text file.

   RECEIVEPOS allows you to copy a line of text from a text file beginning in a specified column.

   They are useful commands because of their flexibility in allowing you to specify a location in the text file.

11. They are useful because they can be used to change the CNF of a rule conclusion during a consultation.

13. The ASCII character $ has a value of 36.

15. The variable called topics, which has the value expert, is a substring of the string VP-Expert, which is stored in the variable called product_name. The substring topics starts in position 3 and has a length of 6 characters.

17. d

19. a

Classification of VP-Expert Keywords

The following list, organized by topic/task, will assist you in determining where a keyword fits into the applications development process. Detailed descriptions of all of these keywords appear in Chapter 9 of the VP-Expert manual and the Addendum for Version 3.0. Classifications are keyed to their VP-Expert instruction type.

Database File Clauses

```
APPEND
GET
INDEX
PUT
SETPOS
```

Graphics Mode Clauses

```
GBCOLOR
GCLS
GCOLOR
GDISPLAY
GLOCATE
MOVETO
PCX
```

Knowledge-Based Programming Clauses

```
ASC
CHAIN
CHR
CONCAT
COUNT
```

```
DATE
ELSE
END
FIND
FOR
FORMAT
GETCH
GETCNF
GETMOUSE
IF
LENGTH
LOADFACTS
LOWER
MENU
MODCNF
MRESET
POP
POS
RESET
RULECHAIN
SAVEFACTS
STEP
SHOWFACT
SORT
SUBSTR
SYSTEM
THEN
TIME
TRUTHTHRESH
UPPER
VPXRUN<file_name>-S
WHILEKNOWN
WHILETRUE
```

Knowledge-Based Programming Statements

```
ACTIONS
ASK
AUTOQUERY
CHOICES
ENDOFF
EXECUTE
HYPERTEXT
PLURAL
RANGE
RULE
RUNTIME
TACK
```

Logic Operations

AND
OR

Printer Device Clause

EJECT

Reserved Word

ALL

Rule Keyword

BECAUSE

Screen Display Clauses

BLINEOFF
BLINEON
CLROFF
CLRON
CLS
COLOR
DISPLAY
ELLIPSE
FILL
GMODE
LINETO
LOCATE
MOUSEOFF
MOUSEON
PRINTOFF
PRINTON
PSET
RECTANGLE
TMODE

Screen Display Statements

BKCOLOR
BUTTON
DBFORM
FORMFIELD
HGUAGE

```
HOTREGION
LBUTTON
METER
VGUAGE
```

Special Symbol

```
! (comment line)
```

System Variables/Values

```
GMODE
MENU_SIZE
MOUSEX
MOUSEY
RECORD_NUM
UNKNOWN
```

Text/External File Clauses

```
BCALL
CALL
CCALL
CLOSE
DELETE
EXISTS
FDISPLAY
RECEIVE
RECEIVEPOS
RENAME
REPORT
SHIPPOS
```

Worksheet Clauses

```
COLUMN
CURR
FIXED
NAMED
ROW
PWKS
SCI
WCLOSE
WFORMAT
WKS
WOPEN
WORKON
WORKSHEET
```

References

Apple, 1987. *Human Interface Guidelines.* Reading, Massachusetts: Addison-Wesley.

Bonnet, A. and C. Dahen. 1988. "Oilwell Data Interpretation Using an Expert System and Pattern Recognition." *Proceedings of IJCAI—88.*

Brown, G.M. 1988. *Human–Computer Interface Design.* Norwood, New Jersey: Ablex.

Buchanan, B. and E. Feigenbaum. 1978. "DENDRAL and METADENDRAL: Their Applications and Dimension." *Artificial Intelligence,* vol. 11.

Buchanan, B. and E. Shortliffe. February 1977. "Production Rules as a Representation in a Knowledge-Based Consultation System," *Artificial Intelligence.*

Buchanan, B. and E. Shortliffe, eds. 1984. *Rule-Based Expert Systems,* 205–301, 571, 596. Menlo Park, California: Addison-Wesley.

"DELTA/CATS-1". The Artificial Intelligence Report, vol. 1, no. 1, January 1984.

Frenzel Jr., L. 1987. *A Crash Course in Artificial Intelligence and Expert Systems.* Indianapolis, Indiana: Howard Sams.

Giarratano, J. and G. Riley. 1989. *Expert Systems: Principles and Programming.* Boston, Massachusetts: PSW-Kent.

Goyal, S. et al. 1985. *COMPASS: An Expert System for Telephone Cable Maintenance.* Waltham, Massachusetts: GTE Laboratories, Inc.

Gresnerietal, R. 1984. "YESMVS: A Continuous Real-Time Expert System." *Proceedings of the American Association for Artificial Intelligence.*

Hendler, J. Ed. 1988. *Expert Systems: The User Interface.* Norwood, New Jersey: Ablex.

"INVEST: An Expert System for Financial Investments." Summer 1988. *IEEE Expert.*

Martin, J. and S. Oxman. 1988. *Building Expert Systems.* Englewood Cliffs, New Jersey: Prentice-Hall.

McDermott, J. January 1982. "Domain Knowledge and the Design Process." *Design Studies,* vol. 3, no. 1.

Pressman, O. 1982. *Software Engineering: A Practical Approach.* New York: McGraw-Hill.

Rubin, T. 1988. *User Interface Design for Computer Systems.* New York: Helsted Press.

Shortliffe, E. and B. Buchanan. 1985. "A Model of Inexact Reasoning" in *Rule-Based Expert Systems.* Reading, Massachusetts: Addison-Wesley.

"Similarity Networks: A Knowledge Representation for Information Analysis and Retrieval." July/August, 1988. *Artificial Intelligence.*

Waterman, D. 1986. *A Guide to Expert Systems.* Reading, Massachusetts: Addison-Wesley.

Wetherbe, J. 1988. *Systems Analysis and Design.* St. Paul, Minnesota: West.

Wright, J. et al. October 1984. "ACE: Going From Prototype to Product with an Expert System." *Conference Proceedings of the Association for Computing Machinery.*

Index

H

I

J

K

N

O

P

Q

R